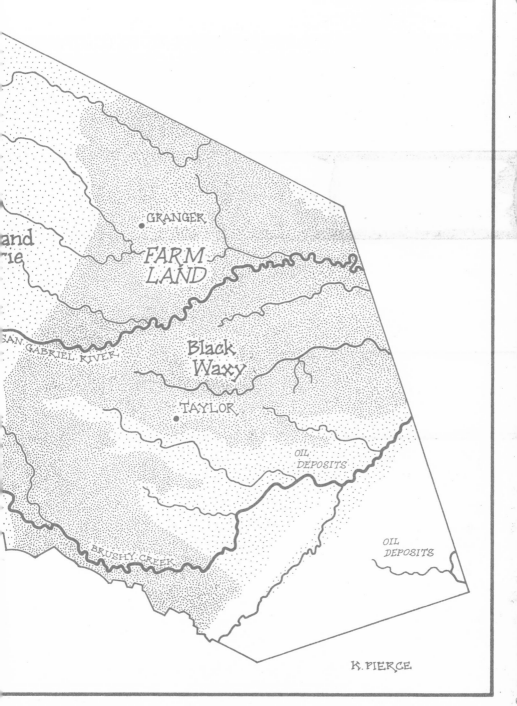

GRANGER

and
_ie_

FARM
LAND

SAN GABRIEL RIVER

Black
Waxy

TAYLOR

OIL
DEPOSITS

BRUSHY CREEK

OIL
DEPOSITS

K. PIERCE

ROAD, RIVER, AND OL' BOY POLITICS

ROAD, RIVER & OL' BOY POLITICS

A Texas County's Path from Farm to Supersuburb

LINDA SCARBROUGH

Texas State Historical Association
Austin

Library of Congress Cataloging-in-Publication Data

Scarbrough, Linda.
    Road, river, and ol' boy politics : a Texas county's path from farm to supersuburb / by Linda Scarbrough.
        p.   cm.
    "Published by the Texas State Historical Association in Cooperation with the Center for Studies in Texas History at the University of Texas at Austin"—T.p. verso.
    Includes bibliographical references (p.   ) and index.
    ISBN 0-87611-202-5 (alk. paper)
    1. Williamson County (Tex.)—History—20th century. 2. Williamson County (Tex.)—Politics and government—20th century. 3. Williamson County (Tex.)—Environmental conditions. 4. Interstate 35—History—20thh century. 5. Express highways—Texas—Williamson County—History—20thcentury. 6. San Gabriel River (Tex.)—Regulation—History—20th century. 7. Dams—Texas—Williamson County—History—20th century. 8. Regional planning—Texas—Williamson County—History—20th century. 9. Suburbs—Texas—Williamson County—History—20th century. 10. Williamson County (Tex.)—Economic conditions—20th century. I. Texas State Historical Association. II. University of Texas at Austin. Center for Studies in Texas History. III. Title.

    F392.W66S27 2005
    976.4'289063—dc22

                                                                2005014928

5 4 3 2 1   05 06 07 08 09

Published by the Texas State Historical Association in Cooperation with the Center for Studies in Texas History at the University of Texas at Austin.

Design by David Timmons.

To "Daddy" Don and "Mama" Clara of the Black Waxy,
who made everything count.

CONTENTS

MAPS

ACKNOWLEDGMENTS

I never could have imagined that so many people would respond so enthusiastically to this project. I cannot possibly thank them all, but I must point out a few whose contributions were beyond measure. Foremost was William H. Goetzmann, professor of American Studies at the University of Texas at Austin, whose vast storehouse of knowledge and experience provided the essential tracings of this work. Before his untimely death, Robert M. Crunden provided his own brand of inspiration during my first months of research. Southwestern University's Walt Herbert kept me going with fried catfish and fine ideas along the way, and my editor, Janice Pinney, rigorously smoothed and improved the final work. Long before I imagined this book, two wonderful editors, Mark Bloom and Paul Burka, encouraged me and critiqued my first feeble attempts to put my thoughts on paper.

Every person I interviewed was gracious about sharing time, memories, and, often, letters and photographs. A few made huge contributions. These include the Fox clan, especially Carol, Marie, Geraldine, and Jim. Without their assistance, I would have been sunk. Also, Loretta Mikulencak and the Honorable J. J. "Jake" Pickle, both of whom died before this could be published, Tom Kouri, Gene Fondren, N. G. "Bunky" Whitlow, "Dot" Labaj Daniel, Opal Wilks, and the Roy Gunn family gave of their time and memories in especially helpful ways. My friend Ron Landes read the manuscript and made great practical suggestions. Newspapers are my business, and I used them extensively. The newspaper archives at the Center for American History at the University of Texas in Austin are superlative; I relied heavily on them. In Williamson County, the *Taylor Daily Press,* the *Round Rock Leader,* and Georgetown's *Williamson County Sun* allowed me to leaf through bound copies, saving my eyesight and sanity. *Texas Highways* editor Jack Lowrey provided similar access to old runs of his magazine, and the *Austin American-Statesman* kindly let me use its morgue.

I have fallen in love with librarians. In every collection I visited, the keepers of the archives went all out for me. Joan Parks at Southwestern University's A. Frank Smith Jr. Library helped me in too many ways to count. Other key librarians included David Chapman at Texas A&M's Cushing Collection, Ben Rogers at the Baylor University Collections of Political Materials, Ralph Elder at the Center for American History, Linda Briscoe at the Harry Ransom Humanities Research Center's photo collection, and Linda Seelke at the Lyndon Baines Johnson Library. When I hit a wall, Tina Houston at the LBJ Library helped me locate Homer Thornberry's papers, which are owned by family members and were not available to the public. Under intense time pressure, Barbara Rust, the National Archives' Fort Worth archivist, managed to pull together everything I needed. And at the Taylor Public Library, Mary Jane Richter was a shining jewel, again and again.

Governments tend to store ancient records in distant, dusty warehouses if they store them at all. In one such Texas Department of Transportation warehouse I culled through highway builder Dewitt Greer's papers to discover early proposed routes for Interstate 35. Texas Department of Transportation employees John Hurt, Helen Havelka, Larry Jackson, Chris Bishop, Jerry Tallus, Anne Cook, and Lee Elkins were always helpful. At the Army Corps of Engineers' headquarters in Fort Worth, Andrew Goss steered me through a confusing maze; at the Brazos River Authority, Mike Bukala let me range with perfect freedom through the Authority's comprehensive and meticulous records. The City of Round Rock, especially Bob Bennett, then the city manager, provided critical information in a timely and courteous fashion. And at the Williamson County Tax Appraisal District, Karen Vanecek made it a joy to seek ancient maps. If mistakes have crept into this book, despite all this help, they are mine alone.

Finally, I thank my family. My grandparents, Auburn and Margaret Stearns, barely escaped death in the 1921 flood that gave rise to Williamson County's reengineered future, giving me a personal stake

Don and Clara Scarbrough bask in the glow of publishing the *Williamson County Sun*'s May 1977 Centennial Edition. *Courtesy* Williamson County Sun, *Georgetown.*

in the saga. The political tales told over the dinner table by my father, Donald Lee Scarbrough, planted early seedlings for this work. Through her history of Williamson County, *Land of Good Water,* my mother, Clara Stearns Scarbrough, laid the foundation and, throughout this project, provided me with encouragement and good advice. Daughter Katherine Alicia Thurmond filled me with delight when I needed it most, and (as always) my husband, Clark Thurmond, provided his special brand of enthusiasm and firm expectation, which kept me from foundering.

INTRODUCTION

Or was that, maybe, an excuse for childishness? What I wanted
was to float my piece of the river again.

JOHN GRAVES, *Goodbye to a River*

W illiamson County, Texas, abuts Austin's northern edge.
Indeed, it *is* north Austin—a powerful generator of residential and economic growth for the nation's second-
fastest-growing city. An agricultural center for a century, Williamson
County became a suburban phenomenon in the late 1980s—one of
the fastest-growing counties in the United States.[1] Between 1970 and
1980, its population went from thirty-seven thousand (where it had
remained essentially frozen since 1900) to seventy-seven thousand. By
1990 it had doubled again to 140,000. At century's end, the county
housed a quarter of a million people and the world's top-grossing
personal computer manufacturer.[2] Fifteen years into the twenty-first
century, Williamson County's population is expected to approach a
million.[3] It is an Edge City in the making.[4]

---

[1] Since 1990 the county has ranked among the ten fastest-growing counties in the
United States.

[2] Dell Computer Corporation, with 16,000 employees in Williamson County in
2002, down from 22,000 the previous year.

[3] U.S. Census, 1900–2000.

[4] Joel Garreau, author of *Edge City: Life on the New Frontier* (New York : Anchor
Books, 1992), coined this phrase to describe new population centers well beyond tradi-

The story of Williamson County's metamorphosis from agrarian backwater to suburban juggernaut reveals a pattern of how several of America's most successful agricultural counties became supersuburbs over the last half of the twentieth century.[5] The twin pillars of this growth surge, most notably in the Dry Sun Belt, were dams and interstate highways funded by the federal government.[6] After World War II, as a deluge of new dams, pipelines, and canals concentrated water in certain portions of the nation's arid zones, and interstate highways linked formerly isolated "oasis" cities, hypergrowth followed. Who decided where to put these massive projects and why? Washington's elite engineers and politicians are often credited (or blamed) for making "top-down" decisions that radically altered local economies and cultures, for better and worse. Powerful state interests often influenced those decisions. But, surprisingly, it was often local "bosses"—savvy rural leaders—in places like Williamson County who conceived many of the projects, altered others, and forged them into reality. Throughout the Dry Sun Belt, dams were usually built to develop existing agricultural economies—not to expand cities. The interstates were supposed to buttress weakening metropolitan central business districts that were losing out to the nation's first round of suburbs developed in the 1920s and 1930s. But the places that acquired dam and interstate highway projects cast off their agricultural identities and became suburban megalopolises on a scale never before contemplated. These new population centers redefined our ideas about cities and became the most potent growth machines of America's second century.[7]

---

tional urban cities that grow to a point at which they can no longer be considered "suburbs," but become employment and cultural centers themselves, competing with the urban center that spawned them.

[5] Outstanding examples include Maricopa County near Phoenix, Boulder County near Denver, Provo and Weber Counties outside Salt Lake City, Bernalillo County near Albuquerque, and Dona Ana County near El Paso.

[6] Throughout this work I use the term Dry Sun Belt in a rather sweeping fashion, to include the western half of the Sun Belt and the arid and semi-arid portions of the Plains States once known as the Great American Desert. In the eastern, or "humid," regions of the United States, the dam-building impulse was not nearly as important to suburban growth as the development of the interstate highway system.

[7] Key sources for this thesis include Marc Reisner's *Cadillac Desert: Water and the*

The pattern replicated itself across the Southwest, with local quirks. In Arizona's Salt River Valley, sprawling Mexican-style *haciendas* nestle among the remnants of orange groves planted in concert with federally funded dams and canals that enabled Greater Phoenix to expand, and inspired Edge Cities like Mesa, Scottsdale (originally Orangedale), and Glendale, all old citrus kingdoms.[8] Near Dallas, thousands of suburban "plantation homes" rise on "Black Waxy" land once famed for its cotton. In Williamson County, Fred Turner's old Charolais ranch gave way to a championship golf course and Texas's first "age-restricted" Sun City planned community. Denver, Salt Lake City, Albuquerque, and El Paso share similar developmental stories. The convergence of four elements was required to produce these places: the promise of abundant new water supplies, easy access to new interstate highways (or an occasional equivalent), relatively inexpensive (usually agricultural) land not far from a modest city, and a determined rural leader.

Transformation of Nature (New York: Penguin Books, 1986) and Donald Worster's *Rivers of Empire: Water, Aridity, and the Growth of the American West* (New York: Pantheon Books, 1985). Another important synthesizer for me was William Cronon's *Nature's Metropolis: Chicago and the Great West* (New York: W.W. Norton, 1991), which made me think more critically about the symbiotic relationship between the development of agrarian areas and cities. Other helpful works included John B. Wright's *Rocky Mountain Divide: Selling and Saving the West* (Austin: University of Texas Press, 1993); Blake Gumprecht's *Los Angeles River: Its Life, Death, and Possible Rebirth* (Baltimore: Johns Hopkins University Press, 1999); Peter Wiley and Robert Gottlieb's *Empires in the Sun: The Rise of the New American West* (New York: Putnam, 1982); Bradford Luckingham's *Phoenix: The History of a Southwestern Metropolis* (Tucson: University of Arizona Press, 1989); Robert A. Caro's *Years of Lyndon Johnson: Volume One, The Path to Power* (New York: Knopf, 1982; cited hereafter as *The Path to Power*); and George Sibley's "The Desert Empire," *Harper's* (Oct. 1977), 49–68. Interviews with Scott Higginson (Aug. 2002), a Phoenix water consultant who served on the Las Vegas city council, and Ken Hull (Aug. 2002), former manager of Del Webb's first Las Vegas Sun City development, were extremely useful. But fundamentally my ideas emerged from my research on Williamson County.

[8] Phoenix's first canals were built by the Hohokam Indians before 600 A.D., rediscovered in 1867 by a prospector named John W. "Jack" Swilling, and repaired and extended by the federal government after a coalition of farmers lobbied hard for Roosevelt Dam, which essentially created Phoenix and its surrounding orange-grove economy. Reisner, *Cadillac Desert,* chapter 8, and Luckingham, *Phoenix,* chapter 1, amplify this sketch, as does Lori K. Baker's "Funneling Water to the Desert," *Arizona Highways* (Feb. 2003), 19–22.

Introduction

Fifty years ago, even thirty, few imagined Williamson County becoming Austin's fastest-growing flank—certainly not the nation's most sophisticated planners in the Army Corps of Engineers or Texas's regional water and transportation gurus. A handful of good ol' "country boys" in Williamson County did. Incredibly, to a degree few people knew or appreciated, these "boys" made it happen by manipulating the federal bureaucracy and national politicians.

As a child growing up in Williamson County during the 1950s, I loved its sweeping landscapes: the rolling black farm land of the east, where "King Cotton" made Taylor an economic and political powerhouse; and the chalky hills of the west, habitat for ranchers, livestock, and cactus. Driving from Taylor to the county seat in Georgetown, I liked watching for the line where the land totally changed. To my eyes, that line (really, a succession of ridges) looked like a jagged barricade on the western horizon rising, in quick stages, two hundred feet or more above the flat land below. This was the Balcones Escarpment, which marked the division between east and west Williamson County—and between the American South and the American West. Everyone instinctively knew this, though no one spoke of it. On the county's eastern blacklands, it was said, a competent farmer could make a handsome living on topsoil forty to sixty feet thick. "Black gumbo," my mother called it. West of the escarpment the vistas were grand but the soil was thin, the living hardscrabble.

The San Gabriel River slices through Williamson County on its journey south and east to join one of Texas's most important rivers, the Brazos, which empties into the Gulf of Mexico. Long before Europeans arrived, the San Gabriel watered Indian tribes, most recently the Tonkawas, who marked favored pecan "bottoms" with bent pecan saplings, a sort of early-day Michelin rating system indicating a promising combination of abundant springs, shelter, and game. A century after the Tonkawas disappeared, one can still see these markers, now old, arching over the river's native bottomlands.[9] Early

[9] For all practical purposes, Williamson County's Native American tribes had vanished by the time Anglo settlers arrived. Presumably most of them died off, or moved on, after being devastated by measles, smallpox, and other diseases brought by the first Europeans to Mexico and Texas. For an excellent study of this phenomenon, see Alfred W. Crosby, *Ecological Imperialism: The Biological Expansion of Europe, 900–1900* (Cambridge,

Road, River, and Ol' Boy Politics

east-west county roads followed the San Gabriel and its chief tributary, Berry Creek, switching from one bank to the other at low water crossings which served as de facto outdoor parks where people fished, picnicked, partied, canoed, camped, inner-tubed, and romanced. Stories about the county's history—and its families—turned on the river's unruly moments, especially the terrifying floods of 1913, 1921, and 1957.

During my school years in Georgetown, in the fifties and sixties, Austin felt light-years away. A trip to the state capital meant a journey to another world. Dallas and San Antonio might as well have been Paris. Within the county, the only towns that really counted were Georgetown, centrally located, and Taylor, over to the east. Georgetown, the government seat, was older, smaller, and considered "stuck-up" by its eastern rival. Unlike the rest of the county, it voted "dry" and was proud to call itself "city of churches."[10] Southwestern University, a small Methodist college, was the town's leading employer and cultural arbiter. During the Depression, when Southwestern couldn't meet its payroll, Georgetown's merchants kept it afloat, accepting university-issued "script" from professors and staff as payment for groceries and clothing.[11] Georgetown was overwhelmingly Anglo and Protestant, but the ranch patriarchs, who came to town every morning at six to drink coffee and swap stories at the L&M Cafe, were a fairly secular bunch—hard, irreverent, or raffish, depending on the occasion. These included Jay Wolf, "Fat" Kimbro, Roy Gunn, I. M. Hausenfluke, the Hawes brothers, and "Doc" Weir. Being the county seat, Georgetown kept many lawyers occupied, but only one local politician was well known outside Williamson County. He was County Judge Sam V. Stone, whose stature peaked in 1937 when he beat an ambitious young politico named Lyndon Johnson in

---

UK: Cambridge University Press, 1986), 94, 98–99, 196–216. For details about Williamson County's Native Americans, see Clara Stearns Scarbrough, *Land of Good Water: Takachue Pouetsu. A Williamson County, Texas, History* (Georgetown, Tex.: Williamson County Sun, 1973); and Ty Adams, "Archaeologists Hit Gold Mine at Site Near Florence," *Williamson County Sun* (Georgetown), Feb. 25, 2001.

[10] It was also known as "Mistletoe Capital of the World."

[11] Tyler Woods, "Southwestern Almost Says 'Finis'," *Williamson County Sun* (Georgetown), Sept. 30, 1998, p. D-7.

Williamson County, but not in the rest of the Tenth Congressional District, which sent Johnson to the U.S. House of Representatives.

Taylor, only seventeen miles away, was a geographical and social world apart. Rather than ranchers, it drew farmers to town on Saturdays, like its satellite railroad towns, Granger, Bartlett, and Coupland. Taylor worshipped cotton, money, politics, beer, and barbecue. Its people formed a mosaic of immigrant groups and religions—Czechs, Germans, Swedes, Mexicans, blacks, and Anglos largely from the states of the Old South; Catholic, Protestant, and Jewish. It was clannish; fistfights often settled ethnic and racial disputes. Taylorites loved T-bone-steak stag dinner "blowouts" and partying at honky-tonks along "The Line," where most of the city's Mexican- and African-Americans lived.[12] A Taylor attorney, Wilson Fox, ran the Williamson County Democratic Party with such an iron hand that county voters routinely provided margins of victory in tight congressional elections. Their loyalty often provided the winning edge needed in elections involving Lyndon Johnson, Homer Thornberry, and J. J. "Jake" Pickle. After his retirement, Pickle half joked that investigators should have checked into one particular Williamson County voting box along with the infamous South Texas "Box 13" that propelled Lyndon Johnson into the U.S. Senate.[13]

Two decades before Johnson's ascension to the Senate, Taylor's Dan Moody was elected Texas's youngest governor. From that time until the early 1970s, Taylor wielded unusual power in Austin for a city of fewer than ten thousand.[14] One of Moody's closest advisers was

---

[12] "The Line" was the county's premiere party street where, in the words of one Taylor native, "one could buy anything," legal or illegal. Fort Hood made the strip off limits to its soldiers and, as Taylor's economic health deteriorated, so did The Line's. In the mid-1990s a grassroots organization called "Turn Around Taylor" marched against the crack houses that infested the street; Governor George Bush joined the effort, and the National Guard was called in to raze the abandoned buildings that housed the worst offenders. Today the street is largely deserted.

[13] Interview, J. J. "Jake" Pickle, July 3, 1998, Austin. He was referring to the voting box in Granger, Wilson Fox's hometown.

[14] Moody was a crusading county attorney who made his name in 1923 and 1924 by obtaining convictions against several Ku Klux Klan members in Williamson County. The sensational trials broke the Klan's domination of Texas government. Moody won the 1926 governorship of Texas as a reformist candidate, attacking corruption and racism

Richard Critz, another Taylor attorney who would serve on the Texas Supreme Court. One of Critz's daughters married Jake Pickle when Lyndon Johnson was a congressman and Pickle worked as Johnson's chief "leg man"; the other daughter married Thatcher Atkin, who became Georgetown's mayor and a "player" when Pickle entered the House of Representatives. Another key adviser of Governor Moody's was his boyhood friend and former Taylor law partner, Harris Melasky, whose clients included several of Texas's leading wildcatters.[15] Lyndon Johnson, too, leaned on Melasky for advice and financial help.[16] In 1941, when Johnson was making his first bid for the U.S. Senate, he ran out of money for radio advertising. A Johnson aide, John Connally (later Governor Connally), came to Taylor to beg Melasky for help. "How much do you need?" Melasky asked. "Twenty-five should do it," Connally said. Melasky wrote out a check for twenty-five thousand dollars and handed it over. Connally stared at the check, pocketed it, thanked Melasky, and quickly departed. He had been praying for twenty-five *hundred* dollars.[17]

Another of Taylor's political strengths was banker John M. Griffith, a close friend of Dewitt C. Greer's, the Texas Highway Department's commander in chief through the fifties and sixties. Greer conceived and built Texas's primary highway system and orchestrated the building of Texas's portion of the interregional highway system. Greer was a Texas version of Robert Moses, master builder of New York City, and Taylor, through Griffith, whispered in his ear.

I grew up steeped in tales of these characters and their exploits. My father, Donald Lee Scarbrough, edited and published the *Williamson County Sun*. He and my mother, Clara Stearns Scarbrough, came of age in Taylor. They started the *Taylor Times* in 1939, then sold it in 1949 after purchasing the *Sun* in Georgetown.[18] Mother wrote a

---

under Governor Jim "Pa" Ferguson and his wife, Governor Miriam "Ma" Ferguson. See Clara Scarbrough, *Land of Good Water*, 376–377, 380–386, and Ken Anderson, *You Can't Do That, Dan Moody: The Klan Fighting Governor of Texas* (Austin: Eakin Press, 1998).

[15] Anderson, *You Can't Do That, Dan Moody*, 9, 144–149.

[16] Caro, *The Path to Power*, 617.

[17] Interview, Tom Bullion, May 12, 2000. Bullion became Melasky's law partner in the 1950s and served for many years as Taylor's city attorney.

[18] They sold the *Times* to friend Henry B. Fox, who sold it to the *Taylor Daily Press*

definitive history of Williamson County's founding century, *Land of Good Water,* which came out in 1973. Dad relished the complexities of rural power politics and passed on that fascination to me.

In May 1962 I left Georgetown for good, or so I thought. Highway workers were pushing Interstate 35 into Williamson County, promising trips to Austin and Dallas at airplane speeds. Georgetown worried the interstate would "kill" the town with its "bypass" routing. A dam controversy had raged for years. Should the Army Corps of Engineers dam the San Gabriel River at Laneport in the eastern end of Williamson County, flooding more than two hundred "black gumbo" farms owned mostly by Czech-Americans? Or should the dam be built on the site west of Georgetown favored by my father, and by most people in the county (except Wilson Fox, the Democratic strongman)? The contending factions had canceled each other out; it appeared neither dam would get built. I was young. I paid little attention.

Sixteen years later, in 1978, I came home to edit the *Sun.* Much had changed. The Corps of Engineers was building two dams, not one, with a third planned. Army engineers had driven Czech farmers off their land; once robust farm towns in eastern Williamson County were sickly or dying. A few ranchers in west Williamson County had gotten rich—for the times—by selling their land for the new lake. Interstate 35 had touched off an economic boom. Round Rock, a village of two thousand when I left in 1962, was now Austin's first suburb of consequence, host to a Westinghouse Electric turbine manufacturing plant, several high-tech computer industries, and fifteen thousand people. It was Williamson County's leading city. The county population had doubled in size, making it the second-fastest-growing county in Texas.[19]

Through the eighties and early nineties as I directed the *Sun,* many questions nagged at me. Precisely what had thrust the county into overdrive? How had agriculture lost its hold on Williamson County? How had Round Rock, of all places, become the leader of the county's sudden pursuit of modernity? Was Round Rock's economic

---

in the mid-1950s. Don and Clara Scarbrough purchased the *Sun* in 1948 and sold it in 1986 to the author and her husband, Clark Thurmond.

[19] U.S. Census, 1980; Clara Scarbrough, *Land of Good Water,* 345–346.

Road, River, and Ol' Boy Politics

development success the result of Austin's proximity, or was there more to it than that? Had the coming of the interstate automatically rearranged Williamson County's economy and social patterns? Or did the two big dams, and the water they collected, make the difference? Would the shift from agricultural to suburban development have taken place without the dams or the interstate highway? Had the planners of these colossal public works projects, which dwarfed everything in Williamson County, understood what changes their projects would bring? Did the final configurations of these projects reflect decisions by federal bureaucrats who knew little about Williamson County and were focused on a theoretical "big picture"? Or, were the final flourishes on the lakes and the highway put there by "special interests"—local boosters and individuals—as some charged? If so, did it matter?

This book emerged from these questions and my desire to see if the answers might reveal useful lessons for fast-track development areas like Williamson County. In such places, I believe, residents risk becoming affluent ciphers, lacking that most fundamental American virtue—civic responsibility. The founding fathers believed that healthy local governments would guide and check a centralized federal government. Strong local governments (that is, communities) still function in this way. But when communities weaken, their ability to influence plummets. As an interstate-highway culture increasingly dominates our nation's economic and cultural life, our sense of "place," so intrinsic to the sense of local community, wavers. What "community" means to America has long been debated. When we talk about community, most of us imply that most people in a given place understand the context of the place they live. To put it another way, a healthy community values and builds on its history—environmental and cultural as well as political and economic. That is what makes its residents feel they belong, and that sense of belonging leads to participation in public affairs.

At the end of the twentieth century, Williamson County was roiled by conflicting responses to its swift growth, which in turn eroded feelings of belonging to any community. While population growth generally translated into economic strength, the county's cities struggled to maintain their identities and political cohesiveness.[20] The county's ability to press its needs in Congress, the Texas Legislature, even to get a hearing from the City of Austin, was hamstrung. A workable synthesis could not be found. Sometimes, local leaders pushing hardest for radical policy shifts were operating, as George W. S. Trow terms it, within the context of no context.[21] But if local leadership is to enhance Williamson County's fundamental state of being—its sense of community and place—its leaders must understand its history.[22] So must the hundreds of thousands of new Williamson County residents who keep moving in. This book was written, at least in part, to help bring that about.

I hope it also will encourage thinking about the importance of place—a large way of saying environment, which I define, broadly, not only through ecological and morphological signposts, but also through cultural, ethnic, and economic characteristics. Other American counties or regional entities, especially those considering large-scale public works projects, might utilize this study, especially as global and national forces increasingly drive our economy while eroding the viability of small, individualistic towns through the loss of locally owned retail and service institutions to the "big boxes" and chains. A tolerable sense of place, I think, includes not only one's feeling for physical landmarks, both environmental and architectural, but also an understanding of a place's economic, ethnic, and cultural context through history. With this in mind, I outline the federal engineering experiments that reinvented Williamson County.

Federal government forces wanted two simple things: a superhighway and a dam. The road was U.S. Interregional Highway 35. It would

---

[20] In many instances, the lion's share of economic growth favored national corporations, which leached profits from local retail or industry, often supplanting locally owned business. In addition, local city and school governments found themselves hard-pressed to keep up with the expense of growth. For much of the last decade, for example, several Williamson County school districts have been forced to build a new elementary school every year to keep up with burgeoning growth.

[21] Two books especially influenced my thinking about this issue of context: George W. S. Trow, *Within the Context of No Context* (New York: Atlantic Monthly Press, 1997); and Neil Postman, *Amusing Ourselves to Death: Public Discourse in the Age of Show Business* (New York: Penguin Books, 1986).

[22] James Kuntsler, *Home From Nowhere: Remaking Our Everyday World for the Twenty-first Century* (New York: Touchstone Book, Simon & Schuster, 1998).

Road, River, and Ol' Boy Politics

eventually link Mexico with Canada, create new suburban markets, transform the trucking and railroad industries, and replace Williamson County's traditional economic, cultural, ecological, demographic, and political patterns with new ones. After a thirty-year battle, one proposed dam became two, making the 112-mile-long San Gabriel one of the shortest non-urban rivers in the United States so rigorously engineered. The dams provided the water that fueled the county's growth binge. But it was the critical combination of interstate and dams that allowed the county to grow so spectacularly. Neither the interstate nor the dams could have done it alone.

Bringing dams to Williamson County required a special brand of local abracadabra. Locating dams, it turned out, was not the "scientific" and apolitical process that Army engineers or United States congressmen liked to profess. Nor were the Texas Highway Department and the U.S. Bureau of Roads immune from judiciously expressed political pressure. Locating dams or interstates to benefit certain interests required the determined intervention of at least one local person who could manipulate the process at either the political or the bureaucratic level. That person might be charming or obnoxious, virtuous or greedy, but he had to be effective or he did not succeed. Men of this type (and this is almost exclusively a story of powerful Anglo males) were crucial to the growth of the Dry Sun Belt's fledgling towns, agricultural empires, and supersuburbs, though historians have tended to minimize their importance.[23]

Eventually I came to understand that Williamson County's story is not unique. Instead, its leitmotiv ripples through America's fertile agricultural valleys, especially those in the Dry Sun Belt. I thought that this pattern deserved close scrutiny—a "thick description," as anthropologist Clifford Geertz put it, of the political, social, economic, and environmental details that pile up until their weight shapes history.[24] While interpreters of the West have proposed various

[23] Local histories, such as those in my seventh and eighth footnotes, have delineated these men's impacts on their fledgling cities, from Jack Swilling of the Salt River Valley to the *Los Angeles Times*'s Otis Chandler, who played a pivotal role in the seizing of Owens Valley's water and in creating modern Los Angeles and Imperial Valley. But as a rule, historians have tended to ignore the local "father" of the story.

[24] Clifford Geertz, *The Interpretation of Cultures: Selected Essays* (New York: Basic Books, 1973), 5–28.

theories on how the modern American West was born—the hydraulic society as espoused by Karl Wittfogel and expanded on by Donald Worster, Marc Reisner's and Alfred Maass's unholy combination of Congressional "pork" and a power-hungry federal bureaucracy, the military industrial complex's embrace of the West during World War II, the creation of supersuburbs by the interstate highway system—the theories barely note the actions of the mostly unknown farmers, ranchers, and small-town businessmen who leveraged radical change for their communities.[25] The French philosopher Henri Lefebvre's notion of how people use space—by farming it (daily use), engineering it (abstract), or making a movie (mythologizing)—helps get our heads around what highway and dam planners do, as opposed to Czech farmers, but it doesn't get us far in Williamson County.[26] One might argue that Williamson County fused dying Old South dreams and Old West longings until the new dams and road eliminated the remnant cultures, which were replaced by a postmodern restless placelessness in which a corporation like Dell, whose workers assemble computers in Williamson County, produces the only "real" space—of the cyber variety. But that would not suffice. While all these theories help explain the modern West, alone they lack completeness. The truth, I believe, is messier.

This work is divided into two parts. In Part I, "The River," I tell the story of the San Gabriel flood of 1921, which by some counts drowned as many as two hundred people; and detail how that flood for fifty years drove the local, regional, and national debate over damming the San Gabriel, finally resulting in the construction of two dams, a configuration long resisted by the Corps of Engineers. In Part II, "The Road," I focus on Interstate 35, from its origins in the German *auto-*

[25] See Worster, *Rivers of Empire*; Reisner, *Cadillac Desert*; Arthur Maass, *Muddy Waters: The Army Engineers and the Nation's Rivers* (Cambridge, Mass.: Harvard University Press, 1951); Gerald D. Nash, *American West Transformed: The Impact of the Second World War* (Muncie: Indiana State Press, 1985); Tom Lewis, *Divided Highways: Building the Interstate Highways, Transforming American Life* (New York: Viking, 1997); and Mark H. Rose, *Interstate Express Highway Politics, 1941–56* (Lawrence: Regents Press of Kansas, 1979).

[26] Henri Lefebvre, *Production of Space*, trans. Donald Nicholson-Smith (Cambridge, Mass.: Blackwell Publishers, 1991), 64.

bahn to its 1968 completion through Williamson County. I show that the interstate's original route would have swept near Taylor, then the county's leading city, but was mysteriously shifted fifteen miles to the west, overturning the county's economic and political power structure; and how a sequence of interchanges meticulously planned at the national level was altered by the relentless lobbying of county property owners and town leaders. An epilogue brings the reader up to date with a survey of the consequences—to individuals, cities, and the region—of the San Gabriel River dams and Interstate 35.

Ultimately this is a story of several "elite" groups struggling for power: the U.S. Army Corps of Engineers and the U.S. Bureau of Roads versus Williamson County's contending "elites" in Taylor, Georgetown, and Round Rock versus Texas's elite rice planters, Dow Chemical, and the Texas Highway Department itself, whose clout could make Lyndon Johnson jump. But Mother Nature held the ultimate trump card, casting flood (or drought) upon plantation owners, engineers, and fieldhands alike, turning their schemes to naught.

Williamson County's bulldozer makeover is a story about the effects of bureaucracy and political maneuvering on the pastoral environment and communities of people. It is a tale of federal power tempered by local desires and of a chain of unintended consequences. For example, the Army engineers intended to dam the San Gabriel River to help farmers—at least, that was how they sold Congress on the idea. But, bittersweet irony, the dams they built literally put the county's most productive farms under water. This work cautions against the "big" solutions that our society so often embraces at a huge cost to our local health.

This "thick description" of Williamson County's road and river projects finds a surprisingly high degree of give and take—horse trading, really—between big government and local forces, elite and otherwise. Time and time again, local firebrands or quiet lobbyists forced federal and state planners and politicians to abandon, redesign, or postpone pet projects. These small-town citizens bent national policy toward what they believed were Williamson County's best interests. Some were "boosters"; some angled for personal gain; several possessed extraordinary vision. I am tempted to label them local heroes.

These rural leaders had one thing in common: they anticipated the future far more accurately, and with more panache, than did the professional planners. In the end, the projects that were built—Interstate 35 and the San Gabriel River dams—represented a convergence of interests among the "bosses" of a small rural county, state and regional power brokers, and the nation's most glorified engineers. Those compromises led directly to Williamson County's astonishing metamorphosis at the end of the twentieth century, one that is mirrored across America's Dry Sun Belt.

LAY OF THE LAND

As one contrasts the civilization of the Great Plains with that
of the eastern timberland, one sees what may be an institu-
tional *fault* . . . roughly following the 98th meridian. At this
fault the ways of life and living changed.

WALTER PRESCOTT WEBB, *The Great Plains*

Seventy million years ago, earthquakes ripped up and down the
belly of Texas, which lay at the bottom of a shallow sea. The
quakes produced a line of faults along which volcanic mountains thrust up and westward, spilling ancient seabeds down the
mountains' eastern scarps. The faulting stopped some twenty-one
million years ago, and the sea retreated southeastward to become the
Gulf of Mexico. The Balcones Escarpment remains to remind us of
those ancient geological upheavals.[1] Rivers running through the hills
and cliffs west of the fault zone have dumped billions of tons of rich
clay silt on the land below, creating a motherlode of fertility in some
places twelve hundred to fifteen hundred feet deep.[2] Thus was created
the Blackland Prairie—"Black Waxy" to early Anglo-American settlers—which pioneering Texas politicans and farmers recognized as
their most important natural resource. "Nearly every foot of its area is
susceptible to a high state of cultivation, constituting one of the most
extensive continuous agricultural regions in the United States," a

---

[1] Some geologists believe this process stopped twenty-four million years ago. See
Austin American-Statesman, Sept. 26, 1999, p. K-3.

[2] *Austin American-Statesman,* Sept. 26, 1999, p. K-3.

scholar wrote in 1889.[3] A farmer's wife said it more pungently: "You can make a better living by accident on the blackland than you can by trying on sandy soil."[4]

Long before farmers arrived, a tiny band of shipwrecked Spanish sailors led by Álvar Núñez Cabeza de Vaca escaped from their Karankawa captors and struck northwest in 1535 from the Gulf of Mexico, meandering across Texas in an effort to reach their compatriots.[5] After marching for days across a rolling grassy prairie, they saw a serpentine escarpment, the first high country the Spaniards had seen in Texas. Its dusky green rim, which stretched three hundred miles across the western horizon, reminded them of a theater balcony, *balcones* in Spanish. And so it was called the Balcones Escarpment.[6]

Nor could Cabeza de Vaca's men and subsequent explorers fail to notice a thick band of rich black soil hugging the eastern side of the rift's spine. This fabulous earth—the Blackland—thinned out, after thirty miles or so, into a less fertile and much larger bench of sandy-land prairie.[7] West of the Balcones Escarpment they found a massive shelf of honeycombed limestone, covered by the thinnest layer of caliche soil and stretching a hundred and fifty miles, forming what we know today as the Edwards Plateau. The ancient faulting that created the escarpment produced fissures through which thousands of springs seeped, feeding rivers that had cut deep canyons through the upland side of the escarpment. East of the escarpment, the rivers sprawled

---

[3] Robert T. Hill, "Roads and Material for their Construction in the Black Prairie Region of Texas," *Bulletin of the University of Texas* (Dec. 1889), 18.

[4] Neil Foley, *The White Scourge: Mexicans, Blacks, and Poor Whites in Texas Cotton Culture* (Berkeley: University of California Press, 1997), 32. The quotation is from folklorist William Owens, retelling his mother's old saying.

[5] T. R. Fehrenbach, *Lone Star: A History of Texas and the Texans* (New York: American Legacy Press, 1983), 23; A. Garland Adair and Ellen Bohlender Coats, *Texas: Its History* (Philadelphia: John C. Winston Co., 1954), 20–22.

[6] Clara Stearns Scarbrough, *Land of Good Water* (Austin: Williamson County Sun, 1973), 6, 8. The Balcones Escarpment was first named on Roemer's map in 1847, but apparently that had been the usage for many years. Scarbrough cites E. H. Sellards and C. L. Baker, *Geology of Texas* (2 vols.; Austin: University of Texas Press, 1934), II, 400–401.

[7] Foley, *The White Scourge*, 16; Scarbrough, *Land of Good Water*, 3, 6; Tom Fowler, "Geography Primer," *Williamson County Sun* (Georgetown) Sesquicentennial Edition, Sept. 30, 1998, p. A-4.

into a maze of streams that threaded through the rich fertile lowlands to the coast. Spain staked its first Texas missions on the Balcones at what became the city of San Antonio, tapping the San Pedro springs to feed a system of *asequias,* or canals, which watered the colonists' first farms and orchards.[8]

By chance, a good part of the Balcones Escarpment roughly follows the 98th meridian, where, historian Walter Webb wrote, "the region of assured rainfall ends and the arid region begins."[9] West of the Balcones lies poor soil and little water—ranch country typical of the American West. East of the Balcones is Texas's Farm Belt. West of the Balcones, one cannot farm successfully without irrigation. East of it, one assuredly can—and on one of the richest strips of soil in North America.

The Balcones Escarpment made another geological mark on the topography of Texas: it uncovered a pencil-thin landform called the Austin Chalk, wedged between the Balcones Escarpment and the Blackland Prairie. The Austin Chalk is composed of pressed layers of the hardened skeletons of primeval sea creatures. Their tiny fossilized bodies make a splendid footing for roads. From the time Europeans arrived, they followed the Chalk's firm path, giving rise to Texas's most important inland routes and cities.[10] The dozen or so cattle paths that collectively became known as the Chisholm Trail, the Missouri-Kansas-Texas railroad, State Highway 81, and finally, Interstate 35 followed the Austin Chalk and "made" San Antonio, Austin, Waco,

---

[8] *Final Environmental Statement: Laneport, North Fork and South Fork Lakes, San Gabriel River, Texas* (Fort Worth, Tex.: U.S. Army Engineer District Fort Worth, Texas, 1972), Section II, pp. 1, 9; J. R. Barnes, hydrologist, personal corresponence, Sept. 24, 1973, held by author; "Carved in Stone," *Austin American-Statesman,* Sept. 26, 1999, p. K-3; Fowler, "Geography Primer," *Williamson County Sun* (Georgetown) Sesquicentennial Edition, Sept. 30, 1998, p. A-4; Clara Scarbrough, *Land of Good Water,* 3–10 .

[9] William E. Smythe, *Conquest of Arid America* (New York: Macmillan Co., 1907), 21.

[10] E. Charles Palmer, "Land Use and Cultural Change along the Balcones Escarpment: 1718–1986," in *Balcones Escarpment, Central Texas,* ed. Patrick L. Abbot and C. M. Woodruff Jr. (San Antonio: Geological Society of America, 1986), 153–162; "Carved in Stone," *Austin American-Statesman,* Sept. 26, 1999, p. K-3; Fowler, "Geography Primer," *Williamson County Sun* (Georgetown) Sesquicentennial Edition, Sept. 30, 1998, p. A-5; Clara Scarbrough, *Land of Good Water,* 4

Dallas, and Fort Worth—all creatures of strategic placement along the border between the Balcones Escarpment and Blackland Prairie.[11]

The Blackland Prairie's thick "black gumbo" soils, while dear to farmers, resisted highway builders. Depending on the weather, its clays shrink or expand six to eight inches a year—a road engineer's nightmare. "In winter time and at other seasons of rainfall the rich clay soil is kneaded into a tenacious paste, through which even an empty vehicle can be pulled only with great difficulty. . . . The roads become sloughs of despond," wrote road geologist Robert T. Hill in 1889.[12] Nothing much has changed. One major north-south artery, Highway 95, and the M-K-T railroad linked a strip of farm towns that thrived in their time courtesy of the "black waxy's" cotton and corn yields, but as agriculture faded in economic importance, so did these cities.

Williamson County, a Rhode Island–sized rumpled rectangle, sits north of Travis County, the seat of the Texas Capitol in Austin. Straddling the Balcones Escarpment, Williamson is an 1848 spinoff of Milam County, originally a vast area claimed by the Spanish crown.[13] Like other Texas counties split by the fault line, Williamson County developed in its first century as a model of agricultural economy, mixing cattle ranching on the Hill Country west of the Balcones Escarpment and cotton and corn on its Blackland half.[14]

In the last fifty years, the old agricultural patterns have given way

[11] The Balcones Escarpment burrows underground north of Waco, but Blackland Prairie spreads most broadly around Dallas and Fort Worth, encompassing Collin, Grayson, Hunt, Fannin, and Lamar Counties. See Foley, *The White Scourge*, 16, and T. U. Taylor, "County Roads," *Bulletin of the University of Texas* (Mar. 1890), 5.

[12] Hill, "Roads and Material for their Construction in the Black Prairie Region of Texas," 18; Fowler, "Geography Primer," *Williamson County Sun* (Georgetown) Sesquicentennial Edition, Sept. 30, 1998, pp. A-4–5.

[13] Mrs. Jeff. T. Kemp, "Significance and Origin of the Names of the Rivers and Creeks of Milam County," 1929, TXC-Z Collection (Center for American History, University of Texas at Austin). Eventually, Milam County was carved into thirty-two Texas counties, including Williamson. Clara Scarbrough, *Land of Good Water*, 71, 75, 114–115.

[14] Clara Scarbrough, *Land of Good Water*, 3–5; Fowler, "Geography Primer," *Williamson County Sun* (Georgetown) Sesquicentennial Edition, Sept. 30, 1998, p. A-4.

Road, River, and Ol' Boy Politics

to new ones: surburbia's spread of subdivision-style "ranchettes" has replaced cattle ranching, though agriculture still rules the Blackland, with fields of maize and cotton augmented by small cattle herds grazing on "improved" (that is, seeded) pastures. From a stable agrarian population that reached nearly forty thousand in 1900 and then essentially froze for seventy years, since 1970 Williamson County's population has doubled, redoubled, and undoubtedly will double again. In the year 2000, it was one of the five fastest-growing counties in the United States, with an official 250,000 inhabitants. Projections show it reaching 850,000 by 2015.[15] More important, it increasingly drives key economic, transportation, social, and political trends in the Austin metropolitan market, since the central city is nearly "developed" out.

The county's first towns lined up on two routes: the earliest following the Balcones Escarpment on the chalk, north from Austin to Round Rock to Georgetown to Temple; and the second along the Blackland Prairie, where in 1876 and 1882 railroad lines were laid, linking the Blackland's copious crop harvests to Houston and San Antonio via Bartlett, Granger, and Taylor. The county seat, Georgetown, hugs the Balcones, as does Round Rock, whose population since the 1970s has burgeoned from twenty-eight hundred to seventy-eight thousand to dominate Williamson County.[16] The Blackland sprouted Taylor, the county's financial and political capital for nearly a century, along with Granger, Bartlett, Hutto, Thrall, Thorndale, and Coupland—all connected by two intersecting railroad lines and two intersecting highways, 79 and 95.

The Balcones Escarpment broke Williamson County into two parts, but the San Gabriel River stitched it back together, giving it its two most valuable geological assets: fecund soil and abundant springs. Without the San Gabriel River, the fertile lowland soil would not exist: the river washed eons of accumulated alluvia from the broken fault zone to form the Black Waxy. And without the hinging action of the earthquake epoch, the county's springs, and the river, could not

[15] U.S. Census, 2000.

[16] 1970 Census of the Population, May 1973, Vol. I, Part 45, U.S. Department of Commerce, 44–45, Texas, Table 10; Round Rock Planning and Development Department 2004 figure, City of Round Rock, Round Rock.

have been created. It is all of a piece. The escarpment begat the springs and the river, and these begat the soil that lured would-be farmers from Germany, Sweden, and, most especially, from Moravia and Bohemia. The soil and springs made possible the first Williamson County— agrarian Williamson County—which lasted a century and still dominates the landscape on the county's eastern end. When white men first arrived in Williamson County, the springs numbered in the thousands; as late as 1960, two hundred bubbled and dripped within seven miles of Georgetown.[17] These springs meant good water for prospective farmers, ranchers, and townspeople. It was not for nothing that the now-extinct Tonkawas, who hunted bison in Williamson County long before the Spanish surveyed it, called their home *takachue pouetsu,* land of good water.[18] And long before even the Tonkawa, roughly 13,500 years ago, an ancient people dwelled up and down the face of the Balcones, taking advantage of all those springs. Think of Williamson County as a tasty archeological sandwich, layering one culture upon another.[19] Rains came to the *takachue pouetsu* in quantities that satisfied Clovis people, the nomadic Tonkawa, and the county's earliest European farmers—thirty-two inches on average, a little higher on the eastern Blackland Prairie, lower to the west. There were years of scant rainfall, but sometimes it rained so hard that rivers and creeks spread out for miles, thickening the rich silt of the alluvial plain.[20]

The permeable limestone of the county's western half masks hundreds, perhaps thousands of caves, natural pipelines for underground rivers. This makes for picturesque panoramas of live oak, cacti, and bluebonnet wildflowers—a landscape inspiring the image of mythic Texas—but it is poor country for farming or profitable ranching. Most settlers avoided it if they could, preferring the San Gabriel Valley's rich alluvial delta. It was the county's eastern farmland that spawned wealth. In the twenties, Taylor boosters dubbed their city "the largest inland cotton market in the world," erecting a sign on the

[17] Clara Scarbrough, *Land of Good Water,* 5.

[18] Ibid., 25.

[19] Interview, Sam Gardner, Austin, Apr. 11, 2005. A University of Texas archeological team has been working in northwest Williamson County near Florence for several years.

[20] *Final Environmental Statement: Laneport, North Fork and South Fork Lakes, San Gabriel River, Texas,* Section II, p. 10.

Road, River, and Ol' Boy Politics

City Hall lawn to that effect. They had reason. The city of six thousand sat snugly in the heart of Blackland Prairie. A major switching point for two national rail lines, Taylor linked Dallas, Austin, San Antonio, and Houston. In 1915, oil fields were discovered nearby, and another was struck in 1930. Family fortunes were made and Taylor boomed, giving it political connections and economic power that pulsed into the 1960s and straight to the White House.[21]

To European peasants of the mid-nineteenth century, often on the point of starvation and embittered by centuries of working other people's land, Texas's Blackland Prairie, which they heard about through relatives' letters, newspaper articles, and proselytizing religious leaders, sounded like heaven on earth. Williamson County's Black Waxy attracted German, Wendish, and Swedish families yearning to till soil they owned, but the Europeans who most definitively stamped their imprint on the new land came from Bohemia and Moravia, lands comprising the "Czech" portion of Czechoslovakia after World War I. From the 1860s on, steamship companies ferrying Czechs across the Atlantic Ocean promoted the Blackland Prairie: "Texas soil: black topsoil six feet in depth," gushed one steamship-line poster at its office in Bohemia.[22] (Actually it was an understatement. In some areas of the Black Waxy, the topsoil descends sixty feet.) As rural life worsened in the Hapsburg Empire's westernmost lands, especially between 1880 and 1900, peasants sailed across the Atlantic, landed at Galveston, and rode the new rail lines into several Blackland Prairie "nodes," including Williamson County's Granger, a "seedbed" for immigrant Czech families, and Taylor, where by 1900 the Czech influence was "prominent."[23] Freed from the Hapsburg Empire's oppressive treatment, the Texas Czechs hired themselves out, saved money, and within a few years bought farms of one or two hundred

---

[21] Ruth Mantor, *Our Town: Taylor* (Taylor: First Taylor National Bank, 1983), 7; *Welcome to Taylor* (Taylor: Taylor Daily Press, 1994), 24; Foley, *The White Scourge*, 29–32; "Taylor, the Biggest Little City," *Taylor Daily Democrat*, Jan. 24, 1923, reprinted in *Taylor Times's* Williamson County Centennial issue, Mar. 18, 1948; Clara Scarbrough, *Land of Good Water*, 302–303, 324, 328–330, 458–459.

[22] Clinton Machann and James W. Mendl, *Krásná Amerika: A Study of the Texas Czechs* (Austin: Eakin Press, 1983), 18–22.

[23] Ibid., 48.

In the first half of the twentieth century, Williamson County's wealth came primarily from cotton fields on the "Black Waxy" east of the Balcones Escarpment. *Courtesy* Williamson County Sun, *Georgetown.*

Granger, "The Paved Street Town," surfaced its streets in 1912, the first Texas city smaller than five thousand to do so. This 1931 photograph was taken by Granger's Jno. Trlica, a renowned Texas–Czech photographer. *Courtesy Photography Collection, Harry Ransom Humanities Research Center, University of Texas at Austin.*

The Czech influence in eastern Williamson County was strong. *Nasinec*, the newspaper of the Catholic Czech Church of Texas, started in Taylor in 1916 and continues to be published in Granger. It is the only Czech-language weekly newspaper in the United States, edited by Joe D. Vrabel. *Courtesy of Williamson County Sun, Georgetown.*

acres on the fabled Blackland Prairie. Their farms became models of intensive cultivation, mixing row crops, livestock, grapes, and truck gardening. Every member of the family labored for the collective enterprise; they succeeded beyond their wildest expectations.

In Texas, almost all Czech immigrants came from the *chalupník,* or "cottager" class. In Europe, *chalupníks* owned tiny homes but no land, working landlords' fields for shares of the crop. Locked in debt and hunger by a corrupt feudal system, a *chalupník's* greatest desire was to own land. To a *chalupník,* land ownership would not only lift his family from poverty to economic security, but would also convey upon the family the political and social respectability reserved in Europe for royals and aristocrats. This accruing collection of benefits led immigrating Bohemians and Moravians to an intense sort of "land worship," a "reverence" for the soil that is hard to exaggerate and difficult for modern Americans to comprehend.[24]

[24] Robert L. Skrabanek, "The Influence of Cultural Backgrounds on Farming Patterns in a Czech-American Rural Community," *Southwestern Social Science Quarterly,* 31 (1951), 258–266; Machann and Mendl, *Krásná Amerika,* 74–75.

Twins Anton and Henry Martinka and their older sisters, Mary Koska and Rosemary Martinka, were photographed by Jno. Trlica in 1927. They were children of a prominent Granger farming family. *Courtesy Photography Collection, Harry Ransom Humanities Research Center, University of Texas at Austin.*

On the Blackland, these new Americans gained what they most desired—freedom from the oppression of feudalism, economic security, and the ability to recreate their traditional cultures in a new land. On Williamson County's Blackland Prairie, they made Granger one of the "most Czech" of Texas towns, driving its robust growth after 1882, when it was laid out as a railroad depot stop; founded three villages—Friendship, Machu, and Moravia—all near Granger on the San Gabriel River's northern flank; and became merchants, mechanics, and a potent voting bloc in Taylor, as that city strove to overtake Temple as the premiere city between Waco and Austin.[25]

[25] Machann and Mendl, *Krásná Amerika*, 47–48. Maps of Williamson County's early towns show three separate Czech settlements near Granger, but Clara Scarbrough's *Land of Good Water* reports that Machu and Moravia were the same. See p. 438.

Where they clustered, Czechs founded schools, churches, fraternal organizations, mutual insurance societies, and newspapers.[26] *Nasinec,* the official organ of the Catholic Czech Church of Texas, was printed in Taylor from 1916 until 1937 and then moved to Granger, where it continues to be published. It is America's only remaining Czech-language newspaper.[27] As a distinct people, the Czechs emphasized group solidarity and economic success, and they found in Williamson County's Blackland Prairie their Promised Land. Though waves of Germans, Swedes, Wends, Mexicans, and Anglo-Americans (largely from Tennessee, Alabama, and Illinois) also farmed the Blackland, some accumulating more wealth than their Bohemian and Moravian neighbors, it was the Czechs who most successfully created a culture in Williamson County that combined their traditional social and religious life with their aspirations as new Americans.[28]

Twenty miles to the west, on the other side of the Balcones divide, a few hardy Anglo-Americans (most of them stemming from Irish, English, or Scottish origins) started drifting over the northern ridge of Stephen F. Austin's Mexican land grant colony in the 1830s. Remnants of Tennessee's planter class, they settled along the creeks and river valleys and the Double File Trail, a trail on the Austin Chalk blazed by Delaware Indians in the 1820s. In 1838, Swedish immigrants arrived; a decade later, an extended Illinois Anglo-American family encouraged other Yankee friends and relatives to follow. The wealthiest among them established farms on the Blackland or built stage stops at Brushy Creek (which became Round Rock), Bagdad (which became Leander), Liberty Hill, and Towns Mill, which vanished after a flood. Others scratched out plaintive livings on pockets of fertile river bottom along the San Gabriel River's two western forks and along Brushy and Berry Creeks. Mostly these pioneers barely clung to their land, calling themselves stockmen, farmers, or mustangers.[29] In 1848, Williamson County was carved from Milam County, and

[26] Barbara McCandless, *Equal Before the Lens: Jno. Trlica's photographs of Granger, Texas* (College Station: Texas A&M University Press, 1992), 4.

[27] Clara Scarbrough, *Land of Good Water,* 220, 222.

[28] Machann and Mendl, *Krásná Amerika,* 4–7, 9–18, 48.

[29] Clara Scarbrough, *Land of Good Water,* 72–87.

Georgetown became its county seat—the result of an enterprising land "boomer" with close ties to the Texas Legislature.[30]

After the Civil War, hundreds of Williamson County stockmen, hardened and sharpened by their military experience, joined an entrepreneurial wave that made cattle barons out of a few, land-poor cattle ranchers out of the rest. They did this by rounding up enormous herds of wild longhorn cattle that had multiplied during the war and driving them over dangerous trails to Kansas, where the beeves were shipped to Eastern markets. John W. Snyder and his brother Dudley H. Snyder, who made their fortunes in Williamson County, were prime examples of the type. As a young boy John cut his teeth as a trader on a wagon trip to Round Rock, selling twenty-five bushels of apples he had purchased in Missouri. After acting as an army supply agent during the Civil War, John and his two brothers amassed enormous ranch holdings in Williamson County, Colorado, and Wyoming. Dudley testified before Congress for the cattle industry and John was pictured on a 1886 postcard bearing the portraits of five "Cattle Kings." After his retirement, John Snyder generously supported the then-infant Southwestern University and helped establish Georgetown's First Methodist Church.[31]

After the cattle drives ended in the 1890s and barbed wire fenced the free range, Williamson County drovers like Willis Thomas Avery, George Washington Cluck, and Greely Weir invested profits from the drives in ranches on the big limestone ledge west of the Balcones Escarpment.[32] By that time, old-line stockmen were subdividing their Blackland Prairie holdings to Czech and German farmers, who could wring bigger profits out of the fertile soil by growing cotton than stockmen could raising beef. But the ranches established on the quirky limestone west of the Balcones Escarpment "wore out" rather

---

[30] Ibid., 124–125. The land "boomer" was George Washington Glasscock, who owned virtually all the land that would become the county seat. Georgetown, then, became Glasscock's namesake and made his fortune. Oddly, the man for whom the county was named, Robert McAlpin ("Three-Legged Willie") Williamson, another early land "developer," never lived in Williamson County.

[31] Clara Scarbrough, *Land of Good Water*, 200–203.

[32] John Graves, *Goodbye to a River* (New York: Alfred A. Knopf, 1983), 86, 128, 132, 215–216.

quickly. The cattle stripped the native buffalo grass and bluestem from the thin topsoil, which washed down the San Gabriel River, adding to the deep Blackland Prairie deposits or flowing uselessly on to the gulf.[33] Still, the old cattle barons and their less successful country cousins, ranchers with thousand-acre spreads and a few dimes in their pockets, cut an heroic image within Williamson County's social firmament. Their social standing towered above that of the clannish German Lutherans and Czech Catholics and Brethren who cultivated modest farms to get the most out of every inch of fabulous dirt and were viewed by the Anglo majority as curiosities: tight with their money, clannish in their society, devoutly religious farmers who insisted on educating their children in private church schools and continued speaking German or Czech at home, drinking wine, making plum brandy, and celebrating European folk festivals in fancy dress. Compared to the ranchers—a masculine society of tight-lipped individuals who attended Methodist and Baptist churches in town where drink was forbidden, though not unknown—it all seemed strange and a bit threatening. The Germans and Czechs raised two and three generations of farmers, successfully passing down their old land-worship values to their children and grandchildren while producing a disproportionately large number of doctors, teachers, and lawyers. Outside their ethnic enclaves, they were often admired but rarely emulated. Deep down in their hearts, most people in Williamson County yearned to own a ranch—have a little of the cattle-baron glamour rub off on them.[34]

As Donald Meinig wrote of Central Texas in his pathbreaking cultural geography of Texas, "It was of course a contrast between a strongly rural, militantly independent, mobile, and aggressive people nurtured on the frontier, and a strongly community-minded people drawn from the rigidly ordered countrysides and villages of Europe

---

[33] John W. Snyder biography, "Cattle Kings" postcard, 1886, Vertical Files (Center for American History, University of Texas at Austin); *Testimony Taken by the Select Committee of the United States Senate on the Transportation and Sale of Meat Products* (Washington, D.C.: Government Printing House, 1889), Sterling C. Evans Library, Texas A&M University, College Station. (Copy in author's possession.)

[34] Clara Scarbrough, *Land of Good Water*, 195–213.

West of the Balcones divide, on the Edwards Plateau, Williamson County ranchers raised cattle, sheep, and goats. This 1947 photograph was taken on Roy Gunn's ranch where springs fed the San Gabriel River. This place now lies under Lake Georgetown. *Edward A. Lane Photo, Courtesy Pat Gunn Spencer, Georgetown.*

who came in groups and clung together as an alien minority."[35] He was describing the Anglo ranchers who dominated the Edwards Plateau west of Austin contrasted against the German burghers of Fredericksburg and New Braunfels, but his generalization perfectly fit Williamson County of the late nineteenth and early twentieth centuries. Meinig also observed that Central Texas "displays the full range of intercultural tensions which are so important a part of Texas life."[36] In Williamson County, those intercultural tensions—especially between Czech farmers and Anglo ranchers—played a significant role in the response of the citizenry at mid-twentieth century when two enormous public works projects—a superhighway and a "new" river—were proposed. The farmers and ranchers saw, albiet dimly and in different ways, that these projects would overturn their lives and lay the foundations for a new Williamson County, one they did not really want.

[35] D. W. Meinig, *Imperial Texas: An Interpretive Essay in Cultural Geography* (Austin: University of Texas Press, 1969), 54.

[36] Ibid., 123.

Road, River, and Ol' Boy Politics

In 1947 Roy Gunn sprays four Brahman bulls for heel flies. Edward Lane photographed the rancher at work for an article in *Country Gentleman*. *Courtesy Pat Gunn Spencer, Georgetown.*

In the late forties Roy Gunn herds 650 Angora goats into a stockade for inspection and medical treatment. *Edward A. Lane Photo, Courtesy Pat Gunn Spencer, Georgetown.*

The San Gabriel, Williamson County's only river, connects the county's two distinct parts. It rises in two forks from underground springs just west of the county line in Burnet County, and runs transparent over limestone riverbeds through the canyon country west of the Balcones Escarpment, shaping the cattle ranches and thick stands of cedar and oak that still anchor a landscape that is fast becoming a sea of suburbia. It cuts through limestone cliffs dripping with springs and dotted with wildflowers. Sometimes a flat piece of alluvial soil spreads out into a fan-shaped pasture or truck garden. The San Gabriel's two main forks, the North Fork and the South Fork, converge at Georgetown, absorbing a shorter Middle Fork. Before reaching Georgetown, the river drops a stunning 750 feet from its source forty-nine miles to the west, rivaling the Los Angeles River's journey from the San Gabriel Mountains to Los Angeles for spectacular, flash flood–inducing potential; after passing Georgetown its gradient flattens and the river lolligags its way through sixty-two miles of undulating coastal plain—the Black Waxy.[37] It ends just across the eastern boundary of Williamson County at a hamlet appropriately named San Gabriel, established in 1747 as the Spanish mission San Xavier.[38] From start to finish, the river's elevation falls sharply from fifteen hundred feet above sea level to 550 feet. Crossing the county, it deposits the precious silt that made Taylor and its satellites wealthy for a century. It picks up a dozen tributaries, the most important being Brushy Creek, before entering a confusing jumble of streams and joining the Little River in Milam County to the southeast, which shortly merges with the Brazos River, whose waters flow on to the Texas coast.[39]

Williamson County sits in one of those odd crooks of the world

[37] Tad Friend, "River of Angels," *New Yorker* (Jan. 26, 2004), 43.

[38] Kemp, "Significance and Origin of the Names of the Rivers and Creeks of Milam County." According to this researcher, the San Gabriel River was named San Francisco Xavier in 1716 by Spanish explorers Espinosa and Ramon. "The name San Gabriel seems to be an American corruption of San Xavier," she wrote, adding that Austin's original map, dated 1829, inscribed it "San Javriel," which became San Gabriel.

[39] *Final Environmental Statement: Laneport, North Fork and South Fork Lakes, San Gabriel River, Texas*, Section II, p. 10.

known to geologists as "weather makers."[40] The Balcones Escarpment stands at the western edge of a long smooth plain that tapers down to the Gulf of Mexico. When a tropical storm swirls out of the gulf, it roars for hundreds of miles across that smooth expanse until it meets the Balcones, the first piece of significant topography in Texas. According to a geographer who has made the Balcones his life's work,

> The area along the Balcones Escarpment is the locus of the greatest frequency of flood-producing storms in the United States. The air masses hit the escarpment and rise ever so slightly. It's already unstable, so that's all it takes to set it off. This particular place is a place that can really get you.[41]

Combine this pattern with a low-pressure system holding steady west of the Balcones Escarpment and storms of exceptional strength and duration can form. Such storms' effects are often magnified by the fact that rivers crossing the Balcones fall rapidly (in Williamson County about 550 altitudinal feet), until encountering the coastal plain, where a flooding river can breach its banks in a breathtakingly short time and spread out for miles. Williamson County saw the normally somnolent San Gabriel River behave this way in 1869, 1900, 1913, 1921, and 1957—most especially in 1921, when it killed at least 154 people in a devastating flood.[42]

---

[40] Another such place is Cherrapunji, on the southern slope of the Himalaya Mountains, which on June 14, 1876, got 40.8 inches of rain in one day. This tidbit appeared in the *Taylor Daily Press* on Sept. 13, 1921, a few days after it rained thirty-two inches in twelve hours at nearby Thrall, culminating in 38.2 inches within twenty-four hours—a national record that held until the mid-1990s.

[41] Interview, C. M. Woodruff Jr., July 1974, Austin.

[42] Letter, George C. Hester to Col. Walter J. Bell, Mar. 17, 1958, Senate Case and Project 1958, Box 639 (LBJ Library, Austin). Hester's brief was the last of dozens of reports from individuals and government bodies regarding the impact on Williamson County from floods on the San Gabriel River. The document ran 75 pages long, though it was unnumbered, and constituted Georgetown's strongest argument favoring dams on the western forks of the river. Hester was an outstanding professor of history and political science at Southwestern University, advisor to Texas governors and a U.S. Senate Finance Committee, a former member of the Texas Legislature, and former mayor of Georgetown.

Two decades after the 1921 debacle, an obscure real estate salesman made it his life's quest to dam the San Gabriel. As science and as literature his description of Williamson County's tendency to flood leaves something to be desired, but as political theater, it worked. For thirty years, in various forms, Owen W. Sherrill relentlessly repeated the same thing to whomever would listen—congressmen, senators, chiefs of the Army Corps of Engineers, and one president of the United States:

> With the waters from three rampaging rivers above George-town carrying 28 percent of the water coming from the highlands . . . [at] over 1,300 altitude . . . through Georgetown, 750, a great drop and is equal to twice the damage in its rampaging rush of three rivers. . . . The combined North, Middle and South Gabriels starts the damage and *should be stopped by dams above Georgetown*. This deluge . . . adds to those stiller waters below on the more level lands . . . where . . . the waters keep adding to make a four mile stream.[43]

During the 1921 flood, Sherrill claimed, the San Gabriel River ran twenty miles wide. That was the salesman talking, Sherrill's typical bombast. And yet, it was almost true. In that flood the San Gabriel River became one with its tributaries—Brushy Creek, Alligator Creek, Willis Creek, Pecan Branch, Mustang Creek, Bull Creek, Boggy Creek, Dry Brushy, and Sore Finger among them—and transformed the eastern third of Williamson County into a watery graveyard. On one malignant September night, ninety people drowned in that graveyard within an area roughly fifteen by twenty miles wide.

---

[43] Owen W. Sherrill, "A Supplemental Report to U.S. Army Engineers on Resurvey Dam Sites," circa 1954, Box 59 (Brazos River Authority, Waco).

Part One

THE RIVER

If there is magic on this planet,
it is contained in water.

LOREN EISELEY, *The Immense Journey*

ONE

THE FLOOD

It may well be that the river itself will have the last word,
after all.

FRANK WATERS, *The Colorado*

Out in the Gulf of Mexico, below the southern tip of the
United States, a hurricane stirred. On Wednesday, September
7, 1921, U.S. Weather Service watchers at Brownsville, Texas,
noted a "disturbance." As the storm rolled slowly inland, heading
north across three hundred miles of frypan-flat coastal plain, it first
seemed nothing more than a welcome autumn thunderstorm. Two
and one-sixteenth inches of rain fell at Corpus Christi. But on Thurs-
day, when the storm reached Texas's first line of hills at the Balcones
Escarpment, something unusual happened: an independent low-pressure system, parked over the Balcones and stretching to the Pacific
Ocean, sucked the gulf storm into its orb, magnifying and lengthen-
ing its life by days.[1]

It rained and it rained and it rained. Lightning laced cobwebs of
shimmering electricity across the sky. Thunder ripped the wet air.
Central Texas was awash in flash floods: Austin eventually reported
18.23 inches of rain and San Marcos was "entirely under water," but

---

[1] *San Antonio Express,* Sept. 11–16, 1921, p. 1; "Record Flood Hits Taylor," *Taylor Press,*
Sept. 10, 1921, p. 1; "Unprecented Rainfall Does Great Damage," *Williamson County Sun*
(Georgetown), Sept. 16, 1921, p. 1; Clara Scarbrough, *Land of Good Water,* 370–373.

the storm did its worst in Williamson County. There, Friday dawned a ghastly pea green, but the rain did not stop. Instead it drummed on. "It just kept raining harder and harder, all day long," said Margaret Tegge Stearns, a young mother of three who lived by a cotton gin on the San Gabriel River at Circleville with her husband Auburn.[2]

When it finally did stop raining, the next day, September 10, 1921, tiny Thrall on the eastern rim of Williamson County had registered 38.21 inches of rain in twenty-four hours, a national record that would not be broken for seventy years. Taylor recorded 23.11 inches during the same span.[3] But rain or no rain, flooding streets or no, on Friday morning, September 9, 1921, Taylor was focused on its prime purpose: getting the cotton crop picked, ginned, and shipped. It was money in everybody's pockets. Scanning copies of the *Taylor Daily Press*, businessmen devoured the latest cotton prices: forty-five points up that week. Since harvest time was upon them, farm owners were vying for cotton pickers, mostly Mexican migrants, along with African Americans who lived in Taylor and Granger. Hundreds, probably thousands, of "hands"—men with wives and small children—camped in tents or lived temporarily in shacks called "hand houses" along the San Gabriel's scenic banks, a short walk from the fields where they worked. They were huddled in those tents and hand houses when the rains came.[4]

Taylor's social set adored Ruth Mantor, the fourteen-year-old daughter of attorney H. C. Mantor. Ruth had inherited her father's dry wit and brilliant mind. On Friday evening, Ruth drove through the storm to a friend's house for a slumber party. The girls had hoped

[2] Interview, Margaret Tegge Stearns, June 1974, Taylor. Stearns was Clara Scarbrough's mother and the author's grandmother. Also see "Greatest Rainfall in Twenty Years," *Taylor Press*, Sept. 10, 1921, p. 1.

[3] Clara Scarbrough, *Land of Good Water*, 370–371.

[4] Interview, Billye Fulcher Cannon, Jan. 28, 2000, Rockdale; "Cotton Sells at 16 Cents in Taylor Today," *Taylor Press*, Aug. 29, 1921; Foley, *The White Scourge*, 1, 8, 10, 44–45, 80; Clara Scarbrough, *Land of Good Water*, 373–374; "She Heard 23 Die," *Rockdale Reporter & Messenger*, Jan. 22, 1998, p. 4; Clara Scarbrough, "Glories, Difficulties," *Williamson County Sun* (Georgetown), Sesquicentennial Edition, Sept. 30, 1998, p. 8; "Unprecedented Rainfall Does Great Damage," ibid., Sept. 16, 1921, p. 1; *San Antonio Express*, Sept. 13, 15, 16, 1921.

Ruth Mantor, approximately ten years after she witnessed the 1921 flood in Taylor. *Courtesy Gail Tharp, Austin.*

some boys would drop by, but the boys never materialized. Fifteen inches of rain had fallen on Taylor since early that day, and the rain kept sheeting down, harder than ever. At about 9:30 P.M., the "electric globes" at Mantor's friend's house flickered and died. Not long after, the telephone went dead. Taylor plunged into darkness. Nearly eighty years later, Ruth Mantor perfectly recalled the storm's fury: "It didn't sound like rain," she said. "It was a roar. It sounded like fire hoses turned on." Mustang Creek had spilled into Taylor's streets, swamping the city's electrical works. Donahoe, Willis, Possum, Turkey, Bull, Mustang, Boggy, and Brushy Creeks, and of course the San Gabriel were behaving similarly, expanding their girths until eastern Williamson County's lowlands were transformed into a shallow sea, punctuated by peninsulas and islands. The floods snapped telephone poles and telegraph lines, leaving most of the county isolated and uninformed

Auburn and Margaret "Peg" Stearns, the author's grandparents, set out in a Model "T" in 1913. A similar vehicle saved their lives when they escaped flood waters in 1921. *Courtesy Clara Stearns Scarbrough, Georgetown.*

about what was happening elsewhere. It took three days to restore electrical power in Taylor.[5]

Saturday dawned at the slumber party. "I had the family car, so when the water started running off the yards, I decided to drive out and take a look at Mustang Creek," Ruth Mantor said. "Normally, Mustang Creek is a little tiny rivulet you could have hopped over, but that morning, it was a mile wide. It had washed out all the pavement on Highway 79 and had taken out the railroad bridge. Steel beams were twisted like horseshoes around trees. Shacks and houses floated by. I saw dead cows float by. The water was a funny sort of milk chocolate color."[6]

---

[5] Interview, Ruth Mantor, Oct. 29, 1999, Taylor; "Record Flood Hits Taylor," *Taylor Press*, Sept. 10, 1921, and "Light and Power Again Restored," Sept. 13, 1921. The *Press* explained it was able to print its paper due to the heroics of a Texas Power and Light employee, who hooked up a special power generator for the press run.

[6] Interview, Ruth Mantor, Oct. 29, 1999, Taylor; Don Scarbrough, "Passing Glance," *Williamson County Sun* (Georgetown), Sept. 18, 1991.

The iron trestle bridge over the San Gabriel River at Circleville
was torn from its moorings after the Stearns crossed it. Jno. Trlica
photographed the bridge where it came to rest, three hundred
yards downstream. *Courtesy Photography Collection, Harry Ransom
Humanities Research Center, University of Texas at Austin.*

While Ruth and her friends watched the flooding from the relative safety of Taylor, in Circleville the Stearnses fled for their lives. All day Friday, cotton-gin owner Auburn Stearns had kept a close watch on the San Gabriel, wading every hour through the knee-deep black clay goop from his house to the river's bank. Every hour he returned, reassuring his wife Margaret, "Nothing to be scared about. That river's running clear water." That evening, though, he started on his trek to the river, but returned immediately. The river had surged over its forty-foot banks, crossed a road and was a foot high at the fence line in the Stearnses' front yard.

"Peg, run! Get the children and get in the car. We've got to get out of here fast!" Auburn shouted. It was a narrow escape. "Just minutes more and we wouldn't have made it," Margaret Stearns said decades later. "Pop was driving an old Model T, which was built high off the road. The road was covered with water, two feet deep and rising fast. If we hadn't been in that Model T, we would have been gone for sure." The car stalled on the iron trestle bridge that crossed the San Gabriel, but Auburn got it restarted. "We had the honor of being the last to cross that old bridge," Margaret said. It was swept three hundred yards downstream later that night.[7]

At Friendship, a Czech village east of Granger, eleven-year-old Billye "Bill" Fulcher struggled to control her terror. She and her six brothers and sisters lived with their tenant farmer parents, who rented a three-story house about three-quarters of a mile north of the San Gabriel River. The house perched on a hill. On a normal clear day, from her upstairs window, "Bill" could see the big native pecan trees of the bottomland along the river. The trees sheltered a line of hand houses for farm workers. "That night," she said, "we heard a sound like a freight train. That was the wall of water coming down. Then we heard this loud wailing noise. I was scared to death. My daddy knew what was happening. 'It's somebody caught in the water,' he said. 'But there's nothing anybody can do.' It was voices—human voices. I'll never, ever forget that sound, not as long as I live."[8]

---

[7] Interview, Margaret Tegge Stearns, June 1974, Taylor.

[8] "She Heard 23 Die," *Rockdale Reporter,* Jan. 22, 1998, p. 4; interview, Billye Fulcher Cannon, Jan. 28, 2000, Rockdale.

Road, River, and Ol' Boy Politics

Billye Fulcher, 11, second from left, with sisters Katye and Laura (with doll) and brother J.B. The photo was taken a few months before the 1921 flood. *Courtesy Margaret Fulcher Caffey, Rockdale, Texas.*

Five Mexican families, most likely refugees from the civil wars that had plagued Mexico for a decade, and two unidentified people, had fled their shacks on Jake Bowers's plantation to take shelter in a large stone house on the river bank that locals called the "Big House." Unfortunately, it could not shield them from the San Gabriel River, which that night rose more than forty feet—Billye Fulcher's "wall of

SECOND STREET SERVICE STATION
Magnolia Gasoline, Motor
Oils and Greases
Magnolia Service Guaranteed Maximum
Mileage

Taylor Daily Press

HUDSON, ESSEX AND CHEVROLET
CARS
Sold to Taylorite
Taylor Motor Company

The Combined United Press Wire Service and International News Service brings the News of the World to the Press Readers Daily

VOLUME VIII   TAYLOR, TEXAS, SATURDAY AFTERNOON, SEPTEMBER 10, 1921   NUMBER 230

TEXAS FLOODED—HUNDREDS PERISH

PROPERTY LOSS RUNS INTO THE MILLIONS

SAN ANTONIO REPORTS HEAVY LOSS OF LIFE AS RIVER GOES OUT OF ITS BANK

Mexican Section of the City Has the Greatest Loss

RECORD FLOOD HITS TAYLOR

Homes Wrecked and Bridges Washed Out as Streets and Creeks Turn to Rivers of Water

GABRIEL RIVER FORTY FEET UP STILL RISING

23.42 INCHES RAINFALL UP TO NOON TODAY

CITY COMMISSIONERS ESTIMATE TAYLOR'S STORM DAMAGE

GREATEST RAINFALL IN TWENTY YEARS

Think It Will Exceed $50,000 and Probably Reach $75,000

HUTTO HIT BY WIND STORM

AUSTIN REPORTS SERIOUS FLOOD; COLORADO RISING

Klu Klux Klan Comes to Rescue Taylor Sufferers

Ivey Foster, Secretary United Charities Receives Letter From Invisible Empire With Check for $500.00

Taylor, Texas, Sept. 10, 1921

Mr. Ivey Foster, Secretary and Treasurer,
United Charities, City:

Dear Sir: We are enclosing the sum of $500.00 to be distributed by you among the flood sufferers of Taylor and vicinity as your judgment directs. We suggest that you wait until all the facts are known to extend to real suffering before distributing this fund.

Yours for the benefit of the people affected by the recent flood,

TAYLOR KLAN NO. 117 OF THE INVISIBLE EMPIRE OF THE KNIGHTS OF THE KU KLUX KLAN.

JONAH REPORTED UNDER WATER

BRIDGES ARE DOWN NEAR GEORGETOWN

TELEPHONE CO. AIDS THE PRESS

On September 10, 1921, the *Taylor Daily Press* bannered news of the flood that killed more than two hundred Texans. A story on the Ku Klux Klan's rescue efforts commands the lower left corner of the page. *Courtesy Texas Newspaper Collection, Center for American History, University of Texas at Austin.*

water."[9] Inside, Pablo L. Quintanilla, a leader and clerk of the San Gabriel Mexican Presbyterian Church, Quintanilla's wife and their five children; Mr. and Mrs. Antonio de la Torre and their three children; Mr. and Mrs. Nicanor Gonzales; Mrs. Maria G. de Mendez and her three children; and several other horrified people felt muddy water bubbling up through the floorboards over their feet. Within seconds, water lapped at their knees, then their waists. They scrambled to tabletops, lifted babies to rafters, punched through roof shingles, and climbed to the roof where they clung, screaming for help. The river surged upward, an implacable monster. In all, twenty-three people died at the Big House. Eighteen of them were members of the San Gabriel Mexican Presbyterian Church.[10]

The tragedy of the Quintanillas and their friends multiplied throughout Williamson and Milam Counties. "Hundreds Perish," a banner headline in the *Taylor Daily Press* shouted the next day. In a week, the final toll of Williamson County deaths was set at ninety-two. In Milam County, just across the county line, sixty-four perished, but at least twenty people were never found. Many victims discovered in Milam County were thought to have drifted on the flood's surging waters from Williamson County. Almost all of the dead were Mexican migrant workers, or Mexican-American and African-American field hands trapped where they had camped along the creeks and rivers, as was the harvest practice in those days. What makes the loss especially poignant is how little was known about the victims—then or now. In a very real sense, the dark-skinned fieldhands who died in Williamson

[9] Interview, Billye Fulcher Cannon, Jan. 28, 2000, Rockdale; "Gabriel River Forty Feet Up Still Rising," *Taylor Press,* Sept. 10, 1921, p. 1; "She Heard 23 Die," *Rockdale Reporter,* Jan. 22, 1998, p. 4.

[10] "Flood Victims on Bowers Farm," *Taylor Press*, Sept. 16, 1921, p. 1; interview, Billye Fulcher Cannon, Jan. 28, 2000, Rockdale; "She Heard 23 Die," *Rockdale Reporter,* Jan. 22, 1998, p. 4. Austin Presbyterian Theological Seminary records show that the San Gabriel Mexican Presbyterian Church disappeared from church records after 1921. The first published report of the fate of the little church's parishioners came in the *Taylor Press* from a well-known evangelistic Presbyterian preacher, the Reverend Walter S. Scott. In describing events at the "Big House," I have borrowed from many stories of survivors of the 1921 flood. At dozens of farms in Williamson County, families clung to their rooftops or held on in treetops for two, sometimes three, days until the waters receded. The Quintanillas and their friends were not so fortunate.

Railroad workers pose on railroad tracks near Granger after the San Gabriel River peeled the tracks off their bed. *Courtesy* Williamson County Sun, *Georgetown.*

County represented the "stoop labor" so easily available to Texas planters. When the laborers and their families were swept away, unknown and unnamed, others, just as nameless, showed up to take their places.[11]

Property damage was staggering. A few miles east of Georgetown, the San Gabriel ripped a Katy iron-trestle railroad bridge from its moorings and flipped a locomotive and rail cars as if they were child's toys. At Circleville, the river pried up railroad tracks, contorting them into roller-coasters. Every bridge and low water-crossing in

[11] "County Death Toll Reaches 92," *Taylor Press,* Sept. 15, 1921; "River of Romance Turned into Torrent of Death," *San Antonio Express,* Sept. 14, 1921, p. 1; "Thorndale Reports 61 Dead," *Taylor Press,* Sept. 16, 1921; "Brazos and Nueces Are Rising," *San Antonio Express,* Sept. 15, 1921; "63 Bodies Are Found at Thorndale," *San Antonio Express,* Sept. 16, 1921; Report, George C. Hester to Col. Walter J. Bell, in anticipation of a Mar. 17, 1958, public hearing on the San Gabriel dams, Senate Case and Project 1958, Box 639 (LBJ Library, Austin).

Williamson County was destroyed or severely damaged, along with most county roads. Repairing them required going deeply into debt—something the county had never before contemplated. Officials estimated it would cost a million dollars to repair and replace public roads and bridges, not counting damage to private property. Farmers lost ten thousand square miles of precious topsoil. Hardly any livestock survived. Hundreds of homes had floated downstream or were beyond repair. County government drafted all Williamson County males to the "Herculean task" of burying mangled corpses and reconstructing bridges and roads. The stench of death, many workers said, was worse than they had experienced as soldiers in the trenches during the Great War in Europe.[12]

In the grim weeks after the flood, Williamson County fixed on one major remedy: dam the two main San Gabriel forks in the highlands west of Georgetown so the river could never again spread death and destruction through the county's heartland. It was an expression of collective will, cutting across all segments of the population. Williamson County pursued that mission until it eventually collided with powerful state and federal forces. Because of its experience in the 1921 flood, and its subsequent efforts to tame the river, Williamson County blocked the will of national politicians, the Army Corps of Engineers, and regional planners for thirty years. In the end, a compromise "solution" produced unexpected results. To a surprising extent, today's Williamson County is an outgrowth of the flood of 1921 and the local determination to correct the causes of that disaster.

[12] *Williamson County Sun* (Georgetown), Sept. 16, 1921; Correspondence, Owen W. Sherrill to Brazos River Authority, 1921 flood photographs, undated, Box 59 (Brazos River Authority, Waco); "Repair of Flood Damage . . . Prompt in Williamson County," *Austin Statesman,* undated, circa Sept. 18, 1921, "Williamson County Scrapbook" (Center for American History, University of Texas at Austin); *Taylor Press,* Sept. 13, 1921; Clara Scarbrough, *Land of Good Water,* 371–376; George C. Hester to Col. Walter J. Bell, Mar. 17, 1958, Senate Case and Project 1958, Box 639 (LBJ Library, Austin).

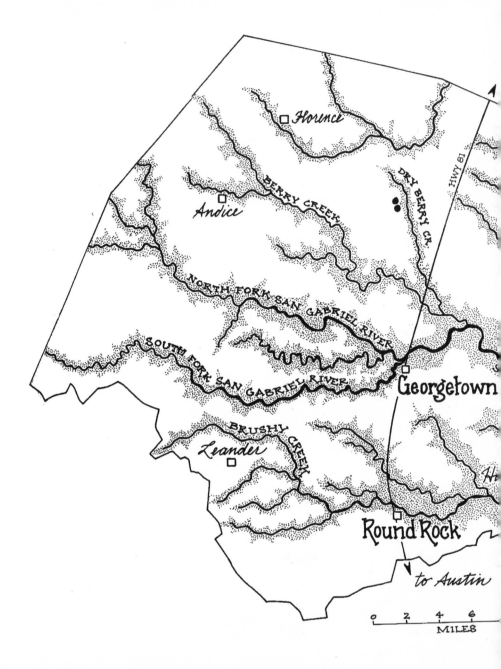

Darkened areas represent flood waters where they swamped Williamson County on the night of September 9, 1921. Every dot represents one person killed. The official death toll for Williamson and Milam Counties was 156, but at least 40 people were never found. Between 200 and 250 people were killed across Texas in the torrential storms of September 1921. *Drawn by Kristen Tucker Pierce.*

Williamson County
1921 Floodwater

□ Area covered by water

• Person killed

Total estimated dead in Williamson and Milam Counties: 156-180

27 died in the "Big House" and on the Bower Place

10 dead in San Gabriel

60 dead, 20 missing in Thorndale

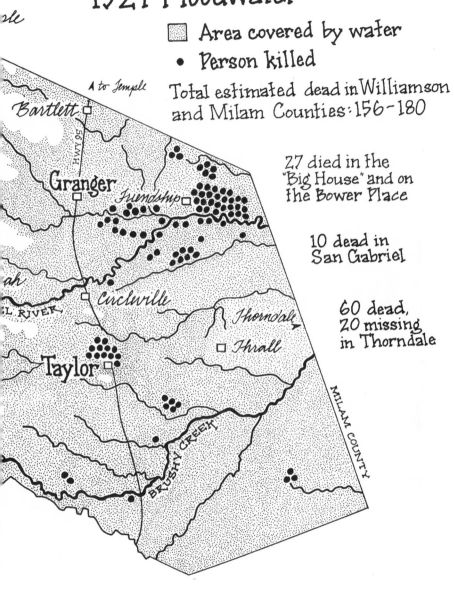

ale

A to *Temple*

Bartlett □

HWY 95

Granger □

Friendship □

Circleville □

Thorndale

□ *Thrall*

Taylor □

L. RIVER

ah

BRUSHY CREEK

MILAM COUNTY

TWO

FOUNDATION & DEVOLUTION

Running water was the great educator of the 1930s, and to a
lesser extent, dust. States could not hope to control erosion,
floods, or dust storms, acting alone.

OTIS L. GRAHAM JR., *Toward a Planned Society*

One fine day not long after the 1921 flood, so the story goes,
Georgetown newspaper editor John M. Sharpe took his childhood friend, Congressman James P. "Buck" Buchanan, on a
carriage ride west of town.[1] Sharpe wanted to break the San Gabriel
River's nasty habit of flooding, a habit violently displayed not only in
1921, but also in 1913, 1900, and 1893.[2] The two men rode four miles
west of the county seat through thick clumps of live oak, juniper, and
sumac until they reached a limestone cliff offering a postcard view of
the San Gabriel's North Fork.

"Here," Sharpe told Buchanan, "is a good place to build our dam."
The story may have been apocryphal, but it was still being repeated
half a century later.[3] Over the years, Buchanan was to become chair-

---

[1] John M. Sharpe Papers, Georgetown Public Library, Georgetown; "Outstanding
Citizens Get Awards at C-C Banquet," *Williamson County Sun* (Georgetown), Feb. 16,
1961.

[2] Letter, George C. Hester to Col. Walter J. Bell, Mar. 17, 1958, Senate Case and Project 1958, Box 639 (LBJ Library, Austin).

[3] Mark Mitchell, "Dam Story," *Sunday Sun* (Georgetown), Oct. 7, 1979. The anec-
dote came from Thatcher Atkin, a former Georgetown mayor during the 1960s, when
the county conflict over damming the San Gabriel was finally resolved.

man of the powerful House Appropriations Committee; Sharpe became a Georgetown patriarch—mayor, postmaster, and confidant of powerful Texans. In Sharpe's mind the dam towered on the North Fork of the San Gabriel River, protecting the county seat and the fertile soil to the east that brought prosperity to industrious farm families and the merchants catering to their needs.

In the early twentieth century, floods plagued Texas. Every few years, citizens of Fort Worth, Austin, and San Antonio watched helplessly as normally lackluster rivers boiled up and destroyed man's hard-won works. As settlers thickened across the rich Black Waxy, floods brought burdensome social and economic costs. Between 1891 and 1932, the Brazos River and its tributaries, including the San Gabriel River, killed 542 people and destroyed property then valued at fifty-four million dollars.[4] The Colorado and Trinity Rivers were almost as unruly. A network of flood-control dams seemed the answer to a problem threatening to cripple the state's economic development.

In 1929, Texas legislators created the Brazos River Conservation and Reclamation District, a pioneering watershed control effort aimed at stopping punishing floods that annually washed tons of precious fertile topsoil into the Gulf of Mexico. A coalition of rice farmers in the lower Brazos basin "took the lead" in shaping the district, it was said, but the idea enjoyed broad support.[5] Unfortunately, the Brazos District had no money, nor did the legislature provide any. It soon became clear that to accomplish its goal of controlling floods, the Brazos District, which later became the Brazos River Authority, needed to build dams that could produce and sell hydroelectric power. But the Depression turned engineering blueprints and high hopes into pipe dreams.

While the Depression battered millions of lives, it produced President Franklin Delano Roosevelt, a fervent believer in public works projects funded by the federal government on a scale never before contemplated. Among the most important of these projects were

---

[4] *Facts* . . . (Mineral Wells, Tex.: Brazos River Authority, 1956), Senate 1956 Case and Project File, Brazos River Authority, Box 1210 (LBJ Library, Austin).

[5] Oral history, William Robert "Bob" Poage, 1985, W. R. Poage Papers, Vol. I, 336 (Baylor Collections of Political Materials, Waco).

dams—to control floods, irrigate parched western farms, and provide water and power for growing cities. Until the thirties, when the biggest high dams on the globe—Hoover, Shasta, Bonneville, and Grand Coulee—were constructed, no significant concrete dam had ever been built in the United States. But when these four colossi reared over the Colorado and Columbia Rivers, a new pattern of dam building was established. Americans listened to Woodie Guthrie's song on the radio:

> That big Grand Coulee 'n Bonneville dam'll
> build a thousand factories f'r Uncle Sam . . .

> Don't like dictators none much myself,
> What I think is the whole world oughta be run by
> E-electricity . . .

All over the country political representatives clamoured for dams—hydropower dams, if possible.[6]

Texas's first hydropower dams were not built on the Brazos but on the Colorado River by a fledgling river authority which became the Lower Colorado River Authority. As Brazos watershed boosters gnashed their teeth in envy, the LCRA built two hydroelectric dams: Buchanan (1935–1937) and Marshall Ford (1936–1940), renamed for Austin mayor Tom Miller upon completion. The entrepreneurial LCRA quickly became a powerful utility company. The Colorado Authority also constructed Inks Dam (1936–1938) and Mansfield Dam (1937–1941), creating a water sportsman's paradise and land developer's dream in the poor chalky hills west of Austin.[7] The LCRA got these dams through the muscle of three powerful Texas congressmen: James "Buck" Buchanan, John Sharpe's old friend who chaired the powerful House Appropriations Committee; Joseph Jefferson Mansfield, chairman of the Rivers and Harbors Committee, which controlled dam projects throughout the United States; and, after Buchanan's death, Congressman Lyndon Baines Johnson of

---

[6] Reisner, *Cadillac Desert*, 158–161.

[7] Robert Cullick (ed.), *State of the River* (Austin: LCRA Corporate Communications, 1993), 12–13.

Road, River, and Ol' Boy Politics

Texas's Tenth District. Eleven days after his 1937 election to Congress, Johnson saved LCRA's Marshall Ford Dam (and its contractors, Brown and Root, Inc., which built its first bridge across the San Gabriel River after the 1921 flood) from certain bankruptcy.[8] While the Brazos River Authority found an effective champion in Congressman W. R. "Bob" Poage, the Waco attorney's fastidious, legalistic approach never could match Johnson's peerless ability to circumvent intractable problems, especially during the desperate early years of the state's watershed districts.[9]

For Williamson County, Sharpe's old dream of damming the San Gabriel moved toward reality on September 26, 1935, when President Roosevelt signed the Emergency Relief Appropriation Act, which gave planning funds to thirteen Brazos watershed dams and many other projects across the United States. Two of the Brazos dams were to be built west of Georgetown on the San Gabriel's North and South Forks—the North Fork dam precisely where John Sharpe had told Congressman Buchanan it should be built.[10] Newspapers bannered the story, "$30,500,000 Brazos River Project Becomes Reality With Signature of President Roosevelt," and Williamson County rejoiced.[11] But twelve months later, the Brazos plan had stalled, mired in a fracas over its board of directors' plan to sell hydroelectric power from Possum Kingdom Dam to a private utility company.[12] Finally, between 1938 and 1941, the Brazos District built Possum Kingdom Dam, which did produce hydropower and help coastal rice farmers

[8] Caro, *The Path to Power*, 373, 379–385, 459–462. Herman and George Brown, who owned Brown and Root, were to become powerhouses in Texas politics and Lyndon Johnson's most important financial supporters.

[9] Oral history, William Robert "Bob" Poage, Nov. 11, 1968, pp. 9–11 (LBJ Library, Austin).

[10] Memo, Franklin D. Roosevelt to Secretary of the Treasury, Sept. 26, 1935, Box 6, H. S. Hilburn File (Brazos River Authority, Waco).

[11] *Temple Daily Telegram*, Sept. 27, 1935, p. 1.

[12] Telegram, Maury Maverick to Poage, Dec. 8, 1936, W. R. Poage Papers, Box 1423, File 24 (Baylor University Collections of Political Materials, Waco); letter, Maverick to Poage, Dec. 11, 1936, ibid.; letter, Poage to Maverick, Dec. 17, 1936, ibid.; Kenneth E. Hendrickson Jr., *The Waters of the Brazos: A History of the Brazos River Authority, 1929–1979* (Waco: Texian Press, 1981), 28–31.

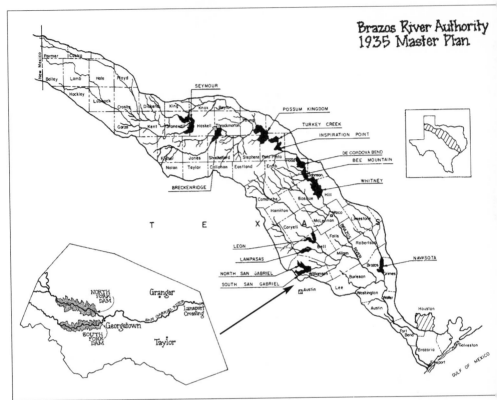

The Brazos River Conservation and Reclamation District's 1935 Master Plan proposed thirteen dams in the Brazos watershed, including two in Williamson County, one on the North Fork and one on the South Fork of the San Gabriel River near Georgetown. *Courtesy Brazos River Authority, Waco. Insert map at bottom left drawn by Kristen Tucker Pierce.*

get water when they needed it. But after Possum Kingdom, no more Brazos Valley dams were built until after World War II.[13]

In the summer of 1945 the federal government gave the Brazos River Authority seventy-five thousand dollars to plan the dam system Roosevelt had authorized ten years earlier. The original thirteen-dam

---

[13] Morris Sheppard Dam, or Possum Kingdom, was the key to the Brazos River Valley's master plan—the BRA's proudest accomplishment but also flawed. Without it, there was no "system" and no money. See Hendrickson, *The Waters of the Brazos*, chapters 5 and 8.

Road, River, and Ol' Boy Politics

plan had been whittled to eight reservoirs, but the San Gabriel's North Fork and South Fork dams remained on the to-do list.[14] In Williamson County there was a sense of inevitability about the San Gabriel dams. Sooner or later, people thought, they would get built; when the next big flood came, the county would be spared. Unfortunately, they were wrong.

The old consensus to curb the river at its upper end, or western forks, was unraveling on three fronts: bureaucratic, political, and financial. In Williamson County few suspected what was happening. Perhaps the most significant unraveling occurred on February 5, 1946, at a secret Austin meeting. Officials from the Army Corps of Engineers and the Bureau of Reclamation, America's two great competitive dam-building agencies, squared off to see who would win the Brazos watershed contracts with their multitude of dams. Both sets of engineers had done cursory surveys. Like the Brazos Water District, the Bureau of Reclamation was keen to hold the San Gabriel's flow above Georgetown, on the river's western forks, releasing floodwaters to irrigate the intensely cultivated land downstream and allowing Black Waxy farmers to develop an agricultural power center. It was a pattern the Bureau had successfully pursued in California, though the California result favored huge farm corporations rather than small family farms such as those in Williamson County. But the theoretical pluses and minuses of the irrigation plan were never debated by the people the plan would have affected, because no one in Williamson County ever knew about the Bureau's scheme.

The Army's engineers, whose mission was flood control, not irrigation, found "no justification" whatsoever for dam sites on the river's forks west of Georgetown. Instead, the Army called for a dam at a tiny river crossing called Laneport in the far eastern reaches of Williamson County—smack in the middle of the Black Waxy and hundreds of extremely productive small farms. In fact, the Army's dam would be built just four miles from Williamson's eastern border with Milam County. Their ideal dam would create a large, shallow lake covering an estimated 250 of the county's thousand-plus Blackland Prairie farms—

----

[14] Hendrickson, *The Waters of the Brazos*, 52.

Texas's most productive nonirrigated farm zone.[15] It was as if a government agency devoted to protecting the nation's oil deposits decided to destroy its most productive oil field, on the theory that other oil fields might be developed through that destruction. But the Army engineers did not really care about farming; they were interested in flood control. They argued that if the point of the Brazos master plan was flood control, as indeed its 1935 enabling act (and subsequent modifications) had specified, a dam at Laneport would control flooding downstream in the Brazos basin with more cost-efficiency than dams on the San Gabriel's North and South Forks. That was probably true, since most of the San Gabriel's tributaries flowed into the main stem below the proposed "fork" dams, but it was not true within Williamson County, where for decades the Brazos River Authority had supported flood-control dams on the North and South Fork of the San Gabriel to protect the farmers of the Black Waxy.

The Corps of Engineers' position nettled the Reclamation Bureau representative at the Austin meeting, who protested that damming Laneport would permanently flood twenty thousand acres of prime agricultural land which the Bureau of Reclamation was prepared to irrigate. Well, countered the Army, there were ten thousand acres of arable land "suitable for irrigation" *below* the Laneport dam site in Milam County. Why didn't the Bureau do something with that?

So the agencies cut a deal. If the Army would agree to make the Laneport dam even bigger than originally contemplated so it could be used to control floods *and* to irrigate, the Bureau would irrigate Milam County's farms and "abandon its potential North and South San Gabriel projects in favor of the Army's Lanesport [*sic*] project."[16] Two months later, the Bureau of Reclamation dropped out of the picture. Presumably it had bigger fish to fry. For all practical purposes, the Army controlled the Brazos basin. For Williamson County's farm

[15] Bridgette Cavanaugh, "Granger Dam," a 1973 University of Texas research paper in which Cavanaugh cites a Texas A&M study that reached this conclusion. The paper was widely circulated at the time and came into the possession of the author in 1974.

[16] Conference, Army Corps of Engineers and U.S. Bureau of Reclamation, Feb. 5, 1946, Austin, Tx., RG 77, Records of the Corps of Engineers, Ft. Worth District, Civil Works Project Files, 1934-1961, Brazos River 001-675, Box 36 (National Archives, Southwest Region, Fort Worth).

life, both economic and social, this decision was to prove as disastrous—and as out of the county's control—as the 1921 flood itself.[17]

Despite brave public talk about the superiority of the Laneport dam site over the upper fork dam sites, the Army engineers appear to have been at odds with themselves. A letter dated May 12, 1948—six months before the Army formally recommended Laneport—reveals disagreement high within the ranks. Col. Henry Hutchings Jr., chief of the Army's Southwestern Division in Fort Worth, wrote Congressman Lyndon Johnson a terse explanation of why a final report on the San Gabriel River was being delayed. "It was found," Colonel Hutchings wrote, "that we differed in some essentials from the opinions expressed by the District Engineer [in Galveston], and it was deemed best to reconcile all those differences prior to submission of my own recommendation."[18] Hutchings did not spell out the details of these "essentials," but clearly they concerned the contest over where the San Gabriel River should best be checked.

The Brazos River Conservation and Reclamation District was struggling with its own internal conflicts. Its manager, R. D. Collins, sympathized with Georgetown and the original plan to dam the San Gabriel's western forks. John Sharpe and F. D. Love, both of Georgetown, were old friends, and both were former directors of the Brazos District's board. The Brazos District's Dallas attorney, John McCall, actively supported Georgetown's cause. On the other side, Congressman W. R. "Bob" Poage of Waco, who had legislated the tools for the Brazos District to develop the Brazos River Valley, staunchly supported a Laneport dam, since Milam County lay below Laneport and was part of Poage's Eleventh Congressional District. And Collins's promising new director, John Howard Fox of Hearne, who had replaced his deceased father, Granger's John Short Fox, leaned toward Laneport because his brother Wilson Fox was lobbying for it so industriously. (On the other hand, Howard Fox's other two brothers,

---

[17] Letter, L. W. Smith to Col. D. W. Griffiths, Department of the Interior, Bureau of Reclamation, Apr. 5, 1946, RG 77, Records of the Corps of Engineers, Ft. Worth District, Civil Works Project Files, 1934–1961, Brazos River 001-675, Box 36, ibid.

[18] Letter, Henry Hutchings Jr. to Lyndon B. Johnson, May 12, 1948, RG 77, Records of the Corps of Engineers, Ft. Worth District, Civil Works Project Files, 1934–1961, Brazos River 001-675, Box 36, ibid.

Henry and Bryan, both of whom lived by the San Gabriel River, vociferously opposed a dam at Laneport, so Howard was torn.)

The big issue for the Brazos Water District regarding Laneport was whether its officials would embrace the Army's regional planning scheme for the Brazos basin. If they did, they were almost forced to support Laneport. In theory, they did. But Collins and his board came from the old school of dam planning, in which dams were erected to meet particular local needs, rather than regional goals. And they could not help empathizing with people they knew and respected, like those in Granger and Georgetown who decried the Laneport plan. When local interests conflicted with those of the entire watershed, as Williamson County's so patently did, the Brazos directors' confusion and pain became acute.

Before the Army Corps of Engineers elbowed the Bureau of Reclamation away from the San Gabriel and determined to dam the river at Laneport, a rags-to-riches Washington, D.C., commodities trader thought he smelled gold near the Black Waxy tenant farm where he grew up. On January 31, 1946—five days *before* Corps and Bureau officials smoked the peace pipe in Austin and two years before the Army plan to dam the San Gabriel at Laneport was revealed to the public—Ralph W. Moore paid nine thousand dollars for just under two hundred acres of San Gabriel River bottomland. It lay close to his birthplace at Friendship—a short walk from the Laneport Crossing—and to the site of the Big House where the Quintanillas lost their lives in 1921.[19]

Moore's purchase was the second sign of trouble for the original plan to dam the San Gabriel where most people in Williamson County believed it should be dammed, on its western forks. Moore's land purchase quickly became fodder for gossip. Nearly everybody in Williamson County knew Moore—a big round man with a moon-shaped face, who nearly always sported a white Stetson and an impeccable white suit. Moore's charm had taken him from his father's tenant farm to the leadership of the powerful Texas Grange, where he

---

[19] Deed Records, 1946–48, County Clerk's Office, Williamson County, Georgetown.

Political cartoon of Washington D.C. "farmer's friend" Ralph Moore, who quipped, "That guy must have used me for a model!" *Reprinted from August 7, 1943,* Liberty.

had befriended politicians and farm promoters across the nation. As the top farm lobbyist in Texas, he created the Texas Corn Carnival, which drew thirty thousand celebrants to the little city of Granger.[20] In 1940 Moore and his wife, Lois, had moved to Washington, where he got rich trading farm futures on the commodities market. He loved telling old Williamson County cronies about hobnobbing on Capitol Hill with "Senator This" or "Senator That." It didn't make sense to the folks back home that Ralph Moore would give up the "high life" to become a gentleman farmer.[21]

---

[20] J. E. Fee, "Corn Becomes King for a Day at Carnival," *Dallas News,* Sept. 29, 1938; "Williamson County Scrapbook" (Center for American History, University of Texas at Austin); Barbara McCandless, *Equal Before the Lens* (College Station: Texas A&M University Press, 1992), 41–46.

[21] Interview, Opal Wilks, Nov. 7, 2000, Taylor.

Texas Grange Master Ralph Moore, third from left, in one of his classic schemes promoting Texas agriculture: miniature cotton bales shipping from Granger to mark the Texas Centennial, May 22, 1936. *Jno. Trlica Photo, Courtesy Photography Collection, Harry Ransom Humanities Research Center, University of Texas at Austin.*

Two years later, as cotton prices fell, Ralph Moore preached corn as the farmer's salvation. In 1938 he initiated the first Texas Corn Carnival in Granger, which was attended by thirty thousand celebrants. In this publicity shot, Granger businessmen dressed like Hollywood farmers ham it up. Moore stands in the center of the back row. *Jno. Trlica Photo, Courtesy Photography Collection, Harry Ransom Humanities Research Center, University of Texas at Austin.*

"Oh, he was something," chuckled former Congressman J. J. "Jake" Pickle, remembering Moore. "He made truckloads of money and lost it, several times. He loved to gamble. Not at the roulette table, you understand, but he took big risks. . . . He was always in the middle of water projects. . . . He was close to the water authorities, maybe close with Poage."[22] (In 1946, when Moore started buying land near Laneport, Bob Poage of Waco had been in Congress for a decade and chaired the House Agricultural Committee. Poage represented Bell and Milam Counties, very near the proposed Laneport dam site in Williamson County, and was a great dam booster.)

On February 22, 1946, Moore bought another 134 acres adjoining the farm he had acquired a few weeks earlier, paying two thousand dollars cash and agreeing to pay off a $5,700 lien on the property. That was just the start. Between January 31, 1946, and June 24, 1948—his final purchase coming five months *before* the Army made public its plan to dam the Gabriel at Laneport—Ralph W. Moore assembled fifteen farm tracts containing 2,472 acres around his native Friendship, hard by a lake that had not yet been officially announced, much less sanctioned, funded, or built. The deeds of sale committed Moore and his wife to pay two hundred thousand dollars for the land—a fortune in 1948.[23] No one can say for certain why Ralph Moore pledged to spend two hundred thousand dollars on agricultural land when farm values were sinking. If he shared his ideas with anybody, the records have long since slipped beyond reach.

There is a theory, of course. People in Williamson County thought Ralph Moore was trying to make a killing. He wanted to create a lakeside development. If the government condemned some of his land for the dam, he would sell at market valuation. If not, he would own lakefront property. About half the owners from whom he bought

[22] Interview, J. J. "Jake" Pickle, Apr. 18, 2000, Austin. Pickle, who was enormously generous with his time and memories for this book, died June 18, 2005.

[23] Deed Records, 1946–1948, County Clerk's Office, Williamson County, Georgetown, Tex. There were thirteen deeds; two of which involved two separate tracts. Land prices fluctuated wildly from $3,832 for 308 acres ($12.44 per acre) to $67,115 for 430 acres ($156 per acre). The average price per acre paid was $80. Most, if not all, landowners lived away from their land; seven of the thirteen lived outside Williamson County.

land lived outside Williamson County; some would have had no idea of a rumored lake.[24] But, the thinking goes, Ralph Moore would have known—*must* have known—that Laneport was a "done deal" to have sunk two hundred thousand dollars on all that land. His speculations had a rank smell to the nostrils of Williamson County residents, a hardy lot long conditioned to the farm odors of dirt and manure. Political "bossism" was a given. "They talk about Duval County," Jake Pickle said, rolling his eyes. "The little old city of Granger used to give [Lyndon Johnson] a vote of about forty or fifty to one. I've often wondered why they didn't investigate Granger instead of just Duval County, because the votes there were just as heavy."[25]

It is unlikely anyone will ever know whether Moore gambled to build a real estate bonanza because someone in the know tipped him about plans to dam the San Gabriel at Laneport, or whether he acted on instinct. It was common knowledge that a dam would be going in *somewhere*, but almost everyone expected it to be west of George-town. But Moore's land acquisition program, preceeding (as it did) public knowledge of the dam's location, fueled feelings of anger and betrayal about Laneport that helped opponents block it for years, and nearly for good.

Suspicions about a supposed "fat cat" deal that had somehow created the Laneport dam plan rankled Williamson County's patriotic World War II vets, high-ranking county officials, Chamber of Commerce boosters, and Czech-American farmers eager to prove themselves good Americans—people who might otherwise have gone along with the project, as Williamson County Democratic Party kingpin Wilson Fox kept assuring his friends in Washington they would. It didn't help Laneport's popularity that Wilson Fox was Moore's personal attorney, that the two spoke frequently on the telephone, and that Fox himself had personal financial reasons for pushing the Laneport dam project with policy makers.[26]

[24] Interview, Loretta Mikulencak, July, 1998, Granger.

[25] Oral History, J. J. ("Jake") Pickle, Section II, p. 25, June 17, 1970 (LBJ Library, Austin). Pickle repeated the same story in an interview with the author.

[26] Interview, Gene Fondren, Sept. 1, 2000, Austin. Wilson Fox, along with the other five surviving Fox children, had inherited land along the San Gabriel from John Short Fox. There was a catch: the land could not be sold except through condemnation.

Williamson County Sun editor Sharpe warned Senator Johnson about rumors of a dam being built at Laneport, rather than west of Georgetown. "I could be mistaken," Sharpe wrote, "but I think I see the hand of the celebrated Ralph Moore in this, he having been reported to have purchased considerable farming land in that area. At any rate I think the matter needs some quiet investigation."[27]

Georgetown farm and ranch realtor Owen W. Sherrill sent a similar missive to Johnson. "Why a Laneport location could even be considered . . . unless, confidentially, the purchase of lands by people, now in Washington, around the Circleville area and near the San Gabriel, which could be condemned at very high prices, would seem possibly to have some bearing (we hope this is in error.)"[28]

There was no error. But despite his involvement in a suspected insider deal, Moore made no fortune at Laneport. Most of the land he bought ended up in the possession of a Granger bank.[29] But he was, as Congressman Pickle put it, a "wheeler-dealer"—and he loved the concept of a lakeside development. Long after he gambled on land near Friendship, Moore purchased a fox-hunting estate in Virginia, dredged three large ponds on it, designed a subdivision so that many of its homes backed up to the water, and named it Warrenton Lakes Estates. That was in the early 1960s. Today, Warrenton real estate agents say it is one of the best places to live in Fauquier County.[30] In 1971, blind and lame from diabetes, Moore died of a heart attack. Seven years later, the lake he had staked—and lost—a fortune on buried his childhood home.[31]

The final unraveling of the twin-fork dam plans occurred on the Potomic River when Lyndon Johnson, a magnetic force in the House

[27] Letter, John Sharpe to Lyndon B. Johnson, Nov. 28, 1947, House Collection 1937–49, Box 314 (LBJ Library, Austin).

[28] Letter, Owen W. Sherrill to Johnson, May 25, 1948, House Collection 1937–49, Rivers, San Gabriel Dam Project 1947–48, Box 314 (LBJ Library, Austin).

[29] Interview, Truett Beard, Apr. 13, 2000, Granger.

[30] Interview, Betty Allen, Nov. 28, 2000, Warrenton, Va.; interview, Opal Wilks, Nov. 6–7, 2000, Taylor.

[31] Obituary, *Fauquier Times Democrat,* undated, Ralph W. Moore Papers, Opal Wilks, Taylor.

of Representatives and brilliant at "delivering the goods" to constituents back home, decided in early May 1948 to run for the U.S. Senate. It was Williamson County's misfortune that Johnson was tied up in a race for his political life that summer, before the Army's position on Laneport had completely jelled. After winning the Democratic primary on August 28 and facing only token opposition in November, Johnson turned away from his old Tenth Congressional District, concentrating instead on state matters. Before the November election, however, Congressman Johnson did pledge—three times on paper—to help Georgetown secure dams on the forks of the western San Gabriel River.

"By now you will have received my letter of the 15th telling you of my deep and abiding interest in . . . the proposed San Gabriel projects near Georgetown. Permit me to say again that if one man can do the job, the Georgetown proposals will be approved at an early date," he wrote realtor Sherrill.[32] But on that same day, Johnson wrote one of his most trusted advisors, Taylor attorney Harris A. Melasky, a tepid letter of support for the Laneport proposition.[33] Johnson could see a collision looming, and he left its solution to his House successor, his intimate friend Homer Thornberry. Had Johnson been focusing on his district, rather than on the state, he might have changed the outcome. The dams on the San Gabriel's forks could have been built and the dam at Laneport dropped. Or all three proposed dams might have been built. Johnson deeply believed in dams, and there was no one better at working the pork barrel than Lyndon Johnson. Still, no one knows what he would have done.

Congressman Poage believed Johnson would have backed Laneport over the Georgetown dams. Twenty years after the fact, Poage would recollect,

> We did have a matter that we worked out while [Johnson] was still a Congressman. . . . We got the agreement on the dams on the San Gabriel just about the time that he made the race for the Senate. . . . Mr. Johnson and I . . . came to an agreement

---

[32] Letter, Johnson to Sherrill, Nov. 16, 1948, House Collection 1937–49, Rivers, San Gabriel Dam Project 1947–48, Box 314 (LBJ Library, Austin).

[33] Letter, Johnson to Harris A. Melasky, Nov. 16, 1948, House File, Box 314, ibid.

Road, River, and Ol' Boy Politics

Congressman Lyndon Johnson visits Granger on July 10, 1946. He stands above and left of the "Granger for Lyndon" sign sporting a white suit and hat. *Jno. Trlica Photo, Courtesy Photography Collection, Harry Ransom Humanities Research Center, University of Texas at Austin.*

that we would support any dams that the Army Engineers approved, and we would support their construction in the order in which the engineers recommended they be constructed. And he always stayed with it. . . . But in all those years, we haven't built anything down there. It has always seemed to me to be clearly the result of the local people not being able to agree to accept that kind of agreement that Lyndon and I made twenty years ago.[34]

Had all this been known at the start of the conflict—the bureaucratic clashes over whether to irrigate or flood the San Gabriel Valley; whether it was worthier to control flooding in Williamson County or in Milam County; Ralph Moore's land speculations and their murky origins; Congressman Johnson's suddenly reordered political universe and Congressman Poage's need to secure dams across the Brazos basin—would the story have played out differently? Would Friend-

[34] Oral history, William Robert "Bob" Poage, Nov. 11, 1968 (LBJ Library, Austin).

ship and Granger still be agricultural strongholds supporting a healthy Czech-American culture? Could irrigation have transformed the Black Waxy from a sustainable economy based on row cropping into a profitable agricultural paradise? Possibly. Had Williamson County forces understood the currents flowing underneath the public debate about Laneport and the North and South Fork dams, Laneport's opponents would have been armed with some heavy artillery. As it was, they conducted an impressive war of attrition. And almost won.

THE FIX

When Uncle Sam waves his hand . . . and says, *"Let there be water!"* we know that the stream will obey his command.

WILLIAM SMYTHE, *The Conquest of Arid America*

They remembered the 1921 flood too well. Williamson County's leading lights—Chamber of Commerce and city officials, county commissioners, farmers, gin owners, farm supply dealers, ranchers, real estate brokers, bankers, cottonseed oil producers, and newspapermen—gathered at Taylor's City Hall at seven o'clock on November 23, 1948, to hear news they expected would alter their lives. Many had served in World War II. Now they were home, intent on grasping a piece of America's predicted postwar prosperity.

Most believed they would hear confirmation of an old plan to master the flood-prone San Gabriel River by building a pair of dams on its two forks west of Georgetown. Just such a plan had been approved since 1935, at least in a sketchy state, by the Brazos River District, the State of Texas, and Congress. But in the spring of 1946, the U.S. Army Corps of Engineers had surveyed the San Gabriel. Unexpectedly, the Army engineers had spent most of their time in the eastern end of the county, where the San Gabriel collects several streams as it lumbers across the broad Black Waxy plain. Exciting tales and distressing rumors had followed in the Army's wake; tonight the county would know precisely where the long-awaited dam, or dams, would land. The county's movers and shakers believed damming the

San Gabriel would serve the county well, stopping or at least diminishing serious floods and storing flood waters for use against drought years. A county publisher expressed the common perception:

> Everybody always assumed that if there ever was a dam, it would be in the western end of the county, near the headwaters of the San Gabriel, to protect the rest of the county from flooding. The east end of the county was that rich, wonderful farm land. No one ever imagined putting a dam there.[1]

As sometimes happens, the dream turned into nightmare. One large landowner remembered the scene. "All the farmers were there, and of course the Corps with its snazzy charts and graphs," he said. "Well, once those farmers realized that the Corps' plan was to stick a twenty-thousand-acre lake in the middle of their best farm land, and further, that the dam wouldn't even protect them from flooding, just people *below* Williamson County, there was this terrible, dead silence. Then most of the farmers just got up and left."[2]

Printed accounts of the two-hour session seem to contradict each other, though they may simply have reflected different pieces of the same story. Newspaper reports highlighted the Army's plan, not the crowd's reaction to it. No official record of the meeting was kept.[3] "Mostly the farmers just yelled," one observer recalled. "They were unanimously against it, and they expressed themselves vociferously."[4] Whether the farmers protested by walking out or shouting or both, they were plainly appalled.

Just as upset, a group of Georgetown businessmen insisted that the Army reconsider its recommendation to build a single dam at Laneport. Owen Sherrill spoke for the group, and by extension for all of Georgetown. Sherrill sold real estate, specializing in ranches and farms, but he was also Texas A&M educated and an early advocate of "scientific" farming. The thought of that perfect black soil, buried under a huge lake, repulsed him. He hammered away at one point:

[1] Interview, Donald Lee Scarbrough, May 1974, Georgetown.

[2] Interview, Henry Fox, May 1974, Circleville.

[3] Col. B. L. Robinson to Col. Henry Hutchings Jr., Jan. 13, 1949, Records of the Army Corps of Engineers Fort Worth District, Civil Works Project Files, 1934–61, Brazos River 001-675, Box 36 E12 (National Archives, Southwest Region, Fort Worth).

[4] Interview, Donald Lee Scarbrough, May 1974, Georgetown.

Road, River, and Ol' Boy Politics

dams must be built west of Georgetown to save the Blackland farms from "rampaging" floods.[5] Sherrill's persistence caused Col. Henry R. "Hank" Norman, chief of the Corps' Galveston Engineering Division, to grudgingly admit that the county "might be protected to a greater extent from flash floods by the construction of the upper dams."[6] On the other hand, Sherrill's delegation diplomatically avoided opposing the Laneport dam—so long as the upper dams were built first. Congressman-elect Homer Thornberry, voted in three weeks earlier, squirmed in his seat, trying to decipher the situation. His friend and mentor, Lyndon B. Johnson, did not attend, having just won a seat in the U.S. Senate. Thornberry had taken over Johnson's old Tenth Congressional District seat. Johnson would be a hard act to follow.

"There's the man to see," Colonel Norman told the Georgetown crowd, pointing at Thornberry. "A loud howl from you folks back here will help get the reservoir in the next appropriation bill."[7] Colonel Norman was playing to the patently obvious desire of the Georgetown contingent for dams on the north and south forks of the San Gabriel, not to the farmers' anguish about a Laneport dam burying their farms. In 1948 it was beyond belief that any group of farmers could stop a dam project, particularly one the federal government was financing. After all, the point of flood control dams, at least in theory, was to serve rural interests. Army engineers expected opposition from anyone whose home lay in the path of a dam project, but those objections were shrugged off. They were part of the price of doing business. The land was being purchased, and it was just too bad if the local tax appraiser routinely undervalued agricultural land. The Corps' backers in Congress generally took the same tack, sorrowfully commiserating with people who were losing their land, but pointing out the benefits for the "greater good." It was also beyond belief that any rational group of people would oppose *any* dam—particularly in Texas, where the flood/drought syndrome reigned: extreme rains and flash floods one year, extreme aridity the next. Indeed, the assumption was exactly the opposite—that "local interests" would turn handsprings to be on the receiving end of a dam, any dam, wherever it was located.

[5] *Taylor Times,* Nov. 25, 1948; *Williamson County Sun* (Georgetown), Nov. 26, 1948.
[6] *Taylor Times,* Nov. 25, 1948.
[7] *Williamson County Sun* (Georgetown), Nov. 26, 1948.

So the ramrod erect Colonel Norman must have been somewhat startled at the crowd's hostility, but not worried.[8] He expected dissent. After all, when the Laneport dam was finished, two hundred or more farms would be flooded; that many families would have to give up their homes. The town of Friendship would end up underwater, along with a couple of smaller hamlets and four cemeteries. But that was part of the bargain with these federal projects—it was what everyone expected. With grave politeness, Colonel Norman fielded the farmers' questions, but he focused most of his attention on the Georgetown group, which seemed more threatening.

From the beginning, the Army brass thought they were seeing a case of pure selfishness in Georgetown's desire for a dam. To a degree, they were right, but they missed the larger, more accurate picture. From the moment Colonel Norman announced a dam would be built at Laneport, no Williamson County outsider adequately calibrated the overwhelming feeling ricochetting around the room—that a Laneport dam spelled disaster for the county. They missed it completely. Hardened by battlefield experience and flushed with engineering expertise, proud of their regional plan, first the Army engineers, then the politicians, misread or discounted that clear signal. In the end, anyone with a stake in the waters of the Brazos paid for that failure: Williamson and Milam Counties; hundreds of farm families who would eventually be forced off their land, taking a rich vein of Blackland Prairie economy with them; the Czech communities of Granger, Friendship, Machu, Moravia, and Taylor; the Brazos River Authority; Congressmen Thornberry and Poage; and Ralph Moore, who bet on Laneport before the game started.

On November 23, 1948, Colonel Norman's engineering lecture turned, perhaps inevitably, to the politics of dams. First he offered Congressman Thornberry up to the crowd, advising the people to "howl." Later that evening, a Texas state water board official, Col. Eugene Spence, echoed Norman's point: "If you people from Georgetown want those two dams bad enough, go to Congress and they'll tell the engineers to build those dams," he said. "That's what Belton did, and look what they got."[9]

<hr />

[8] Ibid., Nov. 26, 1948.

[9] Col. Eugene Spence of the Texas Board of Water Engineers, quoted in the Nov. 26,

Road, River, and Ol' Boy Politics

"They" got Belton Dam—and everybody in the room knew it. In Texas especially, dams were highly political creatures. There was nobody at Taylor City Hall that night unfamiliar with the fact that Senator-elect Lyndon Johnson, who had represented them in Congress for nine years, had inveigled President Roosevelt into completing Marshall Ford Dam, which otherwise would have fallen into bankruptcy, bringing disaster to the fledgling LCRA and ruining its contractors, Brown and Root of Houston, one of Johnson's ardent and (after his help) wealthiest supporters.[10] Johnson bragged about it all the time, and Austin, the state capital, was grateful. So the colonels' advice to use politics to get a dam, though it came from government officials, did not sound strange to those in that room that night, who were, after all, staunch "yellow-dog" Democrats whose county Democratic "machine," run by Ralph Moore's friend Wilson Fox, had faithfully backed Lyndon Johnson and Congressman-elect Thornberry, and could be counted on to do so again if the voters were kept happy.

The men at Taylor City Hall listened carefully to the colonels' advice. Then they relentlessly followed it for two dozen years. They took it in directions unimagined by the Army engineers, sustaining a brilliant political and public relations campaign to overthrow the proposed Laneport dam and substitute two dams on the upper forks of the San Gabriel. Soon the engineers would reverse their argument that politics could influence dam placement. Political push, Corps engineers now asserted, had absolutely no role to play—impossible!—in the Army's scientifically calibrated engineering decisions. Williamson County paid that new argument no mind at all.

The news from Taylor City Hall that November 23 evening did not take Williamson County entirely by surprise. Since March 1946, when the San Gabriel Valley near Laneport was crawling with Army surveyors, some people had suspected the worst. Henry Benjamin Fox, who was editor-owner of the *Granger News* as well as county

---

1948, *Williamson County Sun* (Georgetown). Belton Dam was built before a dam at Waco; logic and need would have suggested the Waco dam be constructed first.

[10] Caro, *The Path to Power,* 459–468; Oral History, Sam Gidean, Oct. 3, 1968 (LBJ Library, Austin).

The Fix

Democratic chairman Wilson Fox's brother, had heard alarm bells. He headlined his exclusive, "Dam Surveying (Some Spell it Differently) Causing Worry":

> Various rumors are circulating, including some which are of the nature of practical jokes. For example, one report has it that a dam will be built near Friendship and will back water up to the level of the bridge at Circleville, which would just about submerge the eastern half of Williamson County.[11]

The "rumor" was exaggerated, but accurate in its essentials. Here it was again, twenty months later, coming out of the mouths of Army engineers whose job it was to stop flooding! What really astounded county leaders was the Army's contemptuous attitude toward the idea of impounding the San Gabriel at its upper forks. The soil was poor there, scratch-thin caliche. Dams west of Georgetown would protect the agricultural wealth of Williamson County against flooding and soil erosion, unlike the Army's plan, which gave the county nothing. Nobody could understand the Army's thinking. They didn't realize that downstream powers—Milam County, Texas rice farmers, and Dow Chemical—were calling the shots.

By November 1948, when Army civil engineers met the people of Williamson County at Taylor City Hall, the notion of damming the upper reaches of the San Gabriel was a still cherished, even sacrosanct, local ambition. In his time, Congressman Buck Buchanan had supported it. Congressman Johnson had supported it for the better part of a decade. The Brazos River District had formally backed it twice since 1935.

The county's historic quest for dams had so completely focused on the San Gabriel's upper western forks it is no wonder that on November 23, 1948, the crowd at Taylor City Hall was struck dumb—and then yelled—when they heard the two dams they had wanted for years would be rolled into one big one at Laneport, which would not only provide no flood protection for the county's richest farm land, but flood it permanently. Worst of all, there seemed to be little hope of changing the Army's mind.

[11] *Granger News,* Mar. 28, 1946.

Road, River, and Ol' Boy Politics

F O U R

THE SALESMAN

He's a man way out there in the blue, riding on a smile and a shoeshine. . . . A salesman is got to dream, boy. It comes with the territory.

ARTHUR MILLER, *Death of a Salesman*

ack in his Galveston office after his first encounter with Williamson County, Colonel Norman relaxed. The worst was over. If the county's reception to the Army's plan to plug the San Gabriel River at Laneport was not exactly warm, it was reasonably polite. The farmers around the dam site hated it, of course, but that was to be expected. And though the Georgetown contingent hated Laneport too, it had mouthed lukewarm support. From Norman's perspective, the Army's Laneport plan was flawless: it gave more effective regional flood protection to the Brazos River basin than dams on the upper San Gabriel could ever achieve; it would cost little more than the two upper San Gabriel dams but would hold twice the water.[1] The politics were simple. Wilson Fox, Williamson County's political mastermind, strongly favored Laneport as did Congressman Poage, a resourceful backer of the Brazos River Conservation and Reclamation District and its plan to build multiple dams, which, of course, were all potentially lucrative Army projects. The San Gabriel

[1] Laneport's superiority as a *regional* flood-control structure stemmed from its placement low on the San Gabriel, at a point below where the river's tributaries dumped their loads in, but above the Little River, which regularly flooded Milam County—which Congressman Poage represented and wanted to protect.

even flowed its last few miles through a lightly populated part of Milam County, in Poage's district. Poage's old counterpart in the House of Representatives, Lyndon Johnson, who as congressman might have insisted on the upper fork San Gabriel dams near Georgetown, had distanced himself from the issue after being elected to the Senate. The new congressman, Homer Thornberry, didn't seem the type to challenge the Army's vaunted engineers.

Colonel Norman judged Laneport's opponents a motley crew—a bunch of inarticulate Czech farmers who didn't want to be uprooted and relocated and a loudmouthed salesman named Sherrill from Georgetown, egging on a Chamber of Commerce that clung to an outmoded plan to dam the river near Georgetown, for reasons no better than civic boosterism. In Norman's view, Sherrill's worry over "rampaging floodwaters" was hyperbolic bunk. The Army's job was to analyze the Brazos River system, design the best *regional* solution for flood control and build it. *That* was the Army's charge, not to fret over Williamson County's fate during a future flood or what the human and economic cost of a Laneport dam might be.

In 1948 the Army Corps of Engineers dominated water policy in America. Founded at the United States Military Academy at West Point in 1802, the Corps had long regarded itself as the Army's most rarified elite. Its earliest cadets married Congressmen's daughters, and Senators appointed friends' sons to the Corps. Breeding had produced a tightly connected inner circle of Army engineers and backers in Congress, politicizing public-works engineering from the start. In 1830 a former West Point superintendent criticized the Corps as "a privileged order of the very worst class . . . a military aristocracy." Over time, reform watered down the "aristocracy," and the Army engineers learned the art of adaptability, designing ironclad battle rams for war and dredging navigable canals from shallow rivers during peace.[2] During World War II they performed brilliantly. In the postwar explosion of American development, Army engineers coasted on their reputation as crackerjack wartime technologists and trolled for work to keep their specialists employed.

---

[2] Todd A. Shallat, *Structures in the Stream: Water, Science, and the Rise of the U.S. Army Corps of Engineers* (Austin: University of Texas Press, 1994), 188, 189.

Road, River, and Ol' Boy Politics

But the Corps' old mistrust of "outsiders"—including other military branches, civilian engineers, and "local interests" that might interfere with their "scientific" works—died hard. Over the years, Army waterway science "reduced river construction to a series of standardized steps," deftly bypassing the human and environmental elements. The Army engineers worshiped at the altar of what they considered science, holding their "faith in problem-solving through quantification, the view of rivers ... as technological systems" rather than as part of an organic ecology that included human beings, wildlife, and plants.[3] The Army engineers' embrace of the technological fix for imponderables like flooding appealed to members of Congress, who discovered after World War II that with big dams proposed to secure the nation's river systems, they could wallow in the joys of "log-rolling," or "pork barrel politics." When "log-rolling" worked properly, congressmen could get "veto-proof legislation" and "something for every region"—a politician's dream.[4]

Congressmen adored the Army engineers. When, sooner or later, constituents back home demanded a solution to a local water problem (flooding, drought, irrigation, "booming" a town with a dependable water supply), the people's representative could deliver up a nice, solid dam. Using what was termed the congressional "courtesy" or "buddy" system, members almost always voted for each other's waterworks legislation, whether the individual project made sense or not. Down the line, if everyone cooperated, every congressman would get something for his or her district. Here is how historian Marc Reisner described it:

> To a degree that is impossible for most people to fathom, water projects are the grease gun that lubricates the nation's legislative machinery. Congress without water projects would be like an engine without oil; it would simply seize up.[5]

In other words, dams were the best federal subsidy most congress-

[3] Donald R. Field, James C. Barron, Burl F. Long (eds.), *Water and Community Development: Social and Economic Perspective* (Ann Arbor, Mich.: Ann Arbor Science, 1974), 1–4, 155–157, 184, 206–207, 280 (quotations).

[4] Shallat, *Structures in the Stream*, 5.

[5] Reisner, *Cadillac Desert*, 308–309.

men could ever hope to bring home. To members of Congress, the Army engineers were like Santa Claus: if you were very, very good (and supported *all* dam projects, regardless of merit), your district would get a juicy plum of a dam. If you were bad (and failed to vote for Corps projects), your district's hope of getting a public works project was slim. William Gianelli described the prevailing situation after President Ronald Reagan appointed him to head the Army engineers: "Over the years, the Corps had been very used to considering itself almost an arm of Congress."[6]

Colonel Norman pondered Owen Sherrill, the man who had arranged the Taylor City Hall meeting through the Georgetown Chamber of Commerce, which Sherill had helped found and still led. Sherrill seemed the most obvious threat to the Army's plans—the sort of "vigorous individual with influence" the Army most disliked, unless, of course, its engineers wanted a project badly enough to work *with* that individual.[7] This was not such a case. For months before the meeting, Sherrill had beleagured Texas politicians and the Army brass with a steady stream of letters expounding on the need for dams *above* Georgetown on the San Gabriel's westernmost forks—dams the Army was loath to build. Sherrill had aimed most of his campaign at Senator Lyndon Johnson. The two had met in 1935 while Sherrill was crisscrossing South Texas for the New Deal's Farm Production Credit Association and Johnson was working for Congressman Richard Kleberg, in theory a mere secretary but in reality a de facto congressman.[8] Apparently Johnson and Sherrill took to each other, for a warm and enduring correspondence ensued. Before the Taylor City Hall meeting, Sherrill had beseeched his old friend "to put your shoulder

---

[6] Robert Gottlieb, *A Life of Its Own: The Politics and Power of Water* (San Diego: Harcourt Brace Jovanovich, 1988), 66.

[7] Robert H. Boyle, John Graves, and T. H. Watkins, *The Water Hustlers* (San Francisco: Sierra Club, 1971), 45; Philip V. Scarpino, *Great River: An Environmental History of the Upper Mississippi, 1890–1950* (Columbia: University of Missouri Press, 1985), 167–169; Robert Kelley Schneiders, *Unruly River: Two Centuries of Change Along the Missouri* (Lawrence: University Press of Kansas, 1999), 256.

[8] Letter, Owen Sherrill to W. R. "Bob" Poage, Nov. 24, 1969, W. R. Poage Papers, Box 692, File 7 (Baylor Collections of Political Materials, Waco).

Road, River, and Ol' Boy Politics

Owen W. Sherrill, "ranch king" and super salesman. *Courtesy Georgetown (Texas) Chamber of Commerce.*

to the wheel in behalf of your old District and help [Georgetown] get the allocation of the two Dams on the San Gabriel."[9]

Sherrill was tough, resilient, and opportunistic. Though not a member of Williamson County's elite class—ranchers, big farmers, and moneymen—he moved easily in those circles. He cultivated political power as faithfully as he attended Georgetown's First Baptist Church, believing the one sure way to advance was to win over Washington's policy makers. He built clout through his adroit use of groups. In 1948, anticipating the Army's preference for the Laneport dam site, Sherrill refashioned the Georgetown Chamber of Commerce into an aggressive fighting force, creating a small committee with a mundane name, Committee on Water Conservation, and running it as if it were a military invasion. Sherrill simply overwhelmed the opposition with requests for technical information, new ideas needing investigation and letters of praise or condemnation. When he thought policy makers were getting tired of hearing from him, he unleashed new voices from his committee, mostly Georgetown's smartest lawyers and financiers. He traveled constantly, often to Capitol Hill to flatter and cajole the Texas congressional delegation.

Sherrill lived and breathed land deals. Whether advertising Crockett Gardens ("A TEXAS LOCATION SUPREME—7 miles NW of Georgetown, Texas, on the beautiful N San Gabriel River with ever-flowing springs, 2 small lakes, traversed by the most beautiful part of the river with fishing, hunting and luxurious living—summer and winter—TOPS"), representing future senator Lloyd M. Bentsen's South Texas ranching interests, or closing a 1958 sale on a five-million-dollar federal center in Fort Worth, Sherrill pursued his calling with the zeal of a street evangelist.[10]

"That man could sell anything. He was the best salesman I ever

___

[9] Letter, Owen Sherrill to Lyndon Johnson, Nov. 6, 1948, House File, Box 314 (LBJ Library, Austin).

[10] Multiple Listing Service, Institute of Farm Brokers, Chicago, Ill., May 10, 1953; letter, Lloyd M. Bentsen to Sherrill, May 25, 1955, Owen Sherrill Papers (Cushing Memorial Library, Texas A&M University, College Station); "Sherrill Scores Again!" *Williamson County Sun* (Georgetown), Aug. 21, 1958, Sherrill Papers, ibid. The multimillion deal in Fort Worth was by no means a fluke: the previous year, Sherrill had brokered a $4.5 million skyscraper sale in Houston.

Road, River, and Ol' Boy Politics

saw," Esther Weir, a long-time Georgetown rancher, would declare decades later.[11]

"Sherrill was 'Ranch King' of Williamson County for many, many years," recalled Austin's Tom Kouri, who made millions of dollars investing in Williamson County property and used Sherrill's expertise from time to time.[12]

"Owen W. Sherrill was effective because he was such a relentless and persistent person," retired Georgetown banker Jay Sloan would say. "He didn't give up on his project. In a lot of cases, people would give in to him just to get him to shut up."[13]

The Depression would have broken a man of weaker spirit. Born in East Texas in 1890, Sherrill graduated in 1910 from Texas A&M University.[14] In 1917 he became Williamson County's first farm extension agent.[15] After the 1921 flood he led a Red Cross team assessing damage around Georgetown, Jonah, and Weir—a stomach-wrenching job that must have strengthened his conviction about damming the river above Georgetown so it could never again swamp the intensely developed farm land below.[16]

After four years as county agent, Sherrill was "drafted into banking," as he put it.[17] In 1922 he went to work for City National Bank in Georgetown and became its president. During eleven years there, Sherrill served for a time as treasurer of the Texas Bankers' Association.[18] In 1931 he bought just the sort of land he later liked selling: 146 acres of prime ranch land south of Georgetown adjacent to State Highway 81, the major north-south route between Austin and Dal-

[11] Interview, Esther Weir, Jan. 19, 2001, Georgetown.

[12] Interview, Tom Kouri, Sept. 22, 2000, Austin.

[13] Interview, Jay Sloan, Sept. 11, 2000, Georgetown.

[14] Interview, David Chapman, Apr. 14, 2000 (Cushing Memorial Library, Texas A&M University, College Station).

[15] *Texas Landsman,* July 1964 (Austin: Texas Real Estate Association). The *Landsman* was the association's trade magazine. It is now the *Texas Realtor Magazine,* a publication of the Texas Association of Realtors.

[16] *Williamson County Sun* (Georgetown), Sept. 16, 1921.

[17] Letter, Owen Sherrill to Poage, Nov. 24, 1969, W. R. Poage Papers, Box 692, File 7 (Baylor Collections of Political Materials, Waco).

[18] "History of Owen Sherrill," not dated, *Austin American-Statesman* morgue. A note says the material was "compiled" by the Texas Real Estate Association.

las.[19] But City National failed in 1933, a casualty of the Depression, and there were hard feelings about Sherrill around town. "He had the bank that went broke in the Thirties and oh, it hurt so many people," remembered a contemporary of Sherrill's. "He had two sons and a wife, and he just walked off. He abandoned them. One of his sons was a cripple. His wife became a recluse and a town 'character.'"[20] Another man, a former postmaster who greatly admired Sherrill, could not forget that episode, either. "It was a sad thing. His first wife became sort of a ragsack Annie."[21] It was a rare thing in those days for such a prominent small-town couple to split up; it naturally sparked a great deal of talk.[22]

Sherrill actually left Georgetown in 1932—before his bank folded. He went to Washington, D.C., where he later recalled telling the U.S. secretary of agriculture, "I am just a country banker from Texas sitting on a keg of dynamite, smoking a cigarette, and wondering when my bank will blow up. I have seen some 60 of my friends turn in their charters for lack of money in the banks and I am wondering when my time will come."[23]

Instead of waiting, he snagged a job helping set up President Herbert Hoover's emergency crop production loan program for Texas, New Mexico, and Arizona. Back in Dallas, where the program was to be headquartered, he wrote, "there were nearly 1000 persons looking for jobs. They were good people from all over Texas . . . . We fenced off the north end of the Baker Hotel wherein we went to work . . . and

[19] Transfer from F. A. Petterson to O. W. Sherrill, Williamson County Clerk, Jan. 29, 1931, and Deed, J. Walter Johnson to O. W. Sherrill, Williamson County Clerk, Jan. 30, 1931, pp. 608–610.

[20] Interview, Ed Evans, Apr. 2000, Georgetown. Sherrill had two sons. "Bo" Sherrill died in the early 1960s in a scuba-diving accident at Lake Travis near Austin. Owen "Jack" Sherrill Jr. is said to have died shortly after his father's death in 1976, though I could not track down proof of a death date.

[21] Interview, J. D. Thomas Jr., Sept. 8, 2000, Georgetown.

[22] The divorce between O. W. Sherrill and Edith Talley Sherrill was filed May 28, 1940, Cause No. 9,752, District Court, Williamson County. Martha Burt Nelson, a close friend and a niece of Sherrill's second wife, Kay, said in 2003 that Edith Sherrill had had a drinking problem, and that for that reason Owen Sherrill never touched alcohol.

[23] Letter, Owen Sherrill to Poage, Nov. 24, 1969, W. R. Poage Papers, Box 692, File 7 (Baylor Collections of Political Materials, Waco). This was an extraordinary "autobiography" of Sherrill's life, written as if Sherrill expected to be the subject of a biography.

Road, River, and Ol' Boy Politics

in 2 years has [*sic*] the lowest cost of operation and the highest collection record in the USA."[24] He moved to Houston, struggled with personal debt, and in 1940 was divorced from his wife Edith. In 1941 he returned to Georgetown with a new wife, Kay, and established the Owen W. Sherrill Agency, "Creator of Ownership."

If Sherrill felt embarrassed about his family history, he kept it hidden. He and Kay, an ambitious and smart business partner, embarked on building a national real estate reputation and pulled it off. He became president of the National Institute of Farm Brokers and of the Texas Real Estate Association's Board of Directors. He earned many honors, including "Texas Realtor of the Year" and "All Time Texas Farm and Land Broker." He served eighteen years as director for the National Association of Real Estate Boards, using his frequent convention trips to cement business and political relationships.[25]

In 1948 Sherrill wanted dams on the upper San Gabriel and he pursued them relentlessly. Before the Army dashed his hopes at Taylor City Hall, Sherrill had tried to woo political leader Wilson Fox to join with Georgetown aiming at getting *three* dams on the San Gabriel. Sherrill wrote Fox:

> I am in hearty accord in getting any and all dams possible on the San Gabriel, and while I do know that heavy political wires have been pulled to get the Laneport Dam, I personally take my hat off to the Powers, and salute as the winner passes . . . I am for you and the Laneport Dam. . . . But we are definitely interested in securing what the Government years ago proposed on the San Gabriel River west of Georgetown. . . . We are going after the Georgetown dams and asking your cooperation.[26]

Sherrill's flattery failed to impress Fox. In fact, Sherrill's strategy of officially supporting Laneport while sniping at it backfired. Laneport's handful of backers, headed by Wilson Fox, maintained that Georgetown's word could not be trusted. The claim had some merit. Within

[24] Ibid.

[25] "History of Owen W. Sherrill," not dated, *Austin American-Statesman* morgue.

[26] Letter, Sherrill to Wilson Fox, Nov. 6, 1948, House Collection 1937–49, Rivers, San Gabriel Dam Project 1947–48, Box 314 (LBJ Library, Austin).

days of the Taylor City Hall meeting, at which "Mr. Sherrill brought the more than 35 [Georgetown] representatives to their feet as a token of their support of [Laneport]," Georgetown changed its position.[27] In a letter to the *Williamson County Sun* Sherrill retreated from Georgetown's "total" support of the previous week (the italics are mine):

> While we are in accord with the location of a dam at Laneport *if such dam would serve a useful purpose to the citizens of Williamson County and if the people whose land will be inundated are in favor of a dam at that location*, I still feel that a dam or dams West of Georgetown will be necessary *before* a Laneport dam could be made to control the rushing floodwaters originating in the higher terrain in the Western portion of the county.[28]

To turn the Army engineers around, Georgetown needed the Brazos River District. *Sun* editor Sharpe had already queried civil engineer John Alexander Norris, the revered creator of the Brazos River Conservation and Reclamation District, about his opinion of Laneport. "The army must have at least *thought* they had a good reason for changing the Gabriel setup," Norris wrote back. "But the soundness of the change has not appeared to me."[29] Sharpe had forwarded Norris's note to Lyndon Johnson.

Now that the Army had actually recommended Laneport, Georgetown wanted Brazos officials to press the Army to restudy the San Gabriel Valley. This would be tricky because the Brazos District needed to have several dams built quickly in order to strengthen its fragile economic base through hydroelectric sales. Any delay would hurt.

The day after the Taylor meeting, Georgetown's Chamber manager wrote to the Brazos water district's manager expressing Georgetown's reservations about Laneport.[30] Sherrill immediately "punched

---

[27] "Georgetown Panel Offers Aid To Secure Reservoir," *Williamson County Sun* (Georgetown), Nov. 26, 1948.

[28] Owen Sherrill, Letter to the Editor, *Williamson County Sun* (Georgetown), Dec. 3, 1948, p. 4.

[29] Letter, John Alexander Norris to John Sharpe, Nov. 7, 1947, House Collection 1937–49, Rivers, San Gabriel Dam Project 1947–48, Box 314 (LBJ Library, Austin).

[30] Letter, Fred Pool to R. D. Collins, Nov. 24, 1948, Box 59 (Brazos River Authority, Waco).

up" one of his "elite" connections, the Brazos River District's general counsel, John D. McCall of Dallas. As a young lawyer, McCall had served as Governor William D. Hobby's secretary and had developed powerful political connections. His sister was married to Georgetown's postmaster, Dr. Hobson Martin, a shy and childless Boy Scout leader, and—an oddity in 1948 Texas—a Republican. McCall favored the dams west of Georgetown, though like most Texans interested in water resources, he believed dams were the state's only salvation and thought it best to build as many as possible at Washington's expense. He had driven from Dallas to attend the Taylor meeting and had been distressed at a "glaring" headline in the *Austin Statesman's* coverage that read, "Georgetown Rebuffed On Dam Plans."[31] McCall sent Chamber manager Fred Pool the clipping, along with his opinion that the "discrepancies" in the "figures and calculations announced by Colonel Norman while he was making his talk" could be used "in support of the resolution which I hoped [sic] will be adopted by the Public Works Committee of Congress authorizing specifically further study of the proposed dams West of Georgetown." He thought Colonel Norman had "rather invited" Congressman Thornberry to present such a resolution. "Please be assured," he added, "that the writer as an employee of the Brazos District, as well as in his personal capacity, will be very happy to cooperate with you and with the good people of Williamson County in furthering these enterprises."[32]

Sherrill leaped on McCall's suggestion, writing Colonel Norman to ask him to send him the resolution form Congressman Thornberry needed to officially request a "reconsideration" of Laneport in favor of dams near Georgetown.[33] The Army engineer stalled. A month later, after repeated queries from Sherrill, Colonel Norman asked the Georgetown realtor to postpone "sending" the resolution (which, of course, Sherrill did not possess) until the House Public Works Committee had approved the Corps' work on the Brazos River and Con-

---

[31] Dave Shanks, "Georgetown Rebuffed," *Austin American-Statesman,* Nov. 24, 1948.

[32] Letter, John D. McCall to Fred Pool, Dec. 13, 1948, Box 59 (Brazos River Authority, Waco).

[33] Letter, Owen Sherrill to Henry R. Norman, Dec. 23, 1948, Brazos River 100-675, Box 36E12, Army Corps of Engineers (National Archives, Southwest Region, Fort Worth).

The Salesman

gress had authorized it. "[A]ny action to secure a review of the conditions on the San Gabriel Rivers [*sic*] before Congressional approval of the Brazos River report may rebound in such a manner as to force the return of the entire comprehensive report for review, thereby delaying development of the entire Brazos watershed," Colonel Norman warned. "The proper time for requesting a resolution . . . is *after* approval of the current report."[34]

Sherrill bowed to Norman's request. He would ask for the resolution at the "proper time," he wrote, adding, "We are again requesting you to furnish us such a resolution at the opportune time, and depending upon you and your united cooperation."[35] It was a tactical mistake from the point of view of Laneport's opponents.

The Army tried to paint Sherrill as a money-grubbing land speculator in his quest for upriver dams west of Georgetown. One division head wrote another, "I suspect, but cannot prove, that Mr. Sherrill is interested in property in the upper reservoirs and that this accounts for his anxiety to secure a review of reports."[36] Actually Sherrill owned no land anywhere along the San Gabriel River, nor would he ever, but he believed in the legitimacy of his cause. He also believed, correctly, that the best way to develop Georgetown (and his business) was to transform the rocky hills of western Williamson County into a lake environment.

Over the next few months, John McCall, the Brazos District's lawyer, shaped Georgetown's political and legal strategy—from the timing of the proposed congressional resolution to the resolution's language and logic. The value of McCall's legal advice cannot be overestimated. The fact that he was advising Sherrill and others in Georgetown in ways that undercut the Army's Laneport plan was duly noted by directors of the Brazos Conservation and Reclamation District, who depended on him for legal and political counsel. McCall's pro bono advice to Georgetown was remarkably detailed. In one letter, for instance, McCall compliments Sherrill on his "excellent

[34] Letter, Henry R. Norman to Owen Sherrill, Dec. 31, 1948, ibid. My emphasis.

[35] Letter, Owen Sherrill to Henry R. Norman, Jan. 4, 1949, ibid.

[36] Letter, B. L. Robinson to Henry Hutchings Jr., Jan. 13, 1949, Army Corps of Engineers Ft. Worth-Dallas, Civil Works Project Files 1934–61, Brazos River, Box 36E12 (National Archives, Southwest Region, Fort Worth).

reasons for impounding in the upper hills as much as possible of the flood waters," but suggests,

> You might consider the advisability of emphasizing the advantages of obtaining as much of the storage as practicable above Georgetown for the threefold reasons you have indicated: (a) lower economic value of lands in the hills; (b) protection of lands between upper reservoirs and Laneport; and (c) reduction of valuable submerged land at Laneport. At the same time you might consider the advantages of eliminating the position that sufficient flood storage will be attained by providing 2950 acres at North San Gabriel, 1600 at South San Gabriel and *2890* at Laneport. This would mean a total of flood storage at the 3 dams of 6440 acres, whereas, the Army engineers now think a total of *11,960* acres is needed. In other words, it might take quite a bit of "selling."[37]

McCall also tossed in an "off the record" tip:

> *Usually*, a stronger case can be made for 'up-stream' people by making an affirmative case for the benefits of the up-stream storage with only *objective* consideration for the lesser amount of lower valley land saved. By this I mean that to the extent that it can be demonstrated that there will be less disturbance of landowners and less total *dollar cost* of land will be used by virtue of the revised plans, the revision will be favored. But the individual reluctance of landowners to give up their land will not be considered as very important. . . . So usually upstream people do not improve their case by making common cause with the owners of the proposed submerged land.[38]

It was an important point that Sherrill, with his Georgetown allies, disregarded. Williamson County citizens, agrarian in outlook and economics, simply did not want to see as many as 250 farm families forced off their land, and with no flood control for the county in the

---

[37] Letter, John D. McCall to Owen Sherrill, Jan. 10, 1949, Box 59 (Brazos River Authority, Waco).
[38] Ibid.

bargain. By March, Sherrill was anxious. Had the Army misled him? "We have certainly cooperated with you, are patiently awaiting the resolution for our Congressman to use in request for consideration of the two dams at Georgetown," Sherrill reminded Colonel Norman. "We are . . . now most kindly asking that you give us the resolution promised."[39]

On March 21 Colonel Norman finally agreed to send Sherrill the magical resolution—but only if Sherrill *insisted.* "[I] am only trying to point out what might happen to the basin-wide plan should the subject review be proposed on a watershed prior to approval of justified and recommended works," Norman wrote, implying that the entire Brazos watershed scheme was doomed if Georgetown persisted in requesting a resurvey. "If you still consider it advisable to ask Congressman Thornberry to initiate such a review now we will be glad to cooperate in furnishing a proposed draft of the review resolution . . . at your request."[40]

By this time, Sherrill had gone public, urging the Williamson County Commissioners Court, headed by the most respected politician in the county, Judge Sam V. Stone, to ask the Army to reconsider the Laneport dam. After hearing protests from "hundreds" of farmers and businessmen, Judge Stone and all four county commissioners forwarded a resolution to Senator Johnson and Congressman Thornberry requesting a "reconsideration" of the Laneport dam plan, "to the end that the results achieved . . . can render a maximum protection to the citizenship of Williamson County . . . and that the best land . . . will not be converted into a shallow lake basin and cause some two or three hundred farmers to be displaced and the danger of further destructive floods in Williamson County will not be lessened."[41] The county officially opposed the Laneport dam. The editor of the *Williamson County Sun* summed up the prevailing sentiment:

> At the outset of the dam activity at a meeting at Taylor, Congressman Thornberry stated that he would like to see the

[39] Letter, Owen Sherrill to Henry R. Norman, Mar. 10, 1949, Army Corps of Engineers, Central Files, Box 36 (National Archives, Southwest Region, Fort Worth).

[40] Letter, Henry R. Norman to Owen Sherrill, Mar. 21, 1949, ibid.

[41] Resolution of the Commissioners Court of Williamson County, Mar. 18, 1949, Senate File 1949–61, Rivers & Harbors 1949–53, Box 847 (LBJ Library, Austin).

Road, River, and Ol' Boy Politics

county get together on what it wanted, and then let him know. There is every evidence, Mr. Congressman, that the county has decided, definitely, that the upper dams should receive primary consideration.[42]

Since November, a small flood of communications from Williamson County had poured into Senator Johnson's office—the county commissioners' resolution, two petitions signed by eighty people and nearly two hundred letters and newspaper articles—all hostile to Laneport Dam. Only one person endorsed it—the owner of the Hare cotton gin, located just below the proposed dam.[43] Johnson was fond of Williamson County, which had always given him huge voting margins. So he picked up the phone and talked to the man he most trusted in Williamson County.

_____

Harris A. Melasky was a soft-spoken, unpretentious attorney with a Taylor practice specializing in oil and gas. He had grown up hard, the son of a Jewish peddler-turned-merchant in a place and time when Jews were viewed with suspicion, if not hostility. Small and wiry, Melasky had played football for the University of Texas.[44] His honesty and political acumen were regarded as stellar, and he had prospered by keeping, with Johnson's help, his oil refinery clients in business despite World War II embargoes. Since then, on several occasions Melasky had saved Johnson's political campaigns from financial disaster.[45] Now Johnson asked Melasky to investigate the Laneport imbroglio. Melasky's investigation, a model of dispassionate fact-finding mixed with political pragmatism, showed why Johnson valued his friend's judgment. "My Dear Lyndon," Melasky began his report,

I find that there is no organized group favoring the construction of [Laneport] dam with the possible exception of the committee of the Taylor Chamber of Commerce. There is

[42] Don Scarbrough, "Dam Public Opinion," "Passing Glance," _Williamson County Sun_ (Georgetown), Apr. 14, 1949.

[43] Letter, George D. Bohlen to Johnson, Mar. 25, 1949, Senate File 1949–61, Rivers & Harbors 1949–53, Laneport (LBJ Library, Austin).

[44] Oral History: J. J. "Jake" Pickle, May 31, 1970, Section I, p. 12 (LBJ Library, Austin).

[45] Ibid., I, p. 13.

considerable organized opposition. One group is composed of the Granger people and the other group is composed of the people at Georgetown who desire the dams built in that area. The people at Taylor, in my opinion . . . are rather disinterested.[46]

Melasky ticked off six "reasons for opposition" to Laneport. At the top of his list were the "small Czech farmers" fighting for their homes and land. There were other arguments against Laneport: it would hurt Granger merchants, who were solidly against it; give Williamson County no flood protection; sink prospects for the dams Georgetown favored; and remove substantial monies from local tax rolls. He proffered no arguments for supporting Laneport. "From my conversation with various parties at both Georgetown and Taylor," Melasky noted,

I stressed the point that unless the dam was built at Laneport there would be no dam built. In nearly every instance, the parties replied that rather than have the dam at Laneport they would prefer that a dam not be built on the San Gabriel River. Frankly, those at Georgetown feel that if they can stop the Laneport dam, they ultimately may be able to get the dams erected at Georgetown.[47]

Melasky's final word, however, was aimed not at the reasonableness of the Laneport or Georgetown dams but on how the issue would play politically. "I doubt if this issue would lose you any substantial number of votes even if the dam should be built at Laneport," he wrote. "On the other hand, you would probably gain votes in other sections of the State which would be benefited by this dam. However, if the dam should not be built, I doubt if it would materially affect you with the voters of Taylor."

In other words, Johnson could get downstream voters and financial backers who were beholden to the rice farming industry and Dow Chemical, two entities pressuring Brazos Water District officials for more reliable flows of water. Let the engineers decide, Melasky

[46] Letter, Harris A. Melasky to Johnson: Apr. 22, 1949, Senate File 1949–61, Rivers & Harbors, Box 847 (LBJ Library, Austin).
[47] Ibid.

advised. If they "deem it in the public interest to erect the dam at Laneport, then certainly you should not oppose it. On the other hand, if the engineers should change their views . . . then I believe you should . . . not attempt to force the construction at Laneport."[48]

Then and there, it appears, Lyndon Johnson made up his mind about the San Gabriel dams. He would steer clear of them. He would sympathize with afflicted parties, provide information and technical assistance when it was sought, but he would not take sides. Instead, as Melasky counseled, along with most members of Congress, he allowed himself to "be guided by the views" of the Army engineers.[49]

Nobody could bring the argument to a conclusion. It was the start of a long war of frustration and growing animosity that exacerbated an old rivalry between Georgetown and Taylor. It didn't help feelings among Taylor's leaders that the new national defense interstate highway route was apparently to be routed through Georgetown, not Taylor. In June Congressman Thornberry, pressed by Granger, Georgetown, and Taylor, requested that Congress pass the Laneport Amendment. He and the chief Army engineer in Galveston agreed to keep news of the amendment secret, except within Army engineering circles.[50] Already Thornberry was sick of Laneport. He was also sick of Owen W. Sherrill. "May I suggest," he cautioned the new Army chief engineer at Galveston,

> that Mr. Norman be very careful in relaying any plans which Colonel Robinson and I discussed to any of the people in Williamson County. I note in your letter you state that Mr. Norman is keeping in close touch with Mr. Sherrill. In my opinion, a great deal of the misunderstanding which has arisen has been due to the fact that there has been too much local discussion.[51]

[48] Ibid.
[49] Ibid.
[50] Letter, Cong. Homer Thornberry to Col. Ellsworth I. Davis, June 27, 1949, Army Corps of Engineers, Central Files, Box 36 (National Archives, Southwest Region, Fort Worth).
[51] Ibid.

Thornberry's Laneport Amendment made no headway in Congress and the Army erected a protective shield around its Laneport project. Sherrill tried to pry maps out of the Army's Galveston office, maps the Army Corps' survey teams should have made after studying dam sites above Georgetown.[52] He never got the maps nor, for that matter, could he get any hard evidence that Corps engineers had even investigated the Georgetown dam sites before choosing Laneport.[53] Meanwhile, the Army blocked new attempts by Thornberry to get Congress to pass a resolution to restudy the San Gabriel River.

Like Sherrill, McCall suspected the Corps of misleading the pro-Georgetown dam folks. The prominent Dallas attorney had asked an Army engineer, "point blank," whether it was not now time for the Georgetown people to "properly ask" for the resolution. The Corps man said no, the resolution had to wait until *after* Congress authorized Laneport. "Very frankly," McCall wrote his brother-in-law in Georgetown, "my recollection is that as soon as the [Flood Control] Committee has acted on the report it would be all right to ask that the resolution be adopted by the Committee." Congressman Tom Pickett, a member of the Flood Control Committee, had also assured McCall this was the correct procedure. "I am definitely committed to Mr. Sherrill and to the Georgetown people to help them . . . get a resolution adopted . . . for further study on these projects," McCall wrote.[54]

But neither the Laneport dam nor its Georgetown alternatives made any real progress. Between 1950 and 1957 and mountains of correspondence, Sherrill, McCall, Thornberry, and Poage attempted, time after time, to break through the San Gabriel River logjam so that *some* dam could be built on the San Gabriel River, to little effect. Sherrill suggested developing hydroelectric power on the upper San Gabriel, an impractical notion which consulting engineers pooh-poohed after McCall asked them to evaluate it.[55] Sherrill also pitched

[52] Letter, Owen Sherrill to Johnson, Jan. 24, 1950, Senate File 1949-61, Rivers & Harbors 1949-53, Box 847 (LBJ Library, Austin).

[53] Letter, Owen Sherrill to Homer Thornberry, Oct. 17, 1952, Box 59 (Brazos River Authority, Waco). I searched for, but did not find, evidence of any Army engineering study of the western forks of the San Gabriel.

[54] Letter, John D. McCall to Dr. Hobson Martin, Mar. 20, 1950, ibid.

[55] Letter, Owen Sherrill to John D. McCall, Sept. 19, 1951, San Gabriel Dams, Box

an idea which resurfaced over the years: combine the planned inter-state highway with dams over the San Gabriel River west of George-town. Dewitt C. Greer, Texas's chief road builder, politely dismissed that prospect.[56] Congress eventually authorized Laneport, along with the other Brazos projects, in the Flood Control Act of 1954, trigger-ing a vigorous letter-writing campaign orchestrated by Sherrill for the long-awaited restudy of the San Gabriel. Congressman Thorn-berry, who by now realized just how unhappy Williamson County was about the proposed Laneport dam, prodded Congress into approving a restudy on July 29, 1955.[57] But the Army got no money to study the San Gabriel, so nothing happened.[58]

It was the river itself that resurrected the San Gabriel dams. Had the San Gabriel's flood-prone behavior not briefly united Williamson County, it is unlikely that any of the dams ever would have cleared Congress, for congressmen were reluctant to appropriate money for civil works projects that split political power bases back home—and Laneport was a prime example of that. But the early fifties brought drought to Texas and changed attitudes about flood control and water supply. Instead of focusing on controlling floods, civic leaders worried about their cities' growing thirst. They still wanted dams, but now they wanted to store flood waters that, during Texas's wet periods, rushed uselessly—wastefully, in the context of the times—into the

---

59 (Brazos River Authority); letter, McCall to Sherrill, Nov. 8, 1951, ibid.; letter, McCall to C. R. Marks, Sept. 3, 1953, ibid.; letter, Marks to McCall, Sept. 16, 1953, ibid.

56 Letter, Dewitt C. Greer to Owen Sherrill, Dec. 22, 1954, I-35/Dam, Box 59 (Bra-zos River Authority); letter, Sherrill to Collins, Jan. 4, 1954, ibid. Two years later, the Williamson County Farm Bureau, led by Martin W. Bergstrom of Georgetown, sent a petition and several letters to Senator Johnson's office promoting the combination of an Inter-Regional Highway and dam at Georgetown, rather than at Laneport. But this effort was no more successful than Sherrill's. See letters, Martin W. Bergstrom to Lyndon B. Johnson, July 28, 1956, and Arthur C. Perry to Bergstrom, July 31, 1956, Senate Case and Project 1956, Box 1230 (LBJ Library, Austin).

57 Public Law 780, 83rd Congress, Flood Control Act of 1954, cited by Senator Johnson in "Resurvey of San Gabriel River," Aug. 1957, Senate Case and Project 1957, Box 1339 (LBJ Library, Austin); Hendrickson, *The Waters of the Brazos,* 122.

58 Letter, Brig. Gen. J. L. Person to Johnson, Mar. 4, 1957, Senate Case and Project 1957, Box 1339 (LBJ Library, Austin).

Gulf of Mexico. McCall first seized on the notion of building dams on the San Gabriel's upper forks for flood control *and* water storage, which the Brazos Water District could then sell to downriver irrigators, or to cities needing more water.[59] It was a novel idea in 1952, when fewer than ten thousand people lived in all of west Williamson County, but one that found growing favor within the Brazos District.

By the summer of 1953, drought had Texas in a death grip. Senator Johnson challenged the nation to underwrite a "coordinated, long-range water program in Texas—a land in which too much sun is followed by too much water." Such an investment, Johnson argued, similar to those the government had already made in California, Arizona, and the Tennessee River Valley, would allow Texas to mine its "vast storehouse of undeveloped resources" and give the nation "elbow room in which to grow."[60] The Corps of Engineers, recognizing a niche it could fill, got Congress to vastly broaden its charge. From being limited to building flood control dams and navigation projects, the Army could now provide municipal and industrial water supplies, as well as enhance fish and wildlife stocks and recreational opportunities.

The drought sucked up Williamson County's water supplies and cracked its vaunted Black Waxy into fissures big enough to swallow chickens and small dogs. By August of 1952, Taylor's wells had dropped to their lowest point ever, due to increasing industrial demand, a growing population, and scant rainfall. Thorndale, the county's fourth-largest town, had to truck its drinking water from Taylor. For the first time, Taylor's city commissioners appealed to Senator Johnson and Congressman Thornberry to authorize the Laneport dam. They needed water.[61]

Meanwhile, the long drought, combined with flash floods, nearly brought the Brazos River District to its knees. One of its biggest cus-

[59] Letter, John D. McCall to Owen Sherrill, Sept. 11, 1951, ibid.; letter, McCall to George C. Change, Sept. 26, 1951, ibid.; letter, McCall to Sherrill, Oct. 10, 1952, ibid.

[60] Speech, Johnson to U.S. Senate, July 29, 1953, Senate Case and Project 1949–61, Box 1186 (LBJ Library, Austin); "Water Supply and the Texas Economy," Department of the Interior, Bureau of Reclamation, July 8, 1953, U.S. Government Printing Office, Washington (Center for American History, University of Texas at Austin).

[61] "City Commissioners Support Revival of Laneport Dam on San Gabriel," *Taylor Times,* Aug. 14, 1952, Box 59 (Brazos River Authority).

tomers, Dow Chemical, demanded more water and the BRA obliged, but in a deal that economically crippled the river authority.[62] But not doing the deal could have caused Dow to "shut down and seriously disrupt the economy of the Gulf Coast."[63] The politically potent rice growers on the Texas Coast also demanded more water released from Brazos reservoirs for their crops. "I think we just about have the rice growers straightened out," Brazos manager R. D. Collins wrote. "However, at the rate they are now drawing on [Lake] Whitney for water, I am afraid that the 50,000 acre feet will soon be exhausted and we will have another crisis on our hands before the rice crop is cured."[64]

The ramifications ricocheted upriver to northwest Texas, where the Brazos River District had built its only hydroelectric plant. Morris Sheppard Dam, commonly called Possom Kingdom, had kept the Brazos District financially afloat by producing electrical power, but in the fifties, it was threatening its creators with bankruptcy. During heavy rains of August 1953, Possum Kingdom Lake backed up and overflowed into Salt Creek, which flooded the sewage works in Graham and polluted the city's water supply. The city lodged a politely worded complaint with the Brazos board. The next summer, in an attempt to boost downstream water supply, the Brazos River District announced it would raise the level of Possum Kingdom Dam by fifteen feet, increasing the lake's water storage capacity by 354,000 acrefeet. Residents of Graham, a well-heeled oil town of about ten thousand, passionately protested. Then, in torrential rainstorms of June and September of 1955, Possum Kingdom Lake surged backwards again. Nearly two hundred Graham homes were badly flooded; its sewage plant and waterworks were swamped. Finally, in late April and early May of 1957, when flooding inundated much of Texas for sixteen days, Graham citizens again found themselves bailing flood waters caused by Possum Kingdom Dam. Graham sued and won a $431,000 judgment. The Brazos River Authority's team of attorneys, led by John McCall, appealed to the Texas Supreme Court, but lost. As the story unfolded, the Brazos River Authority took a pounding from the

[62] Hendrickson, *The Waters of the Brazos*, 48, 66–68, 84.

[63] Memorandum, H. S. Hilburn to Lyndon Johnson, Dec. 1956, Senate Papers 1956 Case and Project File, Box 1210 (LBJ Library, Austin).

[64] Letter, R. D. Collins to J. Howard Fox, Aug. 28, 1952, ibid.

state press for contesting Graham's claims. The fact was, the agency was hurting financially and its lawyers wanted to avoid setting a precedent for settlement. Rowing hard against this legal and public relations disaster, the Brazos directors were attempting to build support for an ambitious Six Dam Plan, which would transform about two hundred miles of the Brazos River Valley into a solid line of lakes between Possum Kingdom and Lake Whitney. But there was stiff opposition to that plan from many quarters. By the end of 1957, the Six Dam Plan was dead.[65]

Against this backdrop, controversy within Williamson County over Laneport and the Georgetown dams looked like a footnote at the renamed Brazos River Authority—that is, until it seemed the BRA might lose its right to service part of its watershed. By 1956 the drought had nearly ruined Williamson County's farmers and ranchers. In February 1957 Taylor sounded the Lower Colorado River Authority on the possibility of building a pipeline to funnel water from Lake Travis on the Colorado River to ease the city's water crunch. This would mean abandoning the Brazos watershed for the Colorado watershed, something that had never been allowed by Texas law. Excitement in Taylor ran high, fueled by breathless press reports. Six months later the Williamson County Commissioners Court created a water control and improvement district that included Taylor, Georgetown, Leander, Round Rock, Hutto, Thrall, and Thorndale. The plan was to acquire water from Lake Travis through the LCRA, or to build a dam at Georgetown with cooperative local funding, or both. Taylor hired two engineering firms to study costs and benefits of several possible options. It appeared that Williamson County was getting ready to divorce the Brazos River Authority and take a new partner, the Lower Colorado River Authority, to meet its water supply needs.[66]

The effect was like a rocket lobbed at the BRA's Mineral Wells

---

[65] Memorandum, Mar. 21, 1955, Walter [Jenkins] to Johnson, Senate Case and Project 1956, Box 1210 (LBJ Library, Austin); Hendrickson, *The Waters of the Brazos*, 84–85. The Brazos River Conservation and Reclamation District changed its name to the Brazos River Authority in the mid-1950s.

[66] Lin Mills, "It Occurs To Me," *Taylor Daily Press,* Feb. 20, 1957; Minutes of BRA directors meeting, Apr. 15, 1957, Minutes, Board of Directors (Brazos River Authority,

headquarters. Brazos officials flung themselves into activities aimed at bringing Williamson County back into its fold. Now the BRA would support dams west of Georgetown, as well as at Laneport, even if the BRA had to partially fund a Georgetown dam itself. The benefit, of course, was that the BRA could sell the impounded water to water-starved municipalities (or to whomever needed it), shutting down the possibility of an LCRA invasion across watersheds. In January 1958 J. H. Kultgen, the BRA's board president, conversed with Col. Walter Wells, the Army's chief regional engineer (who would later become the Brazos Authority's general manager), about the stalled San Gabriel projects and reported, as if the subject had never come up before, that if Georgetown and Taylor citizens "desired" Laneport or the Georgetown dams, "the [Army] engineers will initiate some studies for which they have money to ascertain whether or not the location is feasible."[67] This was a remarkable statement, since Congressman Thornberry had been futilely trying to get a resurvey of the San Gabriel watershed funded for two and a half years.

McCall tried to calm BRA director J. Howard Fox's "anxiety over the situation in which the City of Taylor and another city find them-selves" with respect to obtaining "water from the Colorado River source." Fox was agitated because all of his brothers—Wilson, Bryan, and Henry—were battling over Laneport, and against each other, an alarming state of affairs in east Williamson County, where all three men were highly respected. Emotions were running so high that the opposing Fox brothers had ceased speaking. If the BRA could find the revenue, McCall assured director Howard Fox, it would join with Taylor and Georgetown and build North Fork dam itself "in the

Waco); letter, Owen Sherrill to R. D. Collins, Apr. 29, 1957, Box 59, ibid.; letter, W. S. Gideon to R. D. Collins, Aug. 9, 1957, Box 59, ibid.; note on telephone conversations between Collins, Wilson Fox, and Gideon, Aug. 20, 1957, Box 59, ibid.; memorandum, Roger Tyler to BRA, Aug. 22, 1957, Box 59, ibid.; letter, Collins to Gideon, Aug. 23, 1957, Box 59, ibid.; letter, Cameron Engineering Co. to the Mayor of Taylor, Dec. 27, 1957, Box 59, ibid.; letter, [S. W.] Freese and Nichols Consulting Engineers, Fort Worth, to City of Taylor, Oct. 24, 1958, Box 59, ibid.

[67] Letter, J. H. Kultgen to R. D. Collins, Jan. 29, 1958, Box 59 (Brazos River Author-ity, Waco).

event the Army program is *not* changed to incorporate such a structure."[68] BRA manager Collins forwarded a detailed proposal to the cities of Taylor and Georgetown "on the possibility of the Authority constructing a dam on one of the San Gabriels near Georgetown to supply water for the Cities of Georgetown, Taylor and neighboring communities." The cost was high but not impossibly so. The better economical answer, Collins admitted, "would be for us to hope and pray that the Army eventually decides to build the 2 dams on the San Gabriels near Georgetown and a smaller Laneport Dam."[69] Thus, a decade after the Army engineers picked Laneport, the Brazos River Authority firmly aligned itself on the side of the "local interests" fighting for dams on the upper San Gabriel along with a truncated Laneport—in other words, on Owen Sherrill's side and against the Corps.

Even with the Brazos River Authority behind it, Congress still dragged its heels on funding a resurvey. Congressman Poage, angry at the decade-long delay of delivering promised flood control to Milam County, threatened to bury the resurvey for good. In January 1957, in a series of blistering letters, Poage, once so friendly, attacked Sherrill for his county's intransigence:

> All I have ever asked you for was the assurance that you and the people of Williamson County would accept the decision of a resurvey, whatever it may be. I don't have that assurance. Your letter doesn't give it to me. On the contrary, while I realize you were very careful not to give me an answer one way or another, I think I can clearly imply [*sic*] that you haven't the slightest idea of accepting a decision unless it is the decision you have already made. . . . If Williamson County is determined to fight everything except that which her own people have decided is sound engineering, then we don't need any

---

[68] Letter, John D. McCall to J. Howard Fox, Apr. 28, 1958, ibid. The other city was probably Thorndale, then the fourth-largest city in Williamson County. Round Rock sat conveniently on the pipeline's route, but it was so small it was rarely mentioned until the 1970s.

[69] Letter, R. D. Collins to J. R. Barclay, May 29, 1958 (Brazos River Authority, Waco).

further engineering . . . I would deeply appreciate it if you would give me a straight, frank answer.[70]

Sherrill's "frank" answer offered only part of the assurance Poage sought: "Bob, all we ask for is the same protection instead of damage, just as you are asking for below. To do this Bob, will require dams above Georgetown first. . . . We will accept the Engineer's determination as to THE SIZE OF A DAM AT LANEPORT."[71] Sherrill had Poage over a barrel and Poage knew it. Congress would not budge to fund Laneport until Williamson County convincingly supported it, and this was not happening. Two days after Sherrill sent his "answer," Poage caved in. "I can say without reservation that I am going to do everything I can to help Homer [Thornberry] get the funds needed for a resurvey of the San Gabriel Dam sites," he wrote Sherrill.[72]

Finally, Williamson County presented a solid front. The county government and the Georgetown, Taylor, and Granger governments all called for an immediate resurvey.[73] But even with Poage testily on board and Thornberry pushing, Congress was not inclined to cough up a penny.[74] Then came the flood of 1957.

Fifteen days of hard rain filled the state's dry reservoirs and broke a seven-year drought, and then, on April 24, "torrential" storms brought tornados, flooding, and death to Central Texas. Much of the misery was felt at Georgetown, where the North and South Forks of the San

[70] Letter, W. R. "Bob" Poage to Owen Sherrill, Jan. 17, 1957, Senate Case and Project, 1957 San Gabriel Dams, Box 1339 (LBJ Library, Austin).

[71] Letter, Owen Sherrill to Poage, Feb. 5, 1957, ibid.

[72] Letter, Poage to Owen Sherrill, Feb. 7, 1957, ibid.

[73] Letter, County Judge Sam V. Stone to Poage, Jan. 14, 1957, W. R. Poage Papers (Baylor Collections of Politcial Materials, Waco); Resolution of the City of Georgetown, Jan. 14, 1957, W. R. Poage Papers, Box 692, File 4, ibid.; "Statement in Support of Proposal . . ." by Williamson County Brazos River Association, Georgetown and Taylor Chambers of Commerce, Williamson County Commissioner's Court, Georgetown City Council, Taylor City Commission; "3-Town Request Calls for Gabriel Dams Re-Survey," *Williamson County Sun* (Georgetown), Mar. 7, 1957, Senate Case and Project 1957, Box 1339 (LBJ Library, Austin).

[74] Letter, Owen Sherrill to R. D. Collins, Feb. 25, 1957, W. R. Poage Papers, Box 629, File 3 (Baylor Collections of Political Materials, Waco); Owen Sherrill to Mayor J. T. Atkin, Mar. 11, 1957, ibid.

Gabriel River converged, unleashing a record fifty-foot rise that rolled downstream, washing away farms, bridges, and railroad tracks. No lives were lost in Williamson County, but damage to Georgetown and the Black Waxy farms below the county seat was great.[75]

An Austin newspaperman flying with Ragsdale Flying Service snapped photographs just as the river's crest swallowed a farm house. "Just east of Georgetown, as the heady brown waters swirled around it, a farmhouse shrugged, crumpled and submerged in less than a minute. Only its roof bobbed above the frothy river as it was swept downstream," he reported.[76] A young aviator named Jim Boutwell flew through a thunderstorm to drop leaflets to the crowd assembled at the Highway 81 bridge crossing the San Gabriel River in Georgetown. "STOP THE TRAIN!" the notes ordered. The dramatic air drop was credited with preventing a train disaster on the M-K-T line at the Katy bridge.[77]

Suddenly the climate toward resurveying the San Gabriel River changed. Now it seemed more obvious than ever that dams were needed in the western hills above Georgetown. The river accomplished what nothing else could: it forced Congress's hand. On April 27, three days after the flood roared through Williamson County, Congressman Thornberry released a statement to the press that he would testify before the House Appropriations Subcommittee on Public Works in support of a new study of the San Gabriel River.[78] Two days later, Congressman Poage wrote Sherrill that flood control for Williamson County was essential, though he warned, "with the

---

[75] "Heavy Rains Trigger Floods Over Twister-Wary Centex," "Georgetown Feels San Gabriel Fury," "Destructive Crest Watched from Air," *Austin American,* Apr. 25, 1957, pp. 1, 2, 17, located in the John M. Sharpe Papers, Georgetown Public Library, Georgetown.

[76] Albert Griffith, "Destructive Crest Watched from Air," *Austin American,* Apr. 25, 1957, p. 17.

[77] Robert Lucey, "The Flood of '57," *Williamson County Sun* (Georgetown), Oct. 12, 1994. The same Jim Boutwell flew an airplane around the University of Texas Tower until a sharpshooter could pick off Charles Whitman during a shooting rampage in 1966 in which Whitman killed 21 people. Boutwell later was elected sheriff of Williamson County and served for many years. Governor Ann Richards attended an appreciation dinner for Boutwell when he was stricken with cancer.

[78] Press Release from Cong. Homer Thornberry, Apr. 27, 1957, Senate Case and Project 1957, Box 1339 (LBJ Library, Austin).

present attitude of the [Eisenhower] Administration, I am not at all sure that we are going to achieve much of anything this year."[79]

But Sherrill had covered his bases well, including Lyndon Johnson. The senator's interest in the San Gabriel dams had risen along with the floods and his replies to Sherrill struck an action note, rather than the vague sympathy of the past nine years. "I've been keeping in close touch with Homer," he wrote. "As you know, he is scheduled to appear before the Appropriations Committee of the House on May 13 to urge a resurvey of the flood control requirements of the San Gabriel. I shall follow all of these developments closely and keep in mind your interest."[80]

Two months later, Thornberry's request still awaited action. Sherrill wrote McCall, a close confidant of former governor Dan Moody, Senator Lyndon Johnson, and John Connally who would become governor of Texas. McCall followed up with a note to Johnson. "Our good friend Owen Sherrill of Georgetown wrote me a few days ago that you were optimistic of the prospects of an appropriation for the resurvey of the San Gabriel Rivers," McCall wrote. "This is indeed encouraging. The Brazos River Authority has thought for several years that there should be a resurvey of this area to develop the possibilities of both flood control and water supply for the people and property situated above the proposed Laneport Project."[81]

It was the clincher. Three days later Johnson replied in the carefully coded language politicians use with favored supplicants: "As you may know, the appropriation bill is now pending on the Senate Calendar. I am hopeful to get action on it soon. I am glad to know of your interest in this particular item and will keep in mind your views."[82] Two weeks later, it was done. The Senate earmarked twenty-five thousand dollars for the San Gabriel River resurvey. The Army could proceed. Williamson County rejoiced.[83]

[79] Letter, Poage to Owen Sherrill, Apr. 29, 1957, W. R. Poage Papers, Box 629, File 3 (Baylor Collections of Political Materials, Waco).

[80] Letter, Johnson to Owen Sherrill, May 2, 1957, Senate Case and Project 1957, Box 1339 (LBJ Library, Austin).

[81] Letter, John D. McCall to Johnson, July 22, 1957, ibid.

[82] Letter, Johnson to John D. McCall, July 25, 1957, ibid.

[83] Letter, Don Scarbrough to Johnson, Aug. 9, 1957, ibid.

The San Gabriel River crested at fifty feet during the 1957 flood, reviving plans to dam it. In this photograph the river batters the U.S. Highway 81 bridge in Georgetown. "Northtown"—a church and a few buildings north of the city proper—is in the background. *United Press Photo, Courtesy* Williamson County Sun, *Georgetown.*

It had taken nine years to force the Army Corps of Engineers to re-examine the San Gabriel Valley with an eye toward reducing the size of Laneport Dam or replacing it with dams west of Georgetown. Sher-rill's courtship of the political and water elites finally paid off. Though the power brokers did not always trust him, or even like him, Sherrill's relentless persistence forced them to work with him. In the end, whether they wanted to or not, people gave him what he wanted.

Georgetown teens watch the San Gabriel's South Fork recede. It had topped the
Highway 81 bridge, knocking out iron railings with trees torn from riverbanks by
flood waters. The author stands on the left; the segregated George Washington
Carver School, razed in the 1960s, sits on a bluff above "Blue Hole." *Don Scarbrough Photo, Courtesy* Williamson County Sun, *Georgetown.*

The San Gabriel River sprawls a mile wide east of Georgetown during the 1957 flood, suggesting what the 1921 flood might have looked like. *Don Scarbrough Photo, Courtesy* Williamson County Sun, *Georgetown.*

The Katy Railroad bridge barely escapes being ravaged by the 1957 flood. A pilot dropped leaflets to a crowd in Georgetown ordering, "Stop the Train!" The crowd complied and lives were saved. *United Press Photo, Courtesy* Williamson County Sun, *Georgetown.*

The flood swamped Georgetown's San Gabriel Park, including the Williamson County Rodeo Arena, which filled like a swimming pool, four feet deep. *United Press Photo, Courtesy* Williamson County Sun, *Georgetown.*

HENRY FOX, THE CZECHS, AND "LITTLE" DAMS

The closer we keep our children to the soil the healthier they will be . . .

EDWARD BOK, *Ladies Home Journal*

In 1948, when the Army recommended the Laneport dam, the engineers believed, and Congress took it as a given, that flooding problems were best controlled by big dams on the main stems of rivers. The philosophy had but one serious rival—the small dams movement. As Army engineers hastened to plug the nation's rivers after the war, the Department of Agriculture's Soil Conservation Service mounted a campaign to build tens of thousands of small earthen dams on the upper tributaries of flood-prone rivers. The idea was to prevent flooding at its source, rather than waiting for epic floods to develop and only then trying to stop them. The small dams approach enjoyed several theoretical advantages: collectively, it cost much less than big dams, stopped topsoil from washing downstream, and kept farmers on their land.[1] There was only one major drawback—it provided little water storage for cities wanting to grow.[2]

---

[1] In terms of its impact on small farmers, the "big dam" movement echoed England's enclosure movement, which stripped yeoman farmers of the "common" lands and forced them to take factory jobs.

[2] Elmer T. Peterson, "Big-Dam Foolishness," *Country Gentleman* (May 1952), 26–27, 74–75.

Congressman W. R. "Bob" Poage. *Courtesy Baylor Collections of Political Materials, Waco.*

Ironically, Waco's Congressman W. R. "Bob" Poage, the Laneport dam's biggest House booster from the start, authored the 1952 legislation that launched the "little" or small dams movement. To farmers, Poage was a hero—the "father" of the only legislation that could stop the Army from covering the nation's most fertile river valleys with lakes. Urban interests, fearing destructive floods and needing water, favored the big dam approach, and fell under the sway of the Corps of Engineers. Poage, who had studied geology at Baylor University and had a strong grasp of flood engineering, explained the beauty of the small dams approach:

> Water that soaks into the ground where it falls is an unmixed blessing. It does no one any harm and may do good for many people. It is water that falls and does not sink into the ground that does the damage . . . I think that we have had our attention focussed entirely too closely on the debris-littered torrents pouring down the main stems of our big rivers, and in profitless debate on how to handle that much water—how to *control* floods. We have forgotten, apparently, that somewhere

Henry Fox, the Czechs, and "Little" Dams

back upstream almost every yellow gallon of that boiling flood came out of a tiny branch that any fairly active coon-dog could jump across. Keeping the water back up there in that branch, or persuading it to sink into the ground before it even reaches the branch, is flood *prevention*.[3]

House Resolution 7868 contained serious faults in terms of practicality for farmers who wanted to create watershed districts, but Poage, chairman of the Agriculture Committee's flood prevention subcommittee, believed it better to get something than nothing. He feared that the Army's congressional cheerleaders might kill the small dams legislation if it threatened the Corps' turf. The chief problem with Poage's bill, a Soil Conservation Service official wrote, was that limitations on watershed size and on the volume of impounded water a "small" dam could contain were "so low that it would be a serious handicap to reaching the objective . . . if it did not make it impossible."[4] Poage agreed. "Most of the Committee realized the force of your argument," he wrote, "but all of us finally agreed that from a practical standpoint, of getting the bill passed, we better keep the figures low. . . . I am frankly afraid to undertake to . . . raise these limits . . . because I anticipate that the Public Works Committee is going to be very jealous of us . . . and would, in all probability . . . prejudice the House against us."[5] And so the Department of Agriculture gained the ability, albeit a limited one, to plan and construct networks of small dams to prevent flooding and conserve fertile cultivated soil.

In Williamson County, Poage had many admirers, including Granger newspaperman and Circleville farmer Henry Fox, who in 1952 considered him something of a hero. Poage had a firm grasp of agricultural issues and he truly cared about farmers. He didn't represent Williamson County; indeed, his responsibility to Milam County voters made him favor the Laneport dam, which would give Milam

[3] Press Release, U.S. House of Representatives, Committee on Agriculture, May 15, 1952, Washington, D.C., Box 443, File 1, W. R. Poage Papers (Baylor Collections of Political Materials, Waco).

[4] Letter, Louis P. Merrill to Poage, May 16, 1952, Box 443, File 2, W. R. Poage Papers, ibid. The Soil Conservation Service was part of the Department of Agriculture.

[5] Letter, Poage to Owen Merrill, May 31, 1952, W. R. Poage Papers, ibid.

the ultimate in flood protection at the expense of Williamson County farmers. But after their concurrent elections to Congress in 1937, Poage had worked closely with Lyndon Johnson, and later with Congressman Thornberry, because Blackland Prairie farmers faced similar problems wherever they plowed.

Williamson County made a perfect test site for Poage's small dams program. The Brazos River District and Army engineers had proved that most of the flood waters that periodically piled up across the lowlands of eastern Williamson and western Milam Counties came from the San Gabriel's tributaries—not from its main stem. While shutting down the main stem during a flood would certainly help matters, checking overflows on Brushy, Pecan, Willis, Donahoe, and Berry Creeks was equally important.

The county's chief economic engine—farmers—generally favored the idea of small dams, partly because they believed a carefully designed system of small dams would eliminate the need for the Laneport dam. The Czechs who owned so much of the county's Blackland Prairie were politically conservative but progressive farmers, employing the latest agricultural methods and investing heavily in their land. If the Laneport dam were built, they faced dispossession of their precious Black Waxy, their agricultural livelihoods, their homes and gardens, and their social and religious lives. To them, the small dams program seemed a godsend. But though they penned hundreds of letters to their elected representatives, no one in Washington seemed to heed them. No leader rose to capture the attention of the Washington power brokers. At least not until Henry Fox got interested in small dams and flood prevention.

———

Henry Benjamin Fox grew up in Granger when it was a robust market town, the third most important city in Williamson County. He was the fifth of seven children in a prominent banking family. In the twenties and thirties, Granger Czechs clustered together on the west side of the railroad tracks and the "Americanos" lived on "their" side of town, each group avoiding social and business encounters with the other.[6] But Henry, as Americano as they came, felt at ease with

---

[6] Interview, Betty Hajda and Loretta Mikulencak, Feb. 28, 2001, Granger.

Fox patriarch John Short Fox, left (face obscured), with his sons, left to right: Robert Bryan Fox, Henry Benjamin Fox, John Howard Fox, Walter Fox, and Wilson Harold Fox. They are in front of the Fox home on Fox and San Gabriel Streets in Granger. *Courtesy Carol Fox, Circleville, Texas.*

Czechs. He probably picked it up from his father, John Short Fox, who for many years ran the Czech-owned Granger National Bank.

John Short Fox was a classic American success story. Leaving Lynchburg, Tennessee, after finishing the sixth grade, he rode the train west, carrying a Colt 45 and a nickel in his pocket. On January 1, 1893, he disembarked in Taylor because he was hungry and out of money. He swept floors at a lumber yard and worked his way to the top of the business. He started a competing lumber yard in Granger, went into banking, and lived in a fine house at the intersection of Fox and Gabriel Streets with his wife Fanny and their considerable brood. He was a county school trustee and served on the boards of Brazos River Conservation and Reclamation District and Southwestern University in Georgetown, which gave him an honorary Doctorate of Law. He required that every one of his seven children attend college there. Upon his death in 1944, each child received a gift of three hundred acres of prime bottomland along the San Gabriel River. But under the terms of the bequest, the land could not be sold. "That's why the kids are all scattered up and down that river. He had a long reach," a grandson said. "He was a powerful person, very much the autocrat."[7]

Wilson Harold Fox, the pro-Laneport dam politico, was John Short's eldest child. After serving two terms in the Texas Legislature, Wilson practiced law in Taylor and ran the county's Democratic Party machine for more than two decades. Most people considered him a brilliant behind-the-scenes operator.[8] Robert Bryan Fox, another son, was a gifted orator, taught history and math and became superin-

[7] Interview, Paul Fox, Aug. 31, 2000, Austin; interview, Dr. James "Jim" Fox, Sept. 6, 2000, Austin; interview, Geraldine Fox, Sept. 20, 2000, Granger; "Granger Banker Is Honored By Southwestern U," *Taylor Times*, June 7, 1940; Hendrickson, *The Waters of the Brazos*, viii.

[8] "Wilson H. Fox Services Today," *Austin American-Statesman,* Feb. 11, 1974 (Center for American History, University of Texas at Austin).

Henry B. Fox, known as "The Circleville Philosopher," works in his smokehouse sanctuary. *Courtesy* Williamson County Sun, *Georgetown.*

Henry Fox, the Czechs, and "Little" Dams

tendent of schools in Czech-dominated Friendship.[9] At Southwestern, a third son, John Howard Fox, majored in philosophy and psychology but moved away from Williamson County to Hearne to manage the South Texas Cottonseed Oil Company. In 1947 Governor Beauford H. Jester appointed him to the board of the Brazos River District, over which he presided from 1959 through 1963.[10] The elder Fox's youngest son, Walter, edited the *Bartlett Tribune* and died in 1940 of a heart attack at the age of 30.[11] One daughter, Frances Fox Smither, married a Huntsville man and raised two children.[12] The other, Mary Elizabeth Fox, covered the United Nations as a journalist, then returned to Southwestern to write a master's thesis on Texas roads and direct the college's publicity arm for years.[13]

Only Henry rebelled. The family regarded him as a budding Bolshevik. "I graduated from Southwestern at eleven in the morning and I caught a train at twelve for New York. . . . [But] all I saw up there were writers trying to get rich enough to buy a house and move to the country. I saw a shortcut and figured, why not move to the country and to hell with getting rich?"[14] After a year living in Greenwich Village, he returned to Texas, edited several country newspapers, moved to his inherited three hundred acres at Circleville, and wrote a folksy humor column for *Colliers* magazine. When the magazine folded, he syndicated "The Circleville Philosopher" to about fifty newspapers across the nation with a claimed combined readership of

---

[9] Interview, Geraldine Fox, Sept. 20, 2000, Granger; interview, Paul Fox, Aug. 31, 2000, Austin.

[10] "J. Howard Fox Biographical Material," Box 59 (Brazos River Authority, Waco); "Salvation from Seed," *Houston Chronicle Rotogravure Magazine,* Sept. 22, 1957, Box 59 (Brazos River Authority, Waco).

[11] Interview, Geraldine Fox, Sept. 20, 2000, Granger; telegram, Jno A. Norris to J. S. Fox, Nov. 1, 1937, Box 59 (Brazos River Authority, Waco).

[12] Interview, Marie Fox, Feb. 26, 2001, Circleville.

[13] Biography, Mary Elizabeth Fox, *Austin American-Statesman* morgue, Oct. 1, 1961.

[14] "Book by H. B. Fox 'bamboozles' establishment in small town," *Texas Press Messenger* (trade magazine published in Austin by the Texas Press Association), Apr. 1975; Rick Smith, "Writer-Philosopher Took Shortcut to Beat Rat Race," *Austin American-Statesman,* Mar. 5, 1979; "H. B. Fox Writes Feisty, Folksy Humor," *Daily Texan* (University of Texas at Austin), Oct. 16, 1975 (Center for American History, University of Texas at Austin).

one million.[15] Eventually he published three books of impish humor lightly based on his experiences as a small-town newspaperman. Like Mark Twain and Will Rogers, with whom he was frequently compared, Henry Fox had an uncanny knack for exposing human foolishness, especially of the political variety. (As a young newspaper editor, he had run up against an East Texas congressman who had put every member of his family on the federal payroll, to Fox's disgust. Afterwards he instinctively disliked politicians, period, with the possible exception of Poage.)[16]

At heart his philosophy was a sort of gentrified populism, one that Thomas Jefferson might have recognized. In Henry Fox's ideal world, small-town life based on agrarian values far surpassed the urbanindustrial model as a civilized democratic society. "The nice thing about ranching" he wrote, "is you can ride around looking for cattle and call it work."[17] Or, as he once quipped on the "Today" show, "In New York City, you have problems and no answers. In Circleville, we have answers and no problems."[18] The Army engineers got their first taste of Fox's pugilistic wit in 1949 after they released the first map of the proposed Laneport dam. He ran a large advertisement signed by sixty-eight citizens in his newspaper, the *Granger News*, starting with this question:

IS THIS A GOOD THING?

Take a copy of the map of the proposed Laneport Dam reservoir. Notice where the maximum waterline would be. It would include all land not higher than 530 feet above sea level. This would include an area extending from a point on the San Gabriel River south of Granger almost to the town of Granger

[15] Kent Biffle, "The Sly Old Fox," *Dallas Morning News*, Apr. 28, 1984 (Center for American History, University of Texas at Austin).

[16] Letter, John Short Fox to Johnson, Sept. 19, 1942, House Papers, Box 144 (LBJ Library, Austin). Fox received rave reviews for his political spoofs, *The 2000-mile Turtle* (1975), *Dirty Politics is Fun* (1982), and *Murder in a Small Town—Perhaps* (1984). The first two books were published by Madrona Press; the last by Eakin Press, both of Austin.

[17] Jay Jorden, Associated Press, "Circleville Philosopher Works 'Gentle Fraud'," *Sherman Democrat,* July 31, 1988, p. 1 (Center for American History, University of Texas at Austin).

[18] Interview, Betty Zimmerhanzel, Dec. 8, 2000, Circleville.

itself, would even take in areas NORTH of Granger, would include Friendship.

In short, it would cause the government to buy practically ALL the land east of Granger . . . up to the Laneport Dam.

This land is now producing . . . about

ONE MILLION DOLLARS A YEAR

This would be revenue which Granger would lose, as Granger is the primary beneficiary from this area. It would take the sale of a powerful lot of fish hooks and tackle to off-set such a loss.

Over a full-page reproduction of the Army's map ran the headline:

OFFICIAL ARMY ENGINEER'S MAP OF LANEPORT DAM—
GUARANTEED TO BAFFLE YOU

Underneath the map, which resembled a particularly weird Rorschach test, is this caption:"If you can tell anything about it, here's the map of the proposed Laneport Lake. It's a little hard to figure out, but if you study it long enough and look closely enough you can get some idea of where the shoreline might be, a few minutes before you go crazy.The best way to understand it is to spread the Granger News out in front of you, with the top of this page pointing toward north. Then the Laneport dam will be east of you, or to your right, and you'll be at the south side of the San Gabriel River, looking at the project and wondering whose land is going to get covered up."[19]

Henry Fox could not stop joking, even when he was dead serious. He was almost pathologically shy, but did not care what most people thought of him. He was tailor-made to take on an unpopular dam using a program of small dams sprinkled across Williamson County as an economic, small-scale alternative to a big, unpopular Laneport dam.

While putting out Granger's newspaper, Henry Fox had connected with Czech businessmen who controlled most commerce in that town. Having planted cotton and pecan trees on his fertile river property, he empathized with the Czech families who would lose

[19] Advertisement,"IS THIS A GOOD THING?" and map,"Official Army Engineer's Map of Laneport Dam—Guaranteed to Baffle You," *Granger News,* circa Jan. 1949, House Papers, Box 314 (LBJ Library, Austin).

Road, River, and Ol' Boy Politics

their farms in Friendship, Machu, and Moravia—communities they had created and nurtured—because a dam designed to stop floods elsewhere would permanently flood *them*. To the Czechs it seemed that the Army engineers, the agency of their forced expulsion, were as ruthless as the Hapsburgs had been to their forebears.[20]

The Czechs had vigorously argued their case against Laneport since 1948 but had gotten nowhere. They tried Lyndon Johnson and they tried Homer Thornberry, but the person they turned to most plaintively was Bob Poage, architect of the "small dams" movement. Poage would understand their plight, they thought—all his senses were tuned to agriculture. The Laneport plan affronted common sense, they argued. The remarkable soils of the Black Waxy needed protection, not obliteration. A dam at Laneport would destroy the most productive nonirrigated agricultural land in Texas when flood-control dams could be built on the poor chalk soil west of Georgetown. The Czech farmers could not comprehend the dimwittedness of supposedly smart, "scientific" Army engineers who failed to factor into their cost-effective formulas the economic wealth created by their farms. Comparing *their* soils near Laneport ("rich loam soil of fabulous production") with those of Milam ("sandy lands") and western Williamson County ("chalk rock hills"), Czech protestors expressed stunned disbelief that the culivated Black Waxy should be considered expendable. They were, of course, entirely correct on the matter of soils. The widow of a revered Czech clergyman pleaded with Poage,

It is inconceivable to me . . . that such a program . . . would actually be carried out. Texas has vast areas of poor, sandy and submarginal lands and only a relatively smaller percent is highly productive soil. This land in Granger falls into this class . . . Milam County just below the proposed dam site is inferior sandy, clay soil, that is why it is inconceivable that the good rich land should be sacrificed when only a few miles down the road, the dam could be placed on poor soil—or placed above Georgetown where only *4 families* would be displaced in

[20] Interview, Paul Fox, Aug. 31, 2000, Austin.

Henry Fox, the Czechs, and "Little" Dams

Texas's Czechs largely worshipped either in the Catholic Czech Church or in the Evangelical Unity of the Czech-Moravian Brethren. Jno. Trlica documents "Pioneer Ministers" of the Brethren. *Courtesy Photography Collection, Harry Ransom Humanities Research Center, University of Texas at Austin.*

The village of Friendship was proud of its schools, blanketed in snow in this undated photograph. "The auditorium had a roll-up canvas type stage curtain with a large colorful water wheel scene with many advertisements on it," a local historian wrote. *Courtesy Geraldine Tallas Heisch, Granger, Texas.*

The annual Friendship Fair drew big crowds in the twenties and thirties. Held on Friendship school grounds, the fair featured political speeches, queen coronations, rodeos, airplane rides, and movies in a big tent. In this photograph, local school officials pose under a welcoming sign. Granger Lake now covers Friendship. *Jno. Trlica Photo, Courtesy Geraldine Tallas Heisch, Granger.*

place of the 135 families in the Laneport area—an indication in itself that this must be very productive land.[21]

The writer, Mrs. Hegar, and a Friendship farm wife, Mrs. Henry (Stacy) Labaj, tried to interest Poage in the negative impact the Laneport dam would have on the local Czech culture—but very gingerly. They emphasized not uniquely Czech qualities, but how well the Czech communities measured up as mainstream American. In 1954, the year of *Brown v. Board of Education*, America was moving toward integration of ethnic and racial groups—not toward encouraging diversity. If Friendship had to be sacrificed, Labaj wanted Poage to understand what it was. She wrote about its culture and its quality as a society, but downplayed its Czechness:

> We . . . send our sons and our daughters to universities, to specialized training schools; we take pride in . . . our way of life. We are a community which is organized with a Home Demonstration Club, Mothers' Club, an active Farm Bureau chapter and we have a dividend paying cooperative gin; we have a school of three large buildings . . . and an active church. All these things would be wiped out.[22]

Poage did not get the message. "I think that you probably have not realized that if the Lanesport [*sic*] Dam were built, and you were required to sell your property that it would not be confiscated," he wrote. "On the contrary, you would be paid for everything the Government takes."

Labaj was incensed by his blindered view. She replied, "I regret but feel urged to say you seem not to realize that the passage of the Laneport Dam project is like snuffing out our lives . . . . Ours is really a very fine, modern, active community. . . . We are not a backward group and we know our heritage is worth fighting for and we feel it is the height of poor planning to snatch out of production fine cultivated land such as these 135 farms."[23]

[21] Letter, Mrs. Joseph Hegar to W. R. "Bob" Poage, Nov. 23, 1954, Box 629, File 3, W. R. Poage Papers (Baylor Collections of Political Materials).

[22] Letter, Mrs. Henry A. Labaj to Poage, Aug. 10, 1954, ibid.

[23] Letter, Mrs. Henry A. Labaj to Poage, Nov. 3, 1954, ibid.

These were powerful pleas, but the Czechs of Williamson County lacked leaders who knew how to grab headlines and apply political pressure. Henry Fox knew how to do these things. He loathed the Laneport plan and felt that the San Gabriel River should be controlled upstream, where it would help all of Williamson County, farmers included. That put him and his brother, Bryan, on a collision course with their brother Wilson. Howard, a director of the Brazos River Authority, was left in the uncomfortable position of having to choose between his estranged brothers while steering his water agency toward solvency.

————————

Williamson County's "little dams" movement started in 1954 when the Little River–San Gabriel Soil Conservation District launched a drive to establish an "upstream flood prevention project" on the San Gabriel watershed west of Georgetown. As its directors announced their goal, they also called for "re-consideration" of the proposed Laneport site. Like most people in Williamson County, the district's directors assumed that a new Army survey would recommend dams in the highlands above Georgetown "with supporting adequate upstream flood prevention treatment." If the Army *had* to dam Laneport, surely a smaller dam would do the job, backed by a network of "little" dams.[24] In Georgetown, Bartlett, and Taylor, organizers called "mass meetings" that resulted in petitions, resolutions, and scores of personal letters posted to Washington.[25]

It was a tedious business, establishing water improvement districts under convoluted Department of Agriculture rules, but local farmers and ranchers, led by Henry Fox, made steady progress. In 1956 Congress approved start-up monies and the State of Texas approved the

[24] Letter, Albert Steglich to Poage, July 26, 1954, ibid.

[25] Resolution, Citizens of Williamson County, July 29, 1954, Georgetown, Box 59 (Brazos River Authority, Waco); letter, Louis P. Vitek to R. D. Collins, Aug. 1, 1954, Box 692, File 3, W. R. Poage Papers (Baylor Collections of Political Materials, Waco); letter, Bartlett Home Demonstration Club to Collins, Aug. 11, 1954, Box 59 (Brazos River Authority, Waco); letter, Mrs. Henry A. Labaj to Poage, Aug. 10, 1954, Box 629, File 3, W. R. Poage Papers (Baylor Collections of Political Materials, Waco); letter, Labaj to Poage, Nov. 5, 1954, ibid.; letter, Clarence M. Barnes and 309 others to Senators Johnson and Price Daniel, etc., Sept. 30, 1954, Box 59 (Brazos River Authority, Waco).

Brushy Creek and San Gabriel River water improvement districts.[26] Fox had been elected president of the San Gabriel Water Control and Improvement District. His board of directors included A. C. "Doc" Weir and Donald Irvine of Georgetown, C. V. Hutto of Bertram, and Stacy Labaj's husband Henry of Friendship. On January 14, 1958, the district called an election and passed a 140,000-dollar bond to pay the local share of a five million dollar federal "little dams" flood prevention project for the upper San Gabriel River. The popular vote was overwhelming—four to one in favor.[27]

At this, the Army halted work in the San Gabriel Valley. With Laneport mired in controversy and the small dams program moving forward at a brisk clip, Congressman Thornberry knew the House would not fund the big dam project. From his point of view, the beauty of the resurvey being put "on hold" was that it gave him political "cover" with Owen Sherrill and Wilson Fox—the most outspoken cheerleaders for and against the Laneport dam—while protecting Czech farm families and the Granger business community. He hinted at this in a letter to a Georgetown newspaper publisher. "I have been caught in a dilemma in that I am very anxious to have the report expedited and yet, at the same time, would not want any undue impatience on my part to cause the Engineers not to bring out a recommendation with which all of us would be happy," he cautioned.[28] Whether Thornberry deliberately delayed the resurvey, or whether he simply allowed events to evolve, the net effect was that the Laneport dam was dead. So, too, of course, were the dams at Georgetown.[29]

It was a situation filled with irony. Williamson County residents—

[26] "Brushy Creek Watershed Is Approved by Budget Director," *Taylor Daily Press,* Apr. 23, 1956; "SCS Approves San Gabriel Watersheds," *Taylor Daily Press,* May 9, 1956; letter, Henry Fox to Poage, Feb. 9, 1957, Box 629, File 3, W. R. Poage Papers (Baylor Collections of Political Materials, Waco); letter, Poage to A. C. Weir, May 4, 1957, ibid.

[27] "Bond Vote on Small Dams Slated Tuesday, Jan. 14th," *Williamson County Sun* (Georgetown), Jan. 8, 1958; "San Gabriel Voters Approve $140,000 Small Dams Bond," ibid., Jan. 16, 1958. Voters in Milam and Burnet Counties, which contained a few miles of the San Gabriel watershed, concurred.

[28] Letter, Cong. Homer Thornberry to Don Scarbrough, July 20, 1961, Don Scarbrough Papers (Georgetown).

[29] I could find no hard evidence to support my theory that the congressman purposefully avoided pushing for a resurvey, alas, because Homer Thornberry's congres-

especially the Black Waxy's Czechs, Owen Sherrill, and Henry Fox— had effectively blocked the Laneport dam but didn't quite realize it. Poage laid it out for a bemused Taylor correspondent:

I have never understood the Taylor attitude, but there has been a great deal of objection to the Lanesport [*sic*] site out of Taylor. The situation is such that Congressman Thornberry could not be expected to urge the Lanesport site. He was, in effect, forced to take some kind of action against this site. Obviously, with the Congressman from the area where the dam was to be built opposing it, we had little chance of getting the dam when there were plenty of other dam sites where everybody was in agreement.[30]

With one of the few Milam County residents who ever wrote Poage about Laneport, Poage was more direct. "We have to get along with the Williamson County people; else, we are not going to get any dam at all," he wrote. "They may not be able to get just what they want constructed, but they can certainly block the construction of anything that a large part of them don't want."[31]

But Williamson County residents didn't *just* want to kill the Laneport dam, they wanted dams west of Georgetown *and* fifty or more small dams on the river's creeks and branches. And they were absolutely convinced that an Army resurvey would recommend precisely that. So once more, the forces favoring dams at Georgetown cranked up the heat on the Army to start resurveying again. The Brazos River Authority backed the effort, fearing it would lose future water sales to the Lower Colorado River Authority if a dam at Georgetown did not get built.[32] Business leaders from Georgetown, Granger, and Taylor staged a well-rehearsed "unity meeting" on

---

sional papers are not available to researchers. But from what Poage and Thornberry wrote on the subject, I am confident this was the case.

[30] Letter, Poage to Dr. Edmond Doak, Sept. 5, 1958, Box 692, File 4, W. R. Poage Papers (Baylor Collections of Political Materials, Waco).

[31] Letter, Poage to Emory B. Camp, Sept. 17, 1958, Box 692, ibid.

[32] Letter, Raymond Holubec to R. D. Collins, Feb. 11, 1958, Box 59 (Brazos River Authority, Waco); Collins to Holubec Feb. 16, 1958, ibid.; John D. McCall to Collins, Feb. 24, 1958, ibid.

Henry Fox, the Czechs, and "Little" Dams   *117*

March 19, 1958, to demonstrate the county's common cause in supporting a resurvey.[33]

The problem was that "resurvey" meant different things to different people. No one knows what Congressman Thornberry thought: he kept the resurvey notion alive but didn't keep the Army working at it. The "little dams" people, Henry Fox and the farmers and ranchers of Williamson County, thought it would strengthen their fifty-dam program, so they backed the resurvey. Wilson Fox and Bob Poage thought the resurvey would favor the Laneport dam, so they backed it. Most Williamson County residents thought a resurvey would, as one newspaper confidently predicted, "find that dams above Georgetown will provide maximum protection and benefits to Williamson County and the areas below this county, Milam County in particular."[34] So they too backed it.

After the unity meeting, the Army resumed work in the San Gabriel Valley, investigating four possible reservoir sites—Laneport, North Fork, South Fork, and Berry Creek. The little dams project was put into deep freeze. "It's impossible to intelligently plan small dams until the location of any big dams is decided, due to the fact [that] some of the proposed small dams would be ... covered up by water held back by the big dams. Consequently, we have to wait," Henry Fox's San Gabriel River Water Control and Improvement District announced.[35]

In July of 1960 the Army's chief engineer in Ft. Worth informed Poage and Brazos River Authority officials that it would recommend two reservoirs, one at Laneport and one on the North Fork of the San Gabriel River.[36] But no official report emerged. A year later, Texas's State Board of Water Engineers released a twenty-year plan for developing water resources, recommending that a dam be built on the

[33] "Citizens Agree on Two Dams above Georgetown, Small One at Laneport," *Taylor Daily Press,* Mar. 5, 1958; Statement by R. D. Collins, BRA manager, at Corps of Engineers hearing, Mar. 19, 1958, Box 59 (Brazos River Authority, Waco).

[34] "County-wide 'Unity' Meet On Gabriel Dams Set Here," *Williamson County Sun* (Georgetown), Feb. 13, 1958.

[35] "San Gabriel Dams Plans Postponed," *Taylor Daily Press,* Jan. 27, 1959; "San Gabriel Water Taxes Killed for '61," *Taylor Daily Press*, Mar. 30, 1960.

[36] Letter, Col. R. P. West to Robert Poage, July 25, 1960, Box 692, File 4, W. R. Poage Papers (Baylor Collections of Politcal Materials, Waco).

Road, River, and Ol' Boy Politics

North Fork of the San Gabriel River and conspicuously omitting Laneport as a prospective dam site.[37] Still the Army remained mum about its plans.

On October 15, 1961, Henry Fox lost patience. In Texas, water improvement districts had to "start action" within ten years of their creation, or they would automatically die, he wrote in a newspaper article. The San Gabriel WCID's life span was nearly half spent. "We would have been long gone on the job if the monkey-wrench of big dams had not been thrown," Fox said. "For two years the project has stood still, waiting for the Army Engineers to make up their minds. . . . Time is running out. . . . The question is, shall we sit around hoping for some fabulous big dams costing unbelievably large sums, or shall we take hold of the small dam project now within our reach?"[38]

Eighteen days later Army engineers came clean. In a report dated November 3, 1961, the Army recommended a three-dam package: one at Laneport and two on the forks of the San Gabriel River west of Georgetown. Laneport would be fifteen percent *larger* than originally planned and it would be built immediately. The Georgetown dams should be built "as second and third units, respectively, at such times that additional water conservation storage is needed."[39]

Williamson County threw a tantrum. A Taylor delegation composed of bank presidents and directors of the enormous and influential Taylor Bedding Company, the Williamson County Farm Bureau, the Williamson County Pecan Growers Association, and the South Texas Cotton Oil Company angrily protested the Army's conclusions with Congressman Thornberry in Austin. Thornberry was shaken by the encounter. "Frankly," he told his visitors, "I wish I had never heard of the Laneport dam. It has caused me more concern than any other one thing in my district."[40] People working for the Georgetown dams

---

[37] "North San Gabriel Dam Site Gets Boost by State Board," *Williamson County Sun* (Georgetown), Aug. 3, 1961.

[38] Henry Fox, "Small Dam Program Urged as Means to Cover Wide Range Water Needs," *Taylor Daily Press*, Oct. 15, 1961.

[39] "Notice of Review of Reports on Brazos River and Tributaries, Texas, Covering San Gabriel River Watershed," U.S. Army Engineer Division, Southwestern Corps of Engineers, Nov. 3, 1961, *Taylor Daily Press*, Taylor City Library, Taylor.

[40] "Taylor Group Opposes Dam at Laneport," *Taylor Daily Press*, Nov. 7, 1961.

were livid. "The Laneport dam will be larger than contemplated four-teen years ago and the dams above Georgetown will be built after the Laneport dam has been completed, and then only at the discretion of the Army Engineers. So, it would seem, not much progress has been made in those hard 14 years of work and worry," the *Williamson County Sun* editorialized.[41] Henry Fox described the expanded lake at Laneport to Granger residents:

> During a San Gabriel River flood, [it would] back water up to the edge of Granger; and all the way up the river over the bot-tom lands to the Old Georgetown Road bridge . . . drowning out the pecan trees, covering up the farm land, the roads etc. Then . . . after authorities . . . ruled it was safe, the water would be drained off and the 'lake' returned to its small size, leaving a vast mud hole in what is now productive, tax-paying land.

He asked readers to back the small dams project, "which is ready to go if the big dam can be stopped."[42]

In an odd twist that further inflamed Williamson County, Con-gressman Poage publicly attacked its residents for breaking their word that they would abide by the results of the Army resurvey. He was really referring to Owen W. Sherrill, but he sowed his accusations broadly. He started with a heated letter addressed to his small dams disciple, Henry Fox, which he released to the Taylor newspaper with-out Fox's knowledge. Poage wrote,

> I did agree, and many of your representative citizens of Williamson County agreed, that they would abide by this decision. I am frankly surprised, embarrassed, and a little per-turbed to find that some of the very people who asked me . . . to stand aside these long years . . . to now be petitioning for a repudiation of that agreement. . . . You had your day in court. You had your appeal. . . . You lost on the engineering facts as they were understood by the Corps of Engineers.[43]

[41] "No Dams above Georgetown—Yet," *Williamson County Sun* (Georgetown), Nov. 9, 1961.

[42] Henry Fox, Letter to the Editor, *Granger News*, Nov. 16, 1961.

[43] Letter, Poage to Henry Fox, Dec. 8, 1961, Box 692, File 4, W. R. Poage Papers (Baylor Collections of Political Materials, Waco).

In a classic rendition of good ol' boy skewering, Fox replied to Poage by writing the editor of the *Taylor Press*:

Since you published a letter from Cong. W. R. Poage to me, under this new system whereby Congressmen write individuals and send carbon copies to the newspapers, I would like to reverse the process. I am writing you a letter and will send a carbon copy to Mr. Poage. I would like to comment further on Mr. Poage's original letter, a copy of which got to you before the original got to me.

I am intrigued with this new theory of government, whereby the people are considered unfair if they fail to back up one Congressman's promise to another Congressman, involving the people's own welfare. It's a fine theory, especially for Congressmen, but that's not the way democracy works. A Congressman can commit himself to a course of action if he wants to, but Mr. Poage is the first Congressman I ever heard of claiming the people are obliged to follow, whether they want to or not.

Mr. Poage wrote: "You had your day in court . . . " This too is an intriguing new thought, that the Army Engineers constitute the people's court, and when they speak, the people must keep quiet and submit. [44]

By the time the correspondence sputtered to a close, four rounds of increasingly harsh letters had been exchanged and reprinted in Williamson, Milam, Bell, and McLennan County newspapers, and Poage looked embattled and foolish. In Poage's final letter, which he promised not to send to the newspapers, he asked Henry Fox an extraordinary question:

In all fairness, and as man to man, what would you do if you were in my place? (Frankly, if I were in your place, I might do exactly what you are doing, but you will note I have not criticized you for doing it. You live in Williamson County. I think you own land in the Laneport basin. I don't blame you for trying to protect your interests. Is it fair for you to blame me

[44] Letter, Henry Fox to Poage, Dec. 12, 1961, ibid.

for trying to protect the people I represent?) I will appreciate it if you will write me one more letter and frankly state what you think I, as the Representative of Milam County, should do.[45]

Fox complied. "I think the *quickest* and fairest way for all the people in the San Gabriel River watershed to get flood protection is through the Soil Conservation Service route of 59 flood-retention dams," he wrote. "This would require an about-face on the big dam proposition, to the extent of informing the Board for Rivers and Harbors that opposition to a Laneport dam is so great that delay in building it may go on for years and therefore the plans should be abandoned. . . . I pledge you my written word that we will be in Red Smith's [SCS] office in Temple the day after the big dam at Laneport is disapproved, initiating immediate action. Everything is in shape to do this."[46]

It could not happen. Poage couldn't do an about-face on the Laneport dam. It would have been political suicide, pleasing Williamson County where no one could vote for him and betraying promises he had made his own constituents in Milam County. Still, one wonders whether Poage might have been tempted to take Fox's advice. After all, he had crafted the small dams legislation that might have saved Williamson County a great deal of pain.

The day after Thanksgiving 1961, Henry Fox and his allies dropped off petitions at banks across the county. They had one week to collect names to meet the deadline for the Board of Engineers for Rivers and Harbors, which would rule in January on the Army's proposal for the San Gabriel River. The petition urged that the Army's Laneport dam proposal be killed, and that it be replaced by small dams on the upper reaches of the San Gabriel River and its tributaries.

Nearly three thousand people signed the petition out of Williamson County's population of thirty-five thousand. If the county had ten thousand registered voters, as is probable, the small

[45] Letter, Poage to Henry Fox, Dec. 19, 1961, ibid.
[46] Letter, Henry Fox to Poage, Dec. 20, 1961, ibid.

dams movement captured one-third of its electorate in one week.[47] It was an impressive piece of grassroots democracy. Petitioners came from every part of the county except Round Rock, which was not directly affected and where petitions were not circulated. Granger signatures alone covered twelve of the petition's forty single-spaced, legal-sized pages. Georgetown residents signed ten pages. Taylor contributed almost as many petitioners, with eight pages, followed by Bartlett, with two pages, Liberty Hill and Florence, with two pages each, and Jarrell, with one page. Burnet County residents, mostly from Bertram, added four pages to the petition. One page consisted of petitioners scattered around the county. Attached to the official petition were letters from the Williamson County Farm Bureau, representing thirteen hundred farm families, and half a dozen individuals. The document was signed by every banker in Taylor, Granger, Georgetown, Bartlett, Florence, Walburg, Schwertner, and Bertram.[48] It was a stunning piece of political theater.

It was not enough. Congressmen Thornberry and Poage, who had pledged to support the Army's findings, no matter how the resurvey came out, held their positions. Though he was pressed to do so, Vice President Johnson did not intervene. On January 24, 1962, the Board of Engineers for Rivers and Harbors approved the Army's work.[49] A big Laneport dam would be built. Georgetown's dams would wait, probably forever. And the "little" dams program for the San Gabriel Valley—Congressman Poage's legislative "baby"—was dead, largely by his own hand.

[47] The estimate is conservative, based on average family size and the poll tax's impact on the county's poor, particularly Mexican-American and African-American residents, few of whom could afford to vote.

[48] "Nearly 3,000 Protest Dam at Laneport, Request Small Upstream Structures Instead," *Williamson County Sun* (Georgetown), Dec. 14, 1961.

[49] Letter, Maj. Gen. William F. Cassidy to Vice President Johnson, Dec. 18, 1961, VP Papers, Public Works, San Gabriel File, Box 115 (LBJ Library, Austin); "Laneport Project Authorization Is Recommended," *Taylor Daily Press*, Jan. 28, 1962; "Dam at Laneport Is Recommended . . . In Spite of Homer, H____, & High Water," *Williamson County Sun* (Georgetown), Feb. 1, 1962.

HILDA'S BOTTOM

It's crazy there in America, beer flows on the floor.

BOHEMIAN DITTY, *Kråsnå Amerika*

A t around five o'clock on October 4, 1962, a long line of buses and limousines rolled through Circleville, heading for Wilson Fox's Riverside Ranch and the premiere stag social affair of Central Texas.[1] The limos carried big-shot politicians and small (though Vice President Lyndon Johnson couldn't make it that year), while buses from Austin, Dripping Springs, Lockhart, Luling, Brenham, Marble Falls, Burnet, Johnson City, Blanco, Giddings, Bastrop, Smithville, and Lexington ferried the Democratic faithful from the far corners of the Tenth Congressional District's ten counties.[2] Partygoers were in high spirits, salivating at the thought of one of Wilson Fox's and Roman Bartosh's celebrated barbecue "blowouts" honoring Congressman Homer Thornberry.

The party, which Wilson Fox had launched in 1949 after Thornberry's first election to Congress, took place every other year at Hilda's Bottom (naughtily named after Wilson's wife Hilda), where scores of burr oak towered over a manicured glade that ended abruptly at a cliff dropping forty feet to the San Gabriel River. At

---

[1] Dave Shanks, "Thornberry Barbecue," unedited reporter's copy, *Austin American-Statesman* morgue, Sept. 28, 1963.

[2] Caro, *The Path to Power*, 391.

Thornberry's stag barbecues, Hilda's Bottom resembled Shakespeare's *A Midsummer Night's Dream*. Amber electric bulbs strung through the treetops brushed revellers with golden light. The crowd was giddy: crops were in, bank notes were paid off, and an election was on its way. Taylor Meat Company owners Van, Joe, Charlie, and Paul Zimmerhanzel flipped one-pound T-bone steaks on a massive stone barbecue pit built into the embankment overhanging the river, while twenty black men, crisply dressed in red bow ties, starched white shirts, and black pants, circulated through the crowd, pitchers of beer in each fist, pouring constantly. Their job was to keep all glasses full.[3]

It was a cheap ticket—two dollars for "the biggest steak in Texas" and all the beer you could drink—and an absolutely "must-do" event.[4] "All the politicians were there," reminisced a former Georgetown postmaster who never drank a drop of spirit. "If you wanted to climb, you went. Our office turned out in force. It was always a lot of fun, with a lot of politicking. And a lot of foolishness."[5]

In 1962 fifteen hundred men—a record—were pumping Congressman Thornberry's hand, eager to jawbone with one of Vice President Lyndon Johnson's closest friends.[6] Many did not know that Thornberry was also a key man in the House of Representatives, where as a ranking member of the powerful Rules Committee, he helped set the terms of engagement for all proposed legislation. For years Thornberry had been the "bridge" between Speaker Sam Ray-

[3] Interview, Dr. James "Jim" Fox, Sept. 6, 2000, Austin. Fox is Wilson Fox's only son and attended most of his father's Riverside Ranch barbecues.

[4] Interview, Congressman J. J. "Jake" Pickle, Apr. 18, 2000, Austin. The ticket price rose to $5 before the barbecues ended. The last one was held in the early seventies for Congressman Pickle. In 1972 Wilson Fox switched to the Republican Party, protesting against an activist and liberal national Democratic Party. He died after suffering a heart attack on February 10, 1974.

[5] Interview, J. D. Thomas Jr., Sept. 8, 2000, Georgetown.

[6] The closeness of their relationship was underscored by the concern Senator Johnson showed when Thornberry underwent gallstone surgery. "For some time I have been worried about your very hectic social life and fearful that the consequences would not be good," Johnson wrote. "Too many ambassadors, too much horse racing, too many health roll calls always exact their price. I'll see you Sunday and tell you just how to handle your business. Love, Lyndon." Senate Papers, Selected Names—Thornberry, July 2, 1957 (LBJ Library, Austin).

Relaxing at a Riverside Ranch Stag Party: Congressman Homer Thornberry, left, and Senator Lyndon Johnson, second from right, with hosts Roman Bartosh and Wilson Fox, right, who wanted a dam at Laneport. *Courtesy Dr. James ("Jim") Fox, Austin.*

burn and Johnson. "Everything flowed through him," said Congressman Jake Pickle of his predecessor.[7]

Thornberry had coasted to victory in 1960, but the coming election was, in Pickle's words, a "death-lock race" with the dams as its centerpiece. Since the previous year, when the Army's resurvey had determined on a larger Laneport Dam, Thornberry had heard nothing but anguished cries from Williamson County, until now a "safe" voting bloc for him. He had seen the typed, single-spaced, inch-thick petition signed by three thousand voters against Laneport Dam; he had talked with every banker and business leader in Taylor, Georgetown, and Granger and found them appalled at the Army's plans; now, most dramatically, the State of Texas had come out against the dam.

Not surprisingly, the Republicans were mounting a serious challenge to Thornberry, backing a golden-tongued radio commentator named Jim Dobbs. Dobbs was a Church of Christ minister with violently negative opinions about the big government policies of Presi-

[7] Interview, J. J. "Jake" Pickle, July 3, 1998, and Apr. 18, 2000, Austin; "Senate Nod Tribute To Texan Homer Thornberry," *Austin American,* July 16, 1963, p. 4.

dent John Kennedy and, by extension, Vice President Johnson and his old friend Thornberry.[8] Thornberry should have been elated about his reception at Riverside Ranch that evening. Congress had been keeping him busy, too busy to spend much time working his district. And though he was the most convivial of mortals, Thornberry lacked the stomach for the bitter harangues he had come to expect from opponents of Laneport Dam, who seemed to lurk behind every tree in Williamson County. Secretly, he dreaded politicking. There was a fundamental shyness, a reserve, about Thornberry that emanated, perhaps, from his growing up the son of two deaf-mute teachers, having to master signing as well as English to communicate.[9] Wilson Fox's barbecue—which pulled

[8] Interview, J. J. "Jake" Pickle, Apr. 18, 2000, Austin.
[9] Interview, Charles Patterson, July 1974, New York City. Patterson represented

Wooing the congressman for Georgetown, circa 1958: Don Scarbrough, left, County Judge Sam Stone, second from left, and Charles Forbes, far right, lobby Homer Thornberry, second from right, at a Southwestern University dinner designed to promote a dam on the North Fork of the San Gabriel. *Evans Studio Photo, Courtesy* Williamson County Sun, *Georgetown.*

Hilda's Bottom

all the powers together—and the money it raised to keep him in office, were a blessing.[10]

Tonight, however, Thornberry did not feel blessed, though he knew he should be celebrating. He knew he should be relaxing with Wilson Fox and chatting up the new general manager of the Brazos River Authority, Col. Walter J. Wells (a retired Corps man), but he couldn't relax. For weeks, Thornberry had been working to get the North Fork and South Fork dams into the 1962 Omnibus Rivers and Harbors Bill (characteristically, Poage had predicted it couldn't be done), and both the House and the Senate had rewarded him.[11] Georgetown's fourteen-year campaign for dams west of the county seat had finally moved forward, thanks to him.[12] But the Laneport Dam viper always seemed ready to strike. Just now, he had heard a story that, apparently, was the talk of the party—overshadowing even the exciting news from Congress.

On the country lane leading to Wilson Fox's ranch that evening, a new road sign had popped up. It warned passersby:

WEAK CONGRESSMAN
Three Miles

Someone (everybody suspected Henry Fox) had switched it with the official Texas Highway Department sign that normally read,

---

Williamson County in the Texas Legislature during the late 1960s. Interview, J. J. "Jake" Pickle, Apr. 18, 2000, Austin; "Homer Thornberry—Biographical Sketch," *Austin American* morgue, Nov. 1962.

[10] Interview, Charles Patterson, July 1974, New York City. Also see Dave Shanks, "Thornberry Barbecue," *Austin Statesman* morgue, Sept. 1963; Sam Wood, "Thornberry Quits; Election Upcoming," Sept. 1963, ibid.; Nat Henderson, "Wilson H. Fox Services Today," *Austin American-Statesman,* Feb. 11, 1974 (Center for American History, University of Texas at Austin).

[11] "Summary of Reports and Data on the 'Laneport' Dam," Williamson County Citizen's Committee to Col. F. P. Koisch, Mar. 13, 1963, W. R. Poage Papers, Box 692, File 5 (Baylor Collections of Political Materials, Waco). The House passed the bill October 3, while the Senate passed it October 4, 1962—the day of Thornberry's party. President Kennedy signed into law the Flood Control Act of 1962, which authorized the San Gabriel North and South Fork and Laneport Reservoirs, on October 23.

[12] Letter, Homer Thornberry to unnamed, Jan. 24, 1963, Box 692, File 5, W. R. Poage Papers (Baylor Collections of Political Materials, Waco).

Road, River, and Ol' Boy Politics

"Weak Bridge, 7 Miles." John Wehby, Taylor's KTAE radio station manager, and Wilson's son Jim furiously ripped down the offending object, so Thornberry hadn't seen it, but the story flew around the party.[13]

Weak congressman!

Thornberry knew he should shrug it off, but the message gnawed at his innards. Actually, Thornberry's officially neutral stance on Laneport had never matched his acts, which had effectively iced the dam at the blueprint stage. Moving with extreme caution, presumably to avoid alienating Wilson Fox, the Corps of Engineers, and its admiring Texas Congressional delegation, Thornberry had kept Laneport Dam nicely chilled at the "study" stage for fourteen years. Few if anyone in Williamson County appreciated Thornberry's maneuvers, which he could not afford to explain. Instead, he was derided by almost every faction. Of course, his was a position that could not be sustained, for as long as Laneport Dam was stalled, so were the dams on the San Gabriel's North and South Forks that Georgetown and fifteen hundred farmers strongly desired.[14]

"Thornberry was torn to pieces," Pickle would say decades later. "The controversy literally made him ill. I can still hear him: 'Jake, oh, oh, they're just giving me hell. They're tearing me apart!' He sensed he might lose the next election. Laneport took a toll on him. He didn't know whether to follow the Corps, or follow the people."[15]

In Williamson County, 1962 had been quite a year. After the small-dams petition failed to sway the House's Rivers and Harbors Committee, and Laneport Dam got the go-ahead at the end of 1961, players on both sides of the controversy were spent—all but Owen Sherrill. In late January, Sherrill flew to Washington, D.C., and some-

---

[13] Interview, Dr. James "Jim" Fox, Sept. 6, 2000, Austin; interview, Gene Fondren, Mar. 9 and Sept. 1, 2000, Austin.

[14] The case for this paragraph might be termed circumstantial, but I am convinced it is so from reading hundreds of letters, notes, and memos on the subject. It is unfortunate that papers from Judge Thornberry's congressional days were not available for study; they might have cleared up lingering confusion about Thornberry's role in the Williamson County dam controversy.

[15] Interview, J. J. "Jake" Pickle, July 3, 1998, Austin.

how gathered three powerful Brazos River Valley congressmen—Thornberry, Poage, and Olin "Tiger" Teague—into one room. On January 23 he talked them into signing a pact stating that they would all work for "simultaneous" construction of dams at North Fork and Laneport, along with construction of a South Fork dam and a group of small earthen dams on the San Gabriel's tributaries "as quickly as feasible."[16] The pact was legally meaningless, but it locked the politicians into a new position and inspired a new bandwagon catchphrase back home: "simultaneous" dams.

Seven days later the Army tried to squash Sherrill's coup. Its Board of Engineers for Rivers and Harbors approved the Corps' plan for sequenced—definitely not simultaneous—dam construction on the San Gabriel, with Laneport first in line. The board rejected the Brazos congressmen's request for simultaneous construction, as if it were nothing more than an irritating moth.[17]

Williamson County tried again. The Commissioners Court and Taylor's City Commission backed dams west of Georgetown on the upper Gabriel forks, instead of the Corps' Laneport Dam. A plot developed. Sherrill wrote his ally, the general manager of the Brazos River Authority, "We will meet at Governor Price Daniel's office next Friday, Feb. 16th at 4:30 P.M. to further discuss San Gabriel dams. I trust Mr. Collins, you will be with us then. We are giving no publicity to this meeting, please."[18] Shortly thereafter, BRA directors unanimously urged Governor Daniel to support "simultaneous" construction of three dams on the San Gabriel River (North Fork, South Fork, and Laneport), and requested control of all the impounded water.[19] In April three hundred protestors from Williamson County, led by County Judge Sam V. Stone, pleaded their case before the Texas Water Commission. In the end, the Commission "disapproved" the

---

[16] Memorandum, conference between Owen W. Sherrill, Congressman W. R. Poage, Congressman Olin "Tiger" Teague, and Congressman Homer Thornberry, Jan. 23, 1962, Washington, D.C., Box 59 (Brazos River Authority, Waco).

[17] Letter, Col. Carl H. Bronn to Poage, Jan. 30, 1962, Box 692, File 5, W. R. Poage Papers (Baylor Collections of Political Materials, Waco).

[18] Letter, Sherrill to Collins, Feb. 9, 1962, Box 59 (Brazos River Authority, Waco).

[19] Letter, Collins to Gov. Daniel, Apr. 24, 1962, Box 59, ibid. The board voted at its regular meeting April 16, 1962.

Army's plan to build Laneport Dam—a highly unusual rejection of federal largesse worth millions of dollars to Texas. Indeed, the State of Texas came up with a substitute plan: it "ordered" the Army to build the three proposed San Gabriel dams "as a unit"—and simultaneously. If that was not possible, then North Fork Dam should be built first, South Fork Dam second, and fifty small dams on the San Gabriel's tributaries third. Then, and only if it were clearly necessary, could the Army build an "adequate" Laneport Dam.[20]

The Waco congressman was furious. After Governor Daniel had, in Poage's words, "circumscribed" the Laneport plan, Poage wrote Henry Fox, anger simmering on the typed page:

> I am not charging you or any other individual with any bad faith. You were not a party to [the Sherrill] agreement, but there was an agreement to which I was a party . . . I think it is a waste of time to try to come to any further agreements with the people of Williamson County. I am not charging bad faith, but nobody has any authority to speak for these people and I am not interested in any further agreements where I will be bound but the folks upstream will not be.[21]

Meanwhile Sherrill crowed, in third person, of his coup to Poage: "You can see where Owen—since County Agent days—45 years ago knew so well that the engineers were so wrong, with a mud silt basin at Laneport. . . . You [might] just as well recognize [Laneport] will be killed again and again, if Georgetown and Williamson County is not protected."[22]

There things stood through the sweltering summer months. But in the fall of 1962, facing the formidable Jim Dobbs for his job, Thornberry called in his chits. On October 4, the day of Fox's stag party at Hilda's Bottom, the U.S. House of Representatives, protecting one of

[20] "An Order approving the feasibility of the North San Gabriel . . .," Texas Water Commission, June 25, 1962, Box 59 (Brazos River Authority, Waco); "Alternate Sequence Proposed," *Taylor Press,* June 26, 1962.

[21] Letter, Poage to H. B. Fox, June 28, 1962, Box 692, File 5, W. R. Poage Papers (Baylor Collections of Political Materials, Waco).

[22] Letter, Owen Sherrill to Poage, Aug. 29, 1962, ibid.

its favorite members, authorized two "upper" San Gabriel dams, one on the North Fork, the other on the South Fork west of Georgetown. Since Laneport had been authorized for years, that made a grand total of three dams authorized within one county on one river little more than a hundred miles long.[23] The estimated price tag for all three dams was 45.4 million dollars—approximately Williamson County's total appraised value.[24]

The timing couldn't have been better for Thornberry. In full-page newspaper ads, Dobbs had been indirectly swiping at the congressman by selling Dobbs's "vigorous" and "decisive" conservatism. The implication was that Thornberry could not get things done for his district.[25] But now Thornberry had proved he could deliver. The House had authorized the North Fork and South Fork dams. There was no money to build them, and the question still loomed of what dam to build first. Still, it was a concrete step, and one that appeared to settle the question of whether Thornberry was an effective congressman.[26]

On November 6 Thornberry drubbed Dobbs three to one throughout the Tenth Congressional District, and only slightly less convincingly in Williamson County. He carried every Williamson County voting box, including Granger's, which piled up an astonishing four-to-one margin.[27] Wilson Fox reveled in those numbers, insisting that the Laneport imbroglio was nothing more than propaganda by "a few die hards."[28] Others credited the Czechs' dogged

[23] The San Gabriel River is 112 miles long, 95 miles of which flows within Williamson County.

[24] The Senate quickly followed suit, and President John Kennedy signed the Flood Control Act on Oct. 23, 1962—fourteen days before the off-year elections. "Good News on the Dam," *Williamson County Sun* (Georgetown), Oct. 18, 1962; telegram, Senator Ralph Yarborough to David Hoster, editor of the *Taylor Times*, Sept. 26, 1962 (*Daily Press*, Taylor).

[25] Advertisement, "Jim Dobbs for Congress," *Williamson County Sun* (Georgetown), Nov. 1, 1962.

[26] "Chance for Upper Dams Brighter," ibid., Oct. 3, 1962; memo, RL to unknown, Sept. 26, 1962, Box 692, File 5, W. R. Poage Papers (Baylor Collections of Political Materials, Waco). "RL" was on the staff of Cong. Poage.

[27] "County Vote Follows Trend Set in Texas," *Williamson County Sun* (Georgetown), Nov. 8, 1962.

[28] Letter, Wilson Fox to Poage, Feb. 25, 1963, Box 692, File 5, W. R. Poage Papers (Baylor Collections of Political Materials, Waco).

fealty to the Democratic Party, along with their visceral distaste for anti-Catholic demagoguery, which Dobbs had liberally aimed at President Kennedy. Still others thought the Czechs could be bossed around—or even bought. "Historically, the east side of the county was always viewed by politicians as susceptible to political bossism, because of the nature of the ethnic communities," said Charles Patterson, a former state representative from Taylor. "It was not unlike what can be seen in areas of New York and Boston."[29] During Lyndon Johnson's congressional days, the Czech vote in some "small rural towns" of the Tenth District was said to have been for sale, though that practice may have ended by Thornberry's time.[30] Whatever the reason for Thornberry's stunning electoral success, most notably in Granger, his victory set the stage for a power realignment in Washington and in Williamson County. At last it seemed that Laneport Dam was a sure thing. Only Henry Fox, "sage of Circleville," vowed to continue fighting.[31]

Thornberry did not relish his triumph for long. In fact, he was tired of politics, coveting a quieter life serving on a judicial bench somewhere, anywhere. His close friend, Speaker Sam Rayburn, passed that wish along to President John F. Kennedy. So when a federal district judge retired in El Paso, Kennedy remembered his promise to Rayburn. Vice President Johnson backed up Rayburn's suggestion and on July 9, 1963, President Kennedy announced the appointment.[32] On September 26—at Wilson Fox's 1963 barbecue—Thornberry announced he would leave Congress in two months, on

[29] Interview, Charles Patterson, Summer 1974, New York, N.Y.

[30] In *The Path to Power* (408, 818), Ed Clark, a close friend of Lyndon Johnson's, told Robert Caro that the Czechs in three or four rural communities of the Tenth District were "for sale," but that the price was high. Clark did not specify which communities, but at mid-century, Granger and Taylor contained the largest Czech communities in Texas, except for Fayetteville, which was not part of the Tenth District. Clark, who served as U.S. ambassador to Australia during Lyndon Johnson's presidency, may have been exaggerating. Former Congressman Pickle says he never heard of any Czech votes being for sale in his district—the Tenth.

[31] H. B. Fox to Don Scarbrough, Letter to the Editor, *Williamson County Sun* (Georgetown), Nov. 8, 1962, p. 3; "Continued Fight Against Laneport Dam is Pledged by Circleville Sage," *Williamson County Sun* (Georgetown), Nov. 15, 1963, p. 1.

[32] "JFK Appoints Thornberry Judge, Keeps Promise to Sam Rayburn," *Austin Amer-*

December 20, to be exact. That would give the Tenth Congressional District time to select a new congressman. Under the towering canopy of burr oak trees, Vice President Johnson spoke of Thornberry's friendship and service in Congress, calling him "truly a workhorse." Governor John Connally praised him, as did Wilson Fox. "After hearing all of these remarks, I am almost tempted to announce my re-election for Congress," Thornberry joked.[33]

The two top Democratic hopefuls did not eat steak that night. They shook hands—every hand they could grasp among Thornberry's fifteen hundred well-wishers. Representative Jack Ritter of the state legislature had been campaigning for weeks. Public relations man Jake Pickle, one of Lyndon Johnson's crack political operatives, told friends he'd make an announcement the next day.[34]

"At the time, the most controversial issue in the Tenth was Laneport," Pickle remembered. "Thornberry got Laneport authorized, but he couldn't get any money for it. It was not an easy job selling a dam on the San Gabriel River in those days. Thornberry told me, 'I'm going to leave that with you. Haw, haw.'"[35]

---

ican, July 10, 1963; "Kennedy Names Thornberry Judge," *Williamson County Sun* (Georgetown), July 11, 1963.

[33] Sam Wood, "Thornberry Quits; Election Upcoming," *Austin American*, Sept. 1963, p. 1.

[34] Ibid.

[35] Interview, J. J. "Jake" Pickle. Apr. 18, 2000, Austin.

SEVEN

"HATCHET MAN" AND TECHNOCRAT

Make no little plans, they have no magic to stir men's blood and probably themselves will not be realized.

DANIEL BURNHAM, Master Planner of Chicago

Could Jake Pickle get elected to Congress? Savvy politicians wondered, Vice President Johnson included. Many felt Pickle's liabilities would cancel out his assets. Pickle had, as he put it, "carried wood and water" for Johnson for two decades, both before and after Johnson's election to Congress; later Pickle worked for two Texas governors. He specialized in the rough, queasy edges of campaigning and public relations, while soaking up charisma and tactics from master politicians.[1] While Johnson represented the Tenth District, he treasured Pickle's weekly letters, bristling with enough intelligence to turn a CIA operative green. Pickle was a natural reporter, an astute reader of realpolitik. He was practical, capable, likeable, and a straight shooter. No one could best him as a storyteller. And—a huge advantage—he knew the Tenth and its powerful players intimately. But Pickle carried one thick scar: his reputation as a "hatchet man" for Lyndon Johnson and Governors Allan Shivers and Price Daniel. As he described it years later,

[1] Jake Pickle and Peggy Pickle, *Jake* (Austin: University of Texas Press, 1997), 83; Oral History, J. J. "Jake" Pickle, Section I, p. 12, May 31, 1970, and Section IV, pp. 32–34, Aug. 25, 1971 (LBJ Library, Austin).

I got to be a hatchet man; that is, I got termed as the man . . .
calling the signals. . . . Well, I'll admit to some forcefulness in
some of those actions because if you've got a job to do and
your boss man the Governor says, "Let's do *this*," you *do* it!
And you don't do it in a pussyfooting manner.[2]

Pickle had taken his time deciding to run for Homer Thornberry's
congressional seat, having spent much of the summer before Thornberry's valedictory barbecue thinking about it. For the first time in his
life, he had a high-paying job at the Texas Employment Commission
that did not depend on other men's political fortunes. Widowed
eleven years earlier when his only child, Peggy, was six, Pickle had
recently remarried Beryl Bolton McCarroll, herself a widow with
two sons. They were enjoying family life on Cherry Lane in Austin.
Pickle knew Beryl would recoil at the very notion of political wife-
hood.[3] He also knew that in some corners of the Tenth, he was more
feared than liked. Vice President Johnson sent no word of encouragement. All Pickle heard from his old boss in Washington was a second-
hand caution: might his former aide be too "cut up" to win?[4] Homer
Thornberry, another close friend, warned Pickle that being congressman was "a mean job."[5]

But Pickle was forty-nine years old and ambitious; he did not
want to spend his peak years running a government agency.[6] In the
political arena, he said decades later, "the brass ring comes around
once. You grab it or you miss it. Very seldom . . . does it float around a
second time."[7] After Governor John Connally's wife Nellie quietly
told Beryl that her husband would never stop second-guessing him-

<hr>

[2] Oral History, J. J. "Jake" Pickle, Section IV, pp. 32–34; Pickle and Pickle, *Jake*, 72–75.

[3] Oral History, J. J. "Jake" Pickle, Section IV, pp. 32, 35, 39–40. In 1942 Pickle married
Ella Nora "Sugar" Critz, daughter of a prominent Granger attorney, Richard Critz, who
became a Texas Supreme Court Judge. Peggy, the Pickles' only child, was three when
Sugar Critz Pickle discovered she had breast cancer. Sugar died in 1952, three years after
her cancer was diagnosed.

[4] Ibid., Section IV, pp. 36–38.

[5] Ibid., Section IV, p. 36.

[6] Pickle and Pickle, *Jake*, 222.

[7] Oral History, J. J. "Jake" Pickle, Section IV, pp. 41–42.

self if he did not at least try for Congress, Beryl agreed. Pickle timed his announcement to follow Thornberry's public statement of his retirement from Congress. The day after Wilson Fox's last barbecue for Congressman Thornberry, Pickle announced he would run.

In the November 9, 1963, special election, Pickle led Republican Jim Dobbs and Jack Ritter, a liberal Democrat from Austin—but not by much. Ritter failed to make the cut, and a December 17 runoff date was set between Pickle and Dobbs.[8] Then on November 22 Lee Harvey Oswald assassinated President John Fitzgerald Kennedy during a Dallas motorcade designed to shore up Democratic votes and raise money for Kennedy's reelection bid in Texas. Within hours President Lyndon Johnson was in charge of a stunned nation, frightened at what appeared to be a conspiracy of right-wing extremists emanating from Dallas, and somehow manipulated by the Communist Soviet Union. Dobbs's ultraconservative campaign, bankrolled by Dallas oil tycoon H. L. Hunt and hostile to big government, Roman Catholicism, minorities, and the deceased President, fizzled.[9] As Pickle later said, Dobbs's views "lost their zing."[10]

During the campaign, Pickle refused to take a position on Laneport Dam.[11] "I inherited the controversy in mid-air collision," he joked over lunch at Austin's Headliners Club, shuddering in mock

[8] In those days, Texas's Democratic Party was torn between a "liberal" wing—which included Ralph Yarborough, Maury Maverick, and Ritter—and the "moderate" or more conservative wing, which consisted of Lyndon Johnson, John B. Connally, and Pickle, among others.

[9] Oral History, J. J. "Jake" Pickle, Section IV, pp. 43–44; interview, J. J. "Jake" Pickle, Apr. 18, 2000, Austin. Hunt was a storied character who is said to have used $30,000 he won in a poker game to buy up an East Texas oil field. By 1935 he was worth $100 million. In the 1940s he wrote a novel, *Alpaca,* which portrayed a utopian society in which the votes of the oldest, wealthiest, and most ambitious in society counted more than everyone else's.

[10] Oral History, J. J. "Jake" Pickle, Section IV, p. 44; "Runoff Decision to K-O GOP in 10 Says Jake," *Williamson County Sun* (Georgetown), Nov. 21, 1963.

[11] Letter, L. D. Hammack to Pickle, Oct. 2, 1963, Box 95-112/10, Pickle Papers (Center for American History, University of Texas at Austin); letter, Joe Provaznik to Pickle, Oct. 3, 1963, ibid.; letter, Mr. and Mrs. R. C. Thomas to Pickle, Oct. 4, 1963, ibid.; letter, John Provaznik to Pickle, undated, ibid.; letter, Felix Matl to Pickle, Oct. 10, 1963, ibid.; letter, Alfred J. Wacker to Pickle, Oct. 4, 1963, ibid.; "Gabriel headed for the Gulf," *Williamson County Sun,* Oct. 31, 1963.

horror.[12] In what turned out to be the political miscalculation of his life, Henry Fox tried to squeeze candidate Pickle into a policy vise over Laneport, earning Pickle's undying dislike. Right after Pickle led the special election, Fox shot off a request to Pickle's Austin home. It was politely worded, but it put Pickle in a no-win situation. "Would you . . . commit yourself to abide by the decision of the majority of the qualified voters in the San Gabriel watershed . . . on how the San Gabriel River should be controlled?" Fox queried. The Circleville philosopher asked Pickle to commit himself to back the results of a popular referendum on Laneport Dam, and he wanted Pickle's answer in nine days, three weeks before the runoff election.[13]

At first Pickle struggled to come up with a diplomatic answer to Henry Fox's challenge; he rewrote his reply three times. While he was revising, his brother-in-law, Georgetown mayor J. Thatcher Atkin, forwarded him a note that Henry Fox had mailed to Atkin.[14] "I didn't lift a hand in the election," Henry Fox wrote Atkin rather grandly,

> but if Pickle declines to agree with the inclosed, it'll be a different story in the run-off. Should Pickle decline to respect the majority opinion of us folks on the San Gabriel through the heart of two counties, a lot can be made out of it, not just here but throughout the 10th District, and we've got the talent and money to do it with.

Atkin scribbled a worried postscript: "Dear Jake: I know how harassed you must feel but I believe it would be in your interest to go along on this . . . "[15]

Fox's threat seems to have toughened Pickle's attitude toward

[12] Interview, J. J. "Jake" Pickle, July 3, 1998, Austin.

[13] Letter, Henry Fox to J. J. "Jake" Pickle, Nov. 11, 1963, Pickle Papers, Box 95-112/10 (Center for American History, University of Texas at Austin).

[14] Atkin's wife was Genevieve Critz Atkin, sister of Ella Nora "Sugar" Critz, Pickle's deceased first wife. Though Pickle remarried in 1960, he remained close to Judge and Mrs. Critz, who had helped raise Peggy, and to Sugar's sister and brother-in-law, Genevieve and Thatcher Atkin.

[15] Letters, H. B. Fox to Thatcher Atkin and Atkin to J. J. "Jake" Pickle, Nov. 15, 1963, Pickle Papers, Box 95-112/10 (Center for American History, University of Texas at Austin).

Henry Fox and the anti–Laneport Dam contingent in Williamson County. "I favor the construction of all three dams on the San Gabriel," Pickle now replied. "I am not against any of the dams. I want to be Congressman of a district that builds dams, that provides water storage, water conservation and flood protection. . . . I am strong for the Georgetown dams. . . . At the same time, I do not wish to oppose any dam, and cannot agree with you that a referendum should be the sole determining factor."[16]

Three days later Kennedy was shot and martyred. Pickle became unbeatable. The Tenth's vote—Pickle two to one—was a show of confidence in President Johnson and in the United States of America as much as a validation of Pickle. It was not a time to rattle government stability.[17]

———

Pickle threw himself into the task of securing every possible dam on the San Gabriel River. Dams were almost a religion with him, as they had been for Lyndon Johnson and most Texas congressmen. Three weeks after being sworn into the House, Pickle heard from Wilson Fox, Laneport Dam's most important backer. The Williamson County Democratic Party chairman offered to host one of his patented barbecues for Pickle, continuing the tradition he had established for Thornberry. "I want something done this year on these dams," Pickle wrote back. "If we miss this chance, we may not make the grade again, and that would be a shame."[18] Pickle knew that the good will enveloping the Johnson administration, and him as a staunch supporter of the president, would soon erode. So he pushed hard to bring the San Gabriel dams onto Congress's appropriations platter. Poage, pessimistic as usual, thought it impossible, but on January 21, 1964, less than a month after Pickle entered Congress, the president handed to Congress a 97.7-billion-dollar budget that

[16] Letter, J. J. "Jake" Pickle to H. B. Fox, Nov. 19, 1963, ibid.

[17] "Unite Behind LBJ Jake Pickle Urges," and "LBJ Could Be Hurt Dec. 17th," *Williamson County Sun* (Georgetown), Dec. 5, 1963; "Everything's Jake! Pickle Wins By 2-to-1," ibid., Dec. 19, 1963.

[18] Letters, Wilson Fox to J. J. "Jake" Pickle, Jan. 10, 1964, and Pickle to Fox, Jan. 15, 1964, Leg/Public Works/San Gabriel, Pickle Papers (Center for American History, University of Texas at Austin).

Congressman J. J. ("Jake") Pickle, 1980.
Courtesy J. J. ("Jake") Pickle, Austin.

Colonel Walter Johnson Wells. *Courtesy*
Brazos River Authority, Waco.

included four hundred thousand dollars for the pre-construction
planning of three "simultaneous" San Gabriel River dams—Owen
Sherrill's old scenario.[19]

Two snags developed. The first, and most serious, came from the
Brazos River Authority, which in September 1962 had brought in a
meticulous Army engineer who had headed the Corps' Fort Worth
District to replace R. D. Collins, whose illness of several years had
hampered his ability to run the show.[20] Collins and John McCall
were now both dead, and with both men gone the old connection
between Georgetown and the Brazos River Authority, once so deftly
milked by Sherrill, was moribund.[21]

[19] "There is no chance whatever of getting any appropriation," Poage wrote Sherrill
on Nov. 5, 1963, Box 692, File 5, W. R. Poage Papers (Baylor Collections of Political
Materials, Waco); memo, unsigned to Pickle, Jan. 1964, ibid.; press release, Department of
the Army, Jan. 21, 1964, Leg/PW/San Gabriel reservoirs, Pickle Papers (Center for
American History, University of Texas at Austin); telegram, Sen. Ralph Yarborough to
Owen Sherrill, Jan. 21, 1964, Box 59 (Brazos River Authority, Waco); "Gabriel Dams Are
$400,000 Closer—Maybe," *Williamson County Sun* (Georgetown), Jan. 23, 1964.

[20] Hendrickson, *The Waters of the Brazos*, 142, 180–181.

[21] "John D. McCall, Attorney, Dies," *Dallas Times Herald*, Mar. 23, 1962.

Road, River, and Ol' Boy Politics

The new general manager, Col. Walter Johnson Wells, was a politically astute pragmatist armed with the vision of basin-wide operation, which he pressed upon his directors in opposition to their previous tendency to analyze projects as individual units.[22] He had closely examined his board of directors' politically inspired 1962 directive, urged by Collins, for simultaneous construction of three dams on the San Gabriel and found a blueprint for economic disaster.[23] Wells's reaction threatened to scuttle the entire San Gabriel dam package.

Meanwhile, a freshly energized opposition movement to the Laneport dam—this one emanating from Taylor's leading business concerns—also created problems for Pickle, who, if he wanted hard cash for the projects, had to convince Congress that his constituents really wanted the dams. Despite considerable pressure, Pickle stuck to his guns. "I am not favoring any one dam or dams over another and I . . . will not change my position," he wrote repeatedly to exasperated Williamson County correspondents.[24]

The BRA had overextended itself financially and was in serious trouble on several fronts, particularly along the Navasota River.[25] Four decades later a longtime BRA official described how things looked from within the agency during the sixties:

Everything built in those days was driven by the drought of the Fifties. Everybody was convinced you had to capture every drop of water. The BRA took the position that it would sign contracts with no source of revenue, with no demand. Nobody needed that much water. The view that they took,

[22] Hendrickson, *The Waters of the Brazos*, 162. I have depended heavily on Hendrickson's thorough history of the Brazos Water Authority for my treatment of Wells.

[23] Memo, "Meeting with Georgetown leaders on the San Gabriel projects," Feb. 17, 1964, Box 59 (Brazos River Authority, Waco).

[24] Letter, J. J. "Jake" Pickle to R. J. Bartosh, Jan. 23, 1964, Leg/PW/San Gabriel, Pickle Papers (Center for American History, University of Texas at Austin).

[25] Hendrickson, *The Waters of the Brazos*, 135–148. Several proposed dam sites on the Navasota were opposed by a coalition of landowners who in many respects resembled the opponents of Laneport, except that valuable deposits of lignite were involved. By 1997 Limestone Dam had been built and two others, Millican and Navasota, had been struck from the BRA's "recommended" reservoirs list. See *Water for Texas: A Consensus-Based Update to the State Water Plan* (Austin: Texas Water Development Board, 1997), 3, 150–151.

and mind, you could never repeat it today, was: "We're going to step out on this high wire with no net. We're a regional agency and we have responsibility for this basin." When they signed those contracts, they had no way to pay.[26]

The agreement to simultaneously construct three San Gabriel dams had been the key to unlocking Congress's authorization of the North and South Fork dams, but it also seemed designed to push the BRA off its high wire. Under the simultaneous-dams scenario, the BRA had to sell water from the San Gabriel dams at prices variously estimated at from fourteen to twenty dollars an acre-foot to repay the federal government for the local share of construction costs.[27] But the top water price the Brazos Authority was actually getting was six dollars an acre-foot. BRA board member Harry Provence, publisher of the *Waco News-Tribune*, put it bluntly in a letter to Pickle:

> We have run our necks into a $24 millions noose on other Brazos tributary reservoirs already, gambling that at $6 an acre foot we could come out even in years ahead when water demand increases in the basin. But the prospective $15 to $17 an acre foot on the Gabriels could not be recovered unless that was the only water left in the state, which won't be the case.[28]

The only way to make *any* dam in the San Gabriel Valley pay, ran the new BRA manager's theory, was to build single-purpose flood-control dams but keep the reservoirs behind them dry, except, of course, during floods. "It is difficult to sell an empty reservoir as an idea, but it is the only idea that makes any economic sense on the Gabriels," Provence asserted. Williamson County would never need as much water as the proposed dams would store, Provence thought, along with other BRA officials and the Army's top planners. A few voices from Williamson County preached that Georgetown and Tay-

[26] Telephone interview, Mike Bukala, Oct. 31, 2000, Waco. Bukala was public relations chief for the Brazos River Authority for many years.

[27] Memo, Walter J. Wells for the Record, Feb. 17, 1964, Box 59 (Brazos River Authority, Waco). Later estimates extended the range from $12 to $23 an acre-foot.

[28] Letter, Harry Provence to J. J. "Jake" Pickle, Feb. 29, 1964, Leg/PW/San Gabriel reservoirs, Pickle Papers (Center for American History, University of Texas at Austin).

lor would grow rapidly, if only they had dependable water supplies. (Interestingly, Owen W. Sherrill and his alter ego, Wilson Fox, each hammered at this point, with Sherrill accurately forecasting Georgetown's future development, and Fox overestimating Taylor's.) Everyone else discounted the combined power of the new Interstate 35 and stored waters to send Williamson County's population skyrocketing.

Wells met with Williamson County leaders, warning that the surface water created by the proposed lakes could be too expensive for them—or anyone—to buy.[29] He didn't say how much the water might cost, but he characterized it as "sky high." By now the local leadership had evolved. In Taylor, Wells dealt almost exclusively with Wilson Fox. Pickle's brother-in-law, Thatcher Atkin, led the Georgetown delegation, which included banker Grogan Lord and his top employee, Jay Sloan, businessman Charles Forbes, and newspaper owner Don Scarbrough.[30]

Previously, Taylor and Georgetown had generally agreed that dams above Georgetown would be better for the county than one large dam at Laneport, but now Wilson Fox fired off numerous letters to Wells and Pickle characterizing the controversy as a long-standing contest between Georgetown and Taylor for the dams' anticipated spoils. Fox was simplifying a complex story. Pickle picked up this view and enlarged it in a letter to Wells: "Georgetown and Taylor continue to shoot 16-inch cannon each week at each other—with me in the middle!"[31] Animosity between the two towns grew. Competing newspaper editorials berated the other city's positions as self-serving or worse. For the first time the *Taylor Press* endorsed Laneport Dam.[32] Meanwhile, from Milam County, which rarely seemed exercised

[29] In the terminology of water-supply developers, "surface" water is water from lakes or reservoirs; "ground" water is well water pumped from underground aquifers.

[30] Letter, Wilson Fox to Walter J. Wells, Feb. 3, 1964, Box 59 (Brazos River Authority, Waco); letter, Wells to Fox, Feb. 6, 1964, ibid.; memo, Wells for the Record, Feb. 17, 1964, ibid.

[31] Letter, J. J. "Jake" Pickle to Walter J. Wells, Mar. 3, 1964, ibid.

[32] Don Scarbrough, "Passing Glance," *Williamson County Sun* (Georgetown), Feb. 13, 1964; "Time to Get the Job Done," *Taylor Daily Press,* Feb. 14, 1964; H. B. Fox, "Would You Really Like To Know What the Laneport Dam Would Do?" *Williamson County Sun* (Georgetown), Feb. 27, 1964; Don Scarbrough, "New Dam Era," ibid.

about the dam controversy, a Soil Conservation Service supervisor attacked Laneport. In Granger, a high school teacher whipped up his agricultural students to protest the dam.[33] And a new anti-Laneport, pro-Georgetown dam petition was floated.[34] Taylor and Georgetown flew separate delegations to Washington to lobby Pickle; both cities feared if one dam were built first, the other, or others, would never get built—not an unreasonable worry, as Congress rarely funded multiple-dam projects, especially on rivers as small as the San Gabriel.[35]

In the midst of this confusing swirl, the Brazos River Authority's Wells recommended that his board formally request the Corps to once again reconsider the size of conservation pools at Laneport, North Fork, and South Fork, "with a view toward reducing the costs of such space and the costs of water yielded therefrom." The board agreed. Wells sent the resolution to the House Appropriations Committee's Subcommittee on Public Works and held his breath.[36]

With Pickle's election, Henry Fox and Owen Sherrill, once so formidable, became persona non grata in Congress. Pickle was determined not to be "torn to pieces" over the San Gabriel River's fate as Homer Thornberry had been. Sherrill's missives to Pickle took on a forlorn tone. "I wish you'd call me collect sometime when you're in Texas," Sherrill wrote. "I've been chairman of the dam committee over 10 years." He also complained to Pickle, "I am just your step child."[37] But Sherrill no longer could "pull wires" as he once had, and Pickle took care to finesse him, especially when Sherrill threatened to destabilize a delicate agreement.[38] Henry Fox was another matter.

[33] Letter, Clarence R. Matula to J. J. "Jake" Pickle, Feb. 28, 1964, Leg/Public Works/San Gabriel, Pickle Papers (Center for American History, University of Texas at Austin); Hubert Gorubec to Pickle, Apr. 21, 1964, ibid.

[34] Petition, Felma Headrick to J. J. "Jake" Pickle, May 1, 1964, Leg/PW/San Gabriel Reservoirs, Pickle Papers (Center for American History, University of Texas at Austin).

[35] Letter, J. J. "Jake" Pickle to Harry Provence, Mar. 24, 1964, Box 59 (Brazos River Authority, Waco).

[36] Memo, J. J. "Jake" Pickle to File, Apr. 29, 1964, Leg/PW/San Gabriel Reservoirs, Pickle Papers (Center for American History, University of Texas at Austin); and Draft of Proposed Resolution, Apr. 9, 1964, ibid.

[37] Letter, Owen Sherrill to J. J. "Jake" Pickle, June 1, 1964, Sept. 23, 1966, ibid.

[38] Sherrill was deeply disappointed that South Fork dam was eliminated from the San Gabriel public works package, but he could not (and would not) stop the process

After Wilson's brother tried to become the hatchet man against Pickle during his first campaign, Pickle would have nothing to do with Henry. When Georgetown newspaperman Scarbrough suggested that Pickle "court" his antagonist, Pickle wrote back,

> I don't see how there is any way I can court Henry Fox . . . . Henry not only wants to needle me, but he obviously wants to destroy me. His letter to the editor . . . was a particularly cruel sort of thing to do to me. Surely he knows that pairing a vote is not being afraid to cast a vote and is oftentime a courtesy rather than a lack of determination[39]

Nearly forty years later, and a decade after Henry Fox's death, Pickle bristled at the mention of his name. "He was an aberration," the longtime congressman said. "Henry Fox was amused at causing conflict, humor, consternation, discomfort. He was a controversial animal. You couldn't put your money in the bank on him. He had fun poking fun at people. That satisfied him."[40]

Pickle wanted to move the San Gabriel dams from the authorization to the construction stage, but Williamson County continued its multipronged resistance to compromise. Within the county, virtually anyone paying attention to such things viewed at least one of the proposed dams as a "special interest" boondoggle—the dam the "other" city wanted. The farm and ranch community stalled against them all,

---

once the North Fork dam shared a construction timetable with Laneport. To reduce Sherrill's influence, Pickle rigorously ignored him, but when Sherrill was honored in 1964 as Texas Realtor of the Year, Pickle inserted Sherrill's biography in the *Congressional Record*. In February 1966, after Pickle praised Sherrill's advertisement for a new development in southwest Georgetown, Sherrill wrote Pickle, "You are sweet. Thanks for the kind words." Owen W. Sherrill Papers (Cushing Memorial Library, Texas A&M University, College Station).

[39] Letter, Don Scarbrough to J. J. "Jake" Pickle, June 29, 1964, Leg/PW/San Gabriel Reservoirs, Pickle Papers (Center for American History, University of Texas at Austin); Pickle to Scarbrough, July 7, 1964, ibid.

[40] Interview, J. J. "Jake" Pickle, Apr. 18, 2000. Antipathy between Pickle and Fox hardened on both sides. In one of Fox's books of political satire, *The 2000-Mile Turtle* (Austin: Madrona Press, 1975), Fox lampooned a fictional congressman he dubbed "Jake Dill," who supported a despised Corps of Engineers dam. The caricature was not a kind one.

still hoping for the small dams program it preferred. "Last week," Pickle wrote his newspaper friend in Waco,

> I spent a couple of hours with Don Scarbrough in George-
> town and got hung again on high-center about these dams. I
> offered to get one or two persons from Georgetown, Taylor,
> and Granger together and lock them up in a room with offi-
> cials of the BRA, Texas Water Commission and myself, throw
> away the key until we agreed on something. As usual, I came
> away with no definite answer. I am thinking about making the
> same proposal to the Taylor folks.[41]

Pickle's fantasy was part joke, part hint of the strong-arm "compromise" he later devised and put over with the help of Wells. But at
this point, Pickle was starting to wonder whether the struggle was
worth it. The idea of building three dams simultaneously "might be a
compromise," Pickle told one BRA director, "yet it will basically take
us away from the original purpose of these dams—namely, the conservation of as much water possible in any given reservoir."[42] This was
an arresting statement, because until now, flood control had always
been the driving force behind damming the San Gabriel.

Pickle's words revealed his unwavering belief in the multiple benefits of dams, as well as what had changed over the project's fifteen-
year life span. The original flood-control dams were conceived as the
best way to serve the agrarian population in Williamson County. But
now the county was starting to grow, spurred by the new interstate
highway which was slicing through the Austin Chalk along the Balcones Escarpment, passing ranch land worth very little, maybe thirty-
five to a hundred dollars an acre, but now potentially fetching much
more.[43] For the first time since Europeans set foot in Williamson
County, it was possible to believe that the rugged caliche limestone

---

[41] Letter, J. J. "Jake" Pickle to Harry Provence, July 24, 1964, Box 59 (Brazos River
Authority, Waco). As well as being the publisher of the Waco newspaper, Provence sat on
the board of directors of the BRA. Like many newspaper publishers of that time, he
exerted a powerful influence on government politics and policy.

[42] Ibid.

[43] Clara Scarbrough, *Land of Good Water*, 345–346.

outcroppings and mesquite pastures of west Williamson County might outpace the fabulous Black Waxy as an economic engine. The thought bubbled up and spread, like ripples on a pond, that water supply might be considerably more helpful to development than flood control. A *Sun* editorial worked this theme:

> Strangely, one of the most compelling reasons why George-town people should want dams located above Georgetown has scarcely been mentioned during the past five or six years of agreements and controversy raging around the subject. That reason is a water supply for this growing community.[44]

On June 16, the House included four hundred thousand dollars for pre-construction planning of the San Gabriel River dams in its Public Works Appropriation Bill.[45] In August Congress sent the bill to the President, who signed it into law.[46] Pickle took pains to credit former congressman Thornberry with getting the funds. But the fact that Pickle so aggressively sought dams across his district, and the fact that President Lyndon Johnson was the most notable constituent of Pickle's Tenth District, probably helped.[47]

A solution to the deadlock between Georgetown and Taylor over the fate of the dams started coming together in 1964 during private talks between Pickle and Walter Wells. Rawleigh Elliott, George-town's new mayor, had called the Brazos River Authority during the summer to find out the actual cost to Georgetown of water from the North Fork reservoir. A BRA engineer researched the matter and reported to Wells that North Fork water could cost as much as twenty

[44] Letter, Buck Hood to J. J. "Jake" Pickle, Apr. 21, 1964, with undated enclosure, "Cities Outgrow," *Williamson County Sun* (Georgetown) editorial, Leg/PW/San Gabriel Reservoirs, Pickle Papers (Center for American History, University of Texas at Austin).

[45] Telegram, Pickle to Don Scarbrough, etc., June 16, 1964, Leg/PW/San Gabriel Reservoirs, Pickle Papers (Center for American History, University of Texas at Austin); and "$400,000 Okayed for Planning Dams," *Williamson County Sun* (Georgetown), June 18, 1964.

[46] Telegram, Pickle to Scarbrough, etc., Aug. 15, 1964, Leg/PW/San Gabriel Reservoirs, Pickle Papers (Center for American History, University of Texas at Austin).

[47] Letter, Pickle to David V. Hoster, Aug. 27, 1964, ibid.

dollars an acre-foot, "but I did not tell Mayor Elliott about this."[48] In the meantime Elliott called on Williamson County's city governments and the BRA to discuss what the various cities might expect in the way of water supply and its cost. Elliott's questions focused the BRA's attention on what turned out to be an embarrassing Corps of Engineers blooper, throwing all previous calculations into question. As the BRA's chief engineer dryly noted,

> The Corps of Engineers apparently did not consider prior water rights of downstream appropriators in determining the yields shown in the San Gabriel Watershed Project, and the [Brazos River] Authority has made no studies in sufficient detail to determine the yields of the reservoirs after honoring such rights.[49]

The Army's failure to factor in the rights of the downstream water-users—primarily Dow Chemical Company and the coastal rice farmers, who used enormous quantities of water—meant that the annual estimated cost of water per acre-foot would rise by six cents over original estimates during the first stage of construction and another five cents during the second and third stages. That meant cost increases of 1.8 million dollars for Laneport, 800,000 dollars for North Fork, and 430,000 dollars for South Fork, or a total cost of 48.5 million dollars for the whole San Gabriel project, rather than 45.5 million. And that meant that cities would undoubtedly pay more for their water than previously thought.[50]

The Army engineers offered a quick solution which, as usual, ignored Williamson County's local needs. They suggested building three flood-control dams in Williamson County that normally would hold no water. In other words, the three reservoirs behind the new dams would be dry—dredged pits—except during floods. (This was what the Army had wanted all along for the North and South Fork dams, which it considered poor projects.) County Judge Sam V. Stone

---

[48] Memo, T. B. Hunter, "Cost of Water on San Gabriel's Project," Aug. 12, 1964, Box 59 (Brazos River Authority, Waco).

[49] Memo, Burke G. Bryan, "Proposed San Gabriel Dams," Aug. 21, 1964, ibid.

[50] Memo, Burke G. Bryan, Aug. 13, 1964, ibid.

was horrified, along with everyone else in Williamson County. Stone implored Pickle to work at "retaining the water conservation feature of these 3 dams."[51]

The crisis might have inspired cooperation. Instead it further antagonized Taylor and Georgetown, the towns most clearly representing the fears and ambitions of the people living on the Black Waxy as opposed to those who had settled along the Balcones Escarpment. Pickle complained to Wilson Fox, "I don't see any sign of anybody weakening, compromising or listening—yet."[52] Following Wilson Fox's advice, Taylor's city commissioners resolved that Laneport be constructed at the "earliest possible date," with "no reduction in the height, size or conservation storage of the dam and reservoir as now proposed."[53]

The words sounded innocuous enough, but they seriously threatened Pickle's chance of getting Congress to appropriate money for the dams. With its resolution, Taylor rejected the simultaneous-construction agreement hammered out three years earlier, which had used the language of cooperation required by congressional etiquette, and with it the BRA's attempt to find a formula to reduce the cost of San Gabriel water. Wells was plainly disgusted. With restrained irony, he wrote Wilson Fox, his strongest supporter,

> We are glad to receive this evidence of sentiment in support of the Laneport project. We are disappointed, however, to note that the City's [Taylor's] resolution indicates opposition to the effort being undertaken by the Authority to achieve a reduction in the local interests' share of the project costs. . . . As you know, general agreement of the varied local interests . . . would be extremely helpful in getting reservoirs under construction on the San Gabriel River.[54]

[51] Letter, Sam V. Stone to J. J. "Jake" Pickle, Aug. 18, 1964, Leg/PW/San Gabriel Reservoirs, Pickle Papers (Center for American History, University of Texas at Austin).

[52] Letter, J. J. "Jake" Pickle to Wilson Fox, Aug. 4, 1964, BBQ, Pickle Papers (Center for American History, University of Texas at Austin).

[53] Letter, Wilson Fox to Walter J. Wells, Aug. 28, 1964, with attached Taylor Commission "Resolution," Aug. 26, 1964, Box 59 (Brazos River Authority, Waco).

[54] Letter, Walter J. Wells to Wilson Fox, Sept. 1, 1964, ibid.

In a report to BRA directors, including Wilson and Henry Fox's brother, J. Howard Fox of Hearne, Wells expressed his exasperation at Taylor's surprising change of position. "It looks as though we still have a lot of work ahead of us on the San Gabriels. It is interesting to note that even when people come out in favor of something they have to express opposition to something else at the same time."[55] Ten days later Taylor's Chamber of Commerce reiterated the city's official stance on Laneport, and the *Taylor Press* followed suit, pointing out that Laneport water would cost "only" an estimated twelve dollars an acre-foot, compared to North Fork and South Fork water, at sixteen and twenty-three dollars an acre-foot, respectively. The only reasonable thing to do, the *Press* argued, was to build Laneport immediately.[56] Wells was livid. "Somebody has to pay for the local interests' share of the project costs," he lectured Wilson Fox,

> and the costs even at the Laneport project alone are such that the cost per acre-foot of water is much higher than for any other reservoir in the basin. . . . No one other than the [Brazos] Authority has indicated a willingness to sign up to pay the local interests' share of the cost. If the people are willing to have the Authority assume this responsibility, it seems they should be willing to support the Authority's efforts.[57]

In early September, Mayor Elliott invited officials from Georgetown, Taylor, Granger, Thrall, Circleville, and Round Rock to hear Wells on the subject of projected water costs. Wells used the opportunity to take the competing dam forces to task, suggesting that the opposition simply "erase" itself.[58] The *Taylor Press* reported,

[55] Memo, Walter J. Wells to J. Howard Fox, etc., Sept. 1, 1964, Box 59, ibid.

[56] Resolution, Taylor Chamber of Commerce, Sept. 10, 1964, Box 692, File 5, W. R. Poage Papers (Baylor Collections of Political Materials, Waco); "One or Nothing," *Taylor Daily Press*, Sept. 17, 1964, ibid.

[57] Letter, Walter J. Wells to Wilson Fox, Sept. 24, 1964, Box 59 (Brazos River Authority, Waco).

[58] Letter, R. S. Elliott to Walter J. Wells, Sept. 2, 1964, Box 59 (Brazos River Authority, Waco); Wells to J. J. "Jake" Pickle, Apr. 6, 1966, ibid. Bartlett, Hutto, and Thorndale representatives were invited but did not attend. It is interesting to note that Round Rock, for the very first time, was included in talks about water planning. Until 1964 Round Rock was never mentioned in policy and planning papers related to water supply from

The BRA manager told the civic leaders in sharp language that delays on San Gabriel have been due to a lack of coordinated meeting of the minds of the local people in Williamson County. When there is dissension, said Colonel Wells, Congress approves appropriations for other projects where local citizens are in agreement.[59]

Of course, this was precisely what Thornberry and Poage had been saying for years. But now, with pre-construction planning under way, and the realization that the new interstate highway might stimulate growth, the county needed to decide. If Georgetown and Taylor leaders persisted in pecking the life out of each other's pet projects, Congress would balk. Nor would Congress spend a penny for the San Gabriel River dams unless the BRA stuck by its pledge to repay the local cost of construction. But unless the BRA could maneuver the Army into bringing water costs down to a level that cities could afford, the project was dead. Neither Wells nor Pickle wanted that scenario.

Wells's published comments galvanized Laneport's beleaguered opponents—now mostly farmers whose land would be condemned and sympathetic agricultural interests, as well as a few financial and business interests in Taylor and Granger whose enterprises would suffer with the loss of their agricultural base. Shortly after Mayor Elliott's meeting, Joe Zimmerhanzel, owner of Taylor Meat Company, who with his brothers would grill steaks for Wilson Fox's "Pickle Barbecue Blowout" later that fall, spelled out why he thought Pickle should oppose the Laneport dam. "I just can't help but feel you'd be in favor of dams at Georgetown if you knew this area like I do," he wrote. Then he continued,

> Through my store here at Circleville last year I handled over 700,000 pounds of pecans, and most of this would be gone if the Laneport dam is built. The 18,000 acres of land that would

the San Gabriel River. In 1964 its population was about 2,300, and it was on the path of an interstate highway that was under construction. It was also the nearest city of any size to Austin, a distance of twenty miles. By 1980, 15,000 people lived in Round Rock proper and it was by far the largest water-user in Williamson County; in 2005 it was a budding Edge City—a suburban center so powerful it could compete with Austin.

[59] "Brazos Unit Backs Gabriel Project," *Taylor Daily Press*, Sept. 11, 1964.

be covered is easily producing over $1,000,000 a year in cotton, corn, maize, pecans and cattle, and I don't see how Taylor could ever attract enough industry through a Laneport to equal that. . . . If they build the dam at Laneport, then this business which I've spent over 20 years building up will sure be for sale.[60]

A few days before Pickle's barbecue was to take place, a full-page advertisement ran in the *Taylor Press*. Signed by 320 people from Taylor, Granger, and Circleville, it was addressed to the Chamber of Commerce, the City Commission, and the people of Taylor. The owners of all three Taylor banks signed the ad, as did the owners of Taylor's most important industry, Taylor Bedding Manufacturing. The ad posed this question:

WHY NOT INDUSTRY & AGRICULTURE BOTH?
This year the 16,000 acres of blackland which would be covered by the proposed Laneport Dam have produced over $1,250,000 in revenue . . .

If Taylor had an opportunity to get an industry which would release one and a quarter million dollars into the economy of this area annually, it would be justifiably proud and pleased, yet here some of you are arguing you should wipe out such an industry by building the Laneport Dam.
Let's agree. Taylor in the future may need more water. Why not get that water and any industry it may bring, and still retain those 16,000 acres of good blackland? WHY NOT HAVE BOTH INDUSTRY AND AGRICULTURE?

Two big dams at Georgetown will store enough water for all the towns in Williamson County, even if they triple in size. Taylor could get all the water it needs by letting it flow by gravity down the river to Circleville, where it could be pumped by pipe line to town. . . . This would give Taylor all the water it needs for all the industry it can get.[61]

[60] Letter, Joe Zimmerhanzel to J. J. "Jake" Pickle, Sept. 5, 1964, BBQ, Pickle Papers (Center for American History, University of Texas at Austin).

[61] Advertisement, "Why Not Industry & Agriculture Both?" *Taylor Daily Press,* Sept. 22, 1964, Box 692, File 5, W. R. Poage Papers (Baylor Collections of Political Materials, Waco).

Wilson Fox was beside himself. He and Henry hadn't spoken for years because of their differences over the Laneport dam, but this was the limit. Having personally engineered the Taylor City Commission's and Chamber of Commerce's pro-Laneport resolutions, he should not have been astonished that Henry would reply in kind. Wilson Fox raged in a letter to Wells, grossly underestimating the opposition sentiment flooding into congressional offices, "the opposition to the Laneport Dam led entirely by two landowners in the proposed basin have now bombarded our merchants, bankers, with petitions opposing the construction of the Laneport Dam and understand have gotten some rather prominent people to sign same," he wrote. "I do not believe I have ever seen anything opposed so loudly by so few . . . I am sure that the powers that be pay very little attention to petitions. I largely discount this, and hope you will do the same."[62]

Henry Fox sent Wells a pleasant letter and attached the newspaper ad. He gave Wells background on the 1961 election that had overwhelmingly approved bonds for more than fifty earthen dams on the upper San Gabriel and its tributaries. The perfect solution, Fox suggested, was two big dams above Georgetown and a system of small flood-prevention dams. If the Army could be persuaded, presumably by Wells, to abandon Laneport for such a plan, "I'd be willing to bet the top calf from my herd against the top calf from my brother Howard's herd that Williamson County would unite on the plan overwhelmingly."[63]

It was a tongue-in-cheek challenge, referring to Howard Fox who had recently stepped down as president of the Brazos River Authority but was still a director. It might well have worked a few years earlier; clearly, Wells was irritated with Wilson Fox and Taylor. But Henry Fox's scheme failed to address the issue of water sales, which had become critical to the BRA's future existence.

Wells forged ahead, requesting Williamson County municipalities to resolve to support the BRA's efforts to restructure the Army engineers' dam plan so that the water they banked would not be prohibi-

[62] Letter, Wilson Fox to Walter J. Wells, Sept. 21, 1964, Box 59 (Brazos River Authority, Waco).
[63] Letter, H. B. Fox to Walter J. Wells, Sept. 22, 1964, ibid.

tively expensive. Wells gave every incorporated city in Williamson County a boilerplate resolution and asked the cities to send them to the Corps of Engineers, the Texas Water Commission, the Brazos River Authority, and to Congressman Pickle.[64] To his board of directors Wells commented, "It looks like the pot is still boiling."[65]

Pickle came under intense political pressure from both ends of Williamson County, but he never wavered. His response was markedly different from his predecessor's. Thornberry had agonized whenever controversy over the San Gabriel dams surfaced. He had tried to appease all factions. Pickle just kept pressing forward. "I hate to see this kind of public agitation because it is bound to come to the attention of the Corps of Engineers and the House and Senate Appropriations Committees," Pickle wrote Wilson Fox, referring to Henry Fox's full-page ad. "It should have no direct bearing on their decision but human nature being what it is, sometimes it does."[66] Pickle had fought fiercer battles than Laneport Dam (the loss of his first wife, "Sugar," to cancer, most especially), and he did not flinch.

As the date of Pickle's first Hilda's Bottom barbecue at Wilson Fox's Riverside Ranch approached, Pickle's letters to Fox grew warmer. "When we first decided to go ahead with the barbeque, I had some reluctance," he admitted, "but you folks know how to do things right and big. I'm very grateful to you for your effort and direction in this regard."[67] From then on the Pickle–Wilson Fox relationship grew more trusting, with the smoky flavors of Fox's wildly popular barbecue blowout and his pet project, Laneport Dam, merging into a quiet understanding between the two men.

On September 29, 1964, Hilda's Bottom once again rang with the sound of politicos and their courtiers having a good time. "Oh, man, I remember that party," Pickle said nearly forty years later. "There weren't any better parties than Wilson Fox's—they were *that good*. There was always plenty of beer. . . . It was billed as the biggest steak

[64] Letter, Walter J. Wells to R. F. Holubec, Sept. 16, 1964, ibid.

[65] Letter, Walter J. Wells to W. E. Borger, R. F. Holubec, J. Howard Fox, etc., Sept. 24, 1964, ibid.

[66] Letter, J. J. "Jake" Pickle to Wilson Fox, Sept. 25, 1964, BBQ, Pickle Papers (Center for American History, University of Texas at Austin).

[67] Ibid.

in Texas, but nobody cared. It was a chance for these community leaders across the district—all strictly men—to mix and mingle out under the trees, away from the cares of home. Some ol' boys would have too much to drink and sometimes it got a little wild and wooly. But in general, there were no problems out there. All was in harmony."[68]

Well lubricated after the barbecue, the Georgetown and Granger city councils cranked out the resolutions Wells wanted. Still, it was touch and go. "It was important to resolve," said Pickle. "Granger and Taylor were split on having a dam at all. While I was talking to Taylor, whenever I thought I had something lined up, Georgetown rose up and raised hell. 'You can't build a big dam down there,' they'd yell. 'We'll be ripped apart.' People forget how ravishing the '57 flood was. But what [Georgetown] was really saying was, 'If Taylor and Granger gets one, we get one, too.' "[69]

It was the BRA's Wells who broke the deadlock. Perhaps only a non-politician could have managed it. Just before Christmas 1964, Wells suggested that the Army engineers host a meeting in Austin on January 7, 1965, between three government agencies—the Texas Water Commission, Brazos River Authority, and the Corps—to discuss the San Gabriel projects. No one else was invited. What Wells sought sounded simple, though it had eluded everyone else for fifteen years: "mutual understanding" between the agencies on the amount of conservation space for each dam on the San Gabriel River and (the political sticking point) their order of construction.[70]

Wells was trying to rebuild shattered bridges. Two years earlier, the Texas Water Commission, egged on by Wells's own Brazos River Authority, had rejected the Army's 1958 resurvey recommendation to build Laneport first. The TWC had offered an alternative vision: three dams built simultaneously as an "inseparable" unit. The Texas Water Commission's fall-back position was that North Fork should be built first, South Fork second, and a vastly downsized Laneport third, if at all.

The Corps scoffed at that suggestion. There the matter lay, tattered,

[68] Interview, J. J. "Jake" Pickle, Apr. 18, 2000, Austin.
[69] Interview, J. J. "Jake" Pickle, July 3, 1998, Austin.
[70] Letter, Walter J. Wells to J. A. Cotton, Dec. 15, 1964, Box 59 (Brazos River Authority, Waco).

still contentious and therefore not likely to spur money from Congress. Even if Pickle could push the plan through, the now-revised projected cost of water was so high that the BRA was in the awkward position of having to renege on its old commitment.

Surprisingly, in Austin, the Texas Water Commission and Army representatives gave Wells what he wanted. Perhaps they were worn out by the controversy; perhaps they felt this was their last chance to dam the San Gabriel River. Unofficially, they agreed to kill the South Fork dam project but supported building North Fork and Laneport simultaneously (Sherrill's magic word again) at "optimum yields" that would lower water costs to BRA water users. The skeleton of the final deal was in place.[71]

But Pickle had to make it work politically. He was the only man who could convince Williamson County residents that the dams would work for, not against them. It helped that he honestly believed that, in Texas, the value of conserving water in reservoirs far outweighed any individual sacrifices.

In early April, while attending an Austin banquet, Pickle quizzed Wells about the January meeting. Georgetown's Mayor Elliott, who had learned about the meeting through the grapevine, wanted more details, too. Wells later wrote Pickle:

> I am convinced that there is no single solution which will satisfy everyone in Williamson County. There may be no single solution which will entirely satisfy even a majority of the people. But I do believe that the approach we are exploring offers the best possibilities for getting some reservoirs on the San Gabriels in the near future and meeting the needs and desires of the greatest number of people in Williamson County.[72]

Wells copied this letter to Elliott and to BRA director Raymond F. Holubec of Granger, who owned 350 acres of farm land a few miles

---

[71] "Summary of Conference," Texas Water Commission, Jan. 7, 1965, ibid. The optimum yields were estimated at 25, 20, and 10 cfs for Laneport, North Fork, and South Fork, respectively.

[72] Letter, Walter J. Wells to J. J. "Jake" Pickle, Apr. 10, 1965, Box 59 (Brazos River Authority, Waco).

below the proposed Laneport dam and strongly supported its construction. No one else in Williamson County was in the know.[73]

A week later, Pickle returned to Texas with a copy of Wells's letter tucked under his arm, intending to put it to use during the Easter break from Congress. As Pickle remembers, he came armed with a verbal commitment from Col. Jack W. Fickessen, district engineer for the Corps' Fort Worth District. He had asked Fickessen if the Army would agree to build North Fork and Laneport simultaneously.

"If you can justify them engineer-wise, I can get these Williamson County folks to cooperate," Pickle remembers saying. The Army man agreed, shifting the Corps of Engineers' long-held negative position on the North Fork dam one hundred and eighty degrees.

Over Easter weekend, Pickle took on Taylor and Georgetown. Once he had joked about locking the leaders of competitive towns into a room until they agreed on a plan. He did not quite do that, but he came close. First he tackled Wilson Fox, who had previously stoutly resisted simultaneous construction. Fox succumbed. Pickle does not remember talking to anyone else on the Taylor side of the argument, probably because Wilson Fox *was* the Taylor side of the argument.

Then Pickle turned to the Georgetown crowd. He and an aide stuffed key Georgetown leaders—banker Grogan Lord, newspaperman Scarbrough, and businessmen Charles Forbes among them—into two cars and drove around on back roads, while Pickle made his pitch, first in one car, then in another. Finally Pickle's entourage pulled onto U.S. Highway 81 (Interstate 35 was not yet complete) and stopped at a roadside park halfway between Round Rock and Georgetown.

"We stood under a grove of big old oak trees, and I asked if they would accept the plan. 'If I can do this, you've got to give me your word you won't go back on the deal,' " he remembers saying. They shook hands on it, and the deal was done.[74]

A year passed. The government plodded through the process of

---

[73] "Williamson County Tax Roll," 1964, p. 123; "Williamson County Survey," Tobin Map Co., San Antonio, Texas, pp. 36–37; Map 60741, Williamson County Appraisal District, Feb. 27, 2001, Georgetown.

[74] Interviews, J. J. "Jake" Pickle, July 3, 1998, and Apr. 18, 2000, Austin.

squaring the new plan with the locals and Army engineers. The Corps drew up the plan and the BRA board approved it. A few "die-hards," as Wilson Fox called them, peppered Pickle with objections to Laneport, not knowing that the fight was all but over.[75] An uneasy truce held in Williamson County.

[75] Most potent among these was Owen Sherrill, who never gave up his fight for the South Fork dam, which along with the North Fork, would, in his words, "create the most beautiful and outstanding recreation area USA wise." See letter, Sherrill to Poage, Nov. 24, 1969, File 7, Box 692, W. R. Poage Papers (Baylor Collections of Political Materials, Waco).

Road, River, and Ol' Boy Politics

EIGHT

CLINCHING THE DEAL

The Corps of Engineers is like that marvelous little creature, the beaver, whose instinct tells him every fall to build a dam wherever he finds a trickle of water. Like the Corps, this little animal frequently builds dams he doesn't need, but at least he doesn't ask taxpayers to foot the bill.

SENATOR WALTER MONDALE, *Atlantic*

I t required another year to clinch the deal. On March 25, 1966, Pickle brought officials of the Corps of Engineers, Brazos River Authority, Texas Water Development Board, and Texas Water Rights Commission back to the table in Austin. Again, no one from Williamson County was invited. An estimated fifty-five million federal dollars would flow to the county to build massive federal projects that would completely change the county's complexion, but the subject was still so volatile that local representation was not thought wise. The agencies rubber-stamped the decisions they had informally made a year before: the Laneport and North Fork dams would be built simultaneously and their yields would be reduced so their water "product" would cost less, though specific costs were not delineated. The South Fork dam would not be built.[1] The total estimated price

---

[1] One person who did not attend the March 25 meeting was James A. Cotton, the BRA's new chief consulting engineer. Repeating the Army's view of the previous fifteen years, he wrote Wells that the proposed "approach" was the best available. However, he added, "Personally, I doubt that either the North or South Fork reservoirs are economically justified." See Cotton to Walter J. Wells, Mar. 15, 1966, Box 59 (Brazos River Authority, Waco).

of the San Gabriel dams had jumped twenty percent to 55.5 million dollars—thirty-two million for Laneport and fourteen million for North Fork.[2] Real estate prices, buoyed by speculative investment along the new interregional highway, had risen, inflating the cost of the North Fork dam.[3]

The agencies knew that even with the South Fork dam eliminated from the picture, and even if the other two dams' conservation pools were reduced, Laneport and North Fork water would *still* be "considerably more expensive" than any other Brazos basin reservoir—three times as expensive, to be precise. This worried everybody, since, even in 1966, no one believed Williamson County could ever need enough water to use the new supply. In Austin one BRA man confidently predicted "that there was, and would be in the future, very little demand for the water in the immediate vicinity of the proposed reservoirs."[4]

Apparently, no one questioned that assumption. Texas's chief water planners felt so certain Williamson County would remain a stagnant rural backwater that the Texas Water Development Board representative took the extraordinary step of suggesting that his agency "assume responsibility for making payments to the [federal] Government to the extent and at the time that the [Brazos River] Authority might be unable to do so."[5]

But why should the state water board assume this responsibility? Why did the State of Texas consider the development of Williamson

---

[2] Letter, Col. Jack W. Fickessen to Texas Water Rights Commission, Mar. 31, 1966, Table I (Mar. 25, 1966), "San Gabriel River Preliminary Reservoir Data," Box 692, File 6, W. R. Poage Papers (Baylor Collections of Political Materials, Waco). Though the South Fork dam looked dead, its $9.3 million cost was included in the $55.5 million package total. The Corps rarely removed any authorized public work from its books, figuring that if conditions changed, it would be far easier to build the project without having to go through a difficult re-authorization process. Ultimately, North Fork and Laneport together cost $100 million.

[3] Letter, Walter J. Wells to James A. Cotton, Mar. 14, 1966, Box 59 (Brazos River Authority, Waco).

[4] Walter J. Wells, Memorandum: "Meeting on San Gabriel Projects," Apr. 4, 1966, pp. 1–2. ibid. Walter Wells, J. Howard Fox and Harry H. Moore represented the BRA at the meeting with other water groups and Congressman Pickle.

[5] Ibid., 2.

Road, River, and Ol' Boy Politics

County water to be so important that its top water-policy makers would suggest absorbing the Brazos Authority's bad debt? It was quite true that in 1966, Williamson County couldn't come close to utilizing the water stored by two new dams, even during extreme drought. And yet Wells repeatedly claimed the dams' purpose was "simply to satisfy the desires of the people of Williamson County."[6]

Clearly this was specious logic: strong local contingents had fought both for and against each project, splitting the county for two decades. Texas water development forces wanted the dams for two overriding reasons. Dams at Laneport and North Fork would satisfy the occasional desperate need of powerful "downstream water interests" for vast quantities of fresh water that the BRA could not deliver with its 1966 system of reservoirs. These "interests" counted among Texas's most important industries: Dow Chemical and the coastal rice farmers.[7] Both industries depended on copious amounts of fresh water for survival. In a normal year the Brazos River proved ample, but the fifties had been anything but normal.

During the seven-year drought of the fifties, Dow and the growers had experienced water shortages so dire they threatened the industries' survival. Developing new water reserves with Laneport and North Fork dams—where water seemingly could *not* be utilized locally—was better than digging up buried gold. And, too, state and regional planners acknowledged they could not foretell distant future water needs. The San Gabriel reservoirs' usefulness might not become obvious, as one of Texas's chief water planners noted, until "some 40 to 50 years from now, and the [Brazos River] Authority should consider whether at that time it would be better to be criticized for having too much water available or for not having enough."[8]

_____

In 1966 the Brazos River may have been something of a disappointment in the grand scheme of Texas development. One of Texas's

_____

[6] Minutes, BRA Board of Directors, Apr. 18, 1966, pp. 3831–3832 (Brazos River Authority, Waco).

[7] Letter, Alex Pope Jr. to Lyndon Johnson, Nov. 9, 1956, Senate Case and Project 1956, BRA, Box 1210 (LBJ Library, Austin). The canal companies that supplied water to the rice farmers were also a source of concern to the BRA.

[8] Ibid.

greatest river systems, it had somehow failed to become the agent for wealth that its sister river, the Colorado, had become, especially near Austin and the Highland Lakes. Seasonally the Brazos went berserk, flooding the agriculturally developed lowland plains. Brazoria County, on the coast, billed itself as "the receiving end of Wild River!" and built its own dam to contain the river.[9] After the Texas Revolution, the Brazos Valley had lured many large investors, who established themselves astride its floodplain, where the river crossed the Black Waxy. Cotton planters from the Old South, with slaves to work the fields, arrived first. Some pioneering entrepreneurs believed the Brazos could be made navigable from the Texas coast to Waco, where the Black Waxy terminated; they believed the Brazos's alluvial drifts would provide footing for an agricultural Colossus. The dream of ferrying cotton bales down the Brazos to ships waiting in the Gulf died hard, but after several costly attempts at re-engineering the river, it was abandoned.[10] Nonetheless, the Brazos Valley provided a safe harbor for King Cotton well beyond the turn of the century (longer than in most traditional cotton-producing regions), making fortunes and establishing the most extensive rail system in the United States. But by the late 1920s, King Cotton was playing out.

Meanwhile, from 1900 on, aspiring rice planters were buying up thousand-acre farms southwest of Houston. On a strip of coastal plain roughly twenty to thirty miles wide, railroad promoters flogged rice-growing into existence, despite the need for large capital investments in irrigation canals, pumps, and levees.[11] Most of the planters came from moneyed stock, hiring out their land to menials who did the work. Some Japanese rice farmers were recruited to Texas. In 1911, in

[9] *Welcome to Brazoria County* (Angleton: Brazoria County, 1957), 1 (Center for American History, University of Texas at Austin).

[10] Hendrickson, *The Waters of the Brazos*, 5–10. The historian of the Brazos River Authority traces the fascinating story of attempts to re-engineer the Brazos into a navigable stream.

[11] *The Texas Rice Book* (circa 1900), Houston: Gulf, Western Texas & Pacific Railroad's Southern Pacific Sunset Route, Edward A. Clark Texana Collection (Southwestern University, Georgetown), 41. In 1900 the outlay required to start a rice farm ranged from $50,000 to $300,000, depending on the size of the operation, this promotional booklet asserts.

a fairly typical brochure, one railroad trolled for tenants, claiming, "It is not unusual for tenants to buy their own farms with the results of one year's work."[12] It was an exaggeration, but some serious wealth was built on rice.

By 1966, when the Brazos River Authority was trying to boost dam-building on Brazos tributaries, rice was Texas's number three crop—and a very profitable one at that. Unlike cotton or corn, though, it needed prodigious amounts of fresh water precisely when the farmer called for it, and a simple but extensive irrigation system.[13] A railroad pamphleteer described how it worked when the farmer opened his flood gates: "The rice farmer from this time until harvest begins has only to watch his levees and cry out, 'Give me water, water,' which he keeps up for about seventy days, the usual period of irrigation."[14]

It sounded easy, but it wasn't, especially during drought years when calling out for water would hardly do the trick. In 1955 the rice growers pressured their political representatives, including Senator Lyndon Johnson, against doing *anything* that would interfere with their right to use precious Brazos River water—from building new dams on the river's main stem to raising the height of Possum Kingdom Dam, the Brazos River Authority's only significant source of income. The Brazos Authority's need to develop a new hydropower dam—a cash cow—and its directors' belief in the "multi-bucket" system of river management warred with the state's concerns about keeping the rice growers in business.[15] Something had to give.

Spurred by World War II demands, another industry quickly dwarfed rice agriculture on the Texas coast. In 1940 Dow Chemical officials, pressed by the United States military to produce magnesium for the war effort, built a chemical plant on the mouth of the Brazos

[12] *Texas and Louisiana Rice* (Houston: Passenger Department Sunset Route, Industrial Department, South Pacific Railroad Company, 1911), 19 (Center for American History, University of Texas at Austin).

[13] Marshall R. Godwin and Lonnie L. Jones (eds.), *The Southern Rice Industry* (College Station: Texas A&M University Press, 1970), 5–7.

[14] *Texas and Louisiana Rice*, 19.

[15] Memo, Walter [Jenkins] to Senator Johnson, Mar. 21, 1955, Senate Case and Project 1956, BRA, Box 1210 (LBJ Library, Austin).

River at Freeport. Before long, Dow's Texas Division was its flagship. The Gulf of Mexico provided Dow with the tremendous quantities of sea water its scientists needed to process magnesium. Once the magnesium was bathed in salt water, it needed a fresh-water rinse. The Brazos River, neatly carved into two channels by the citizens of Brazoria County, fit the bill perfectly.[16] A Dow company magazine writer elaborated:

> The relationship of the Dow plants to the Brazos warrants something more than the usual recognition of the importance of a river in the functions of an industrial facility located snugly on its shores. . . . This specification was met by taking ingenious advantage of the quirks of the old and new channels of the Brazos.[17]

When it moved to Texas in the early forties, one of Dow's first orders of business was to secure rights to Brazos River water. In Texas, the state owns all water in its rivers and lakes, but people or corporations may lay claim to use those waters. The state will permit them to use the water so long as no prior claim exists. The most desirable water rights in Texas are the claims made earliest in time and lowest in the river basins, along the Texas coast. Thus, the canal companies that supplied the rice growers with water had the best rights on the Brazos, but Dow held an enviable position, too. In 1942 the Brazos River Conservation and Reclamation District, which became the Brazos River Authority, acting as the state's agent, signed a thirty-year contract with Dow allowing the industry to use up to 150,000 acre-feet of water per year from the Brazos. The water was free. It was a great deal for Dow, granted, no doubt, because of the importance of the industry to Texas. There was a catch: If Dow needed more than 150,000 acre-feet in a year, or if the Brazos's flow could not meet Dow's contractual allotment, Dow could ask the Brazos District to release "emergency" water from Possum Kingdom Dam, four hundred miles upstream. Once that water reached Dow's pumping station, the emergency Possum Kingdom water (mingled, of course,

---

[16] "From Salt Marsh to Chemical Center: The Texas Division," *Dow Diamond* (Oct. 1955), 4, 6, 8, 9–11 (Center for American History, University of Texas at Austin).
    [17] Ibid., 4.

with water from downstream Brazos tributaries), would cost Dow one dollar per acre-foot. Even that was dirt cheap. In addition Dow paid the BRA fifteen thousand dollars a year to seal this "right."[18]

In 1956, near the end of the seven-years' drought, Dow made its first request for "emergency" water. At that point, Dow was faced with only a twenty-day supply of water from the shrunken mouth of the Brazos. Its two private reservoirs were drawn down to mud. Anxious memoranda flew around state offices. "Should Dow run out of water, it would be forced to shut down and disrupt seriously the economy of the Gulf Coast," the president of the Brazos River Authority board of directors wrote. There was no easy fix. Possum Kingdom, the theoretical back-up "bucket" in reserve, was drawn down to the nub, too. Even if every acre-foot of water in Possum Kingdom Lake were released, Dow still wouldn't receive enough water to keep operating. It was a pending disaster, not only for Dow and the Texas economy it fueled, but for the Brazos River Authority. "Such withdrawl from Possum Kingdom," wrote the same BRA official, "might make it impossible for the Authority to fulfill other prior contracts. In this event, the Authority might lose $42,500 a year of income for several years."[19] He did not add that the BRA could be forced to renege on numerous small contracts, potentially spawning dozens of lawsuits.

The solution turned on Lake Belton, a new Corps of Engineers flood-control dam twenty miles north of Georgetown in Bell County. Open only two years, its water supply was untapped, so BRA officials adroitly turned a liability into an asset. They released Lake Belton's reserves to Dow, preserved the BRA's contracts with its other Possum Kingdom water users, and theoretically saved thirty thousand acre-feet of water that would have evaporated on the long trip from Possum Kingdom in north Texas to Dow on the coast. The BRA even made money on the deal.[20] Lake Belton proved the allure of multiple

[18] Contract, "Brazos River Conservation and Reclamation District and the Dow Chemical Company: A Water Supply Contract," Dec. 30, 1942, Box 362 (Brazos River Authority, Waco).

[19] Memo, "Helping a Neighbor in Need," H. S. Hilburn to BRA board of directors, Dec. 1956, Senate Case and Project 1956, BRA, Box 1210 (LBJ Library, Austin).

[20] Ibid., 2.

dams within a major river basin run as one flexible unit. The success of Lake Belton during the 1956 Dow emergency warmed the hearts of Texas water managers, emboldening them to seek more dams and complex integrated projects.

Lake Belton taught another lesson: the desirability of dams on relatively insignificant upstream tributaries where water was *not* in heavy demand. Such reservoirs could create transportable liquid pools for desperate situations in the high-priority downstream world of rice and Dow. As a BRA spokesman put it, "You could go to another bucket that was full—or fuller."[21] This was the sub rosa reason the BRA made a case for two, or even three, dams on the tiny San Gabriel, and why the Texas Water Commission and politicians like Lyndon Johnson and Jake Pickle backed them up. Sympathy for Williamson County and its flooding miseries were the cover story that helped sell the project, but by the sixties, greater forces—Dow and the rice industry—were Texas policy-makers' larger concerns.

But the policy makers missed one trend completely: they failed to imagine the impact Interstate 35 would have on Williamson County. Though national patterns of suburban growth triggered by freeways had been materializing for well over a decade, the planners did not extrapolate those experiences to the new interstate's path between San Antonio and Dallas. Between 1950 and 1960 Williamson County's population had actually declined, from 38,833 to 35,044.[22] But in Williamson County a handful of people saw that the combination of the new interstate and new lake water might trigger desperately needed growth. (After all, the county's population had essentially stalled since the twenties, and its agricultural bounty no longer translated into wealth.) Owen Sherrill was certain of growth.[23] But no one making water policy for Texas expected Williamson County to grow. If anything, the state's professional fortune-tellers suspected it would shrink.

Pickle believed in dams. He had seen what Lyndon Johnson's

[21] Telephone interview, Mike Bukala, Oct. 31, 2000, Waco.

[22] Clara Scarbrough, *Land of Good Water,* 345.

[23] Advertisement, "San Gabriel Hills," Owen Sherrill Agency, Feb. 21, 1966, J. J. "Jake" Pickle Papers (Center for American History, University of Texas at Austin).

string of dams on the Colorado River had done for Austin, and that was all he needed to know.[24] Certain he was correct, Pickle wrapped up two decades of Williamson County's internal battling over San Gabriel dam proposals by imposing on the county a "compromise" solution crafted by government technocrats and approved by a few carefully selected local "leaders." To get that agreement, Pickle deliberately avoided informing or involving those most affected by the proposed dams until the deal was cemented.

Williamson County was brought into the picture on April 29, 1966, in a carefully choreographed affair. Pickle billed the Austin meeting as a chance to hear a "preliminary pre-construction planning and engineering report" on the San Gabriel dams. His invitation list was short: the mayors of Georgetown, Taylor, and Granger along with County Judge Sam Stone. He asked each man to limit his party to a dozen invitees. Congressman Poage would bring "a car load of Cameron people."[25] Corps, BRA, and Texas water-agency officials would be on hand to answer questions. What Pickle, Wells, and Colonel Fickessen feared was a public discussion flaring into argument, so they prepared a press release to be distributed at the meeting. The point was to freeze into official policy the unofficial agreement between the three key government agencies and the half-dozen county leaders Pickle had wooed, while discouraging "local interests" from jousting and derailing the project.[26]

They need not have worried. The mayors and the county judge picked attendees precisely for their polite disinclination to "make waves" while television cameras rolled. Sherrill was the only real dan-

---

[24] This should in no way imply that Pickle had not educated himself about dams; he attended learned conferences and sought information constantly from experts in the field. But he came at the issue with his aims firmly fixed, and then applied his considerable knowledge and skills toward achieving them. See his speech in "Proceedings, Seminar on Management of River Basins," Apr. 5, 1965, sponsored by the University of Texas Center for Research in Water Resources (McKinney Engineering Library, University of Texas at Austin).

[25] Note, J. J. "Jake" Pickle to Poage, Apr. 19, 1966, with attached form letter also dated Apr. 19, Box 692, File 7, W. R. Poage Papers (Baylor Collections of Political Materials, Waco).

[26] Letter, Walter J. Wells to J. J. "Jake" Pickle, Apr. 20, 1966, with attached press release, dated Apr. 29, 1966, Box 59 (Brazos River Authority, Waco).

Clinching the Deal

ger, but he could not be left out. At the meeting, Pickle approached Sherrill with the sort of backhanded flattery that only someone with an ego like Sherrill could enjoy. Sherrill actually boasted about the incident in a letter he wrote to Poage several years later:

> There was a time and a meeting called at Austin by Jake ... when Jake picked me from a back observation seat, set me down on the front row and said, "Owen ____ ____ don't you say a word." This was when North Gabriel and Laneport were placed in the program leaving out the lonesome South Gabriel dam.[27]

Amity reigned. Newspaper reports reflected the meeting's cheery glow. "Gabriel Dams Are Assured," the *Williamson County Sun* headlined, and topped an editorial with, "Yippee! We'll Get That Dam!!"[28] Pickle and Wells congratulated each other, though Wells made sure Pickle got public credit. "This was *your* baby—so I know you were doubly proud," Pickle wrote to Wells.[29]

The perception of unanimity was far stronger than the reality—a smoke and mirrors trick—but it worked. Poage, always the pessimist when it came to Williamson County, fretted that county residents might not "stay hitched" to the deal, remembering how another agreement had fallen apart in 1962. Pickle worried about getting appropriations for two dams on one river in Lyndon Johnson's 1967 budget. But for the first time in twenty years, on the subject of dams, Williamson County stayed mum.[30]

Early that autumn of 1966 Pickle celebrated with twenty-five hundred followers at the Tenth's stag "whingding," as Wilson Fox and Roman Bartosh dubbed it, at Hilda's Bottom. The company feasted

[27] Letter, Owen Sherrill to Poage, Nov. 24, 1969, Box 692, File 7, W. R. Poage Papers (Baylor Collections of Political Materials, Waco).

[28] *Williamson County Sun* (Georgetown), May 5, 1966.

[29] Letter, J. J. "Jake" Pickle to Walter J. Wells, May 2, 1966, Box 59 (Brazos River Authority, Waco).

[30] Letters, J. J. "Jake" Pickle to Walter J. Wells, May 2, 1966, and May 14, 1966, ibid.; Poage to Wells, May 21, 1966, Box 692, File 7, W. R. Poage Papers (Baylor Collections of Political Materials, Waco).

on two thousand steaks and three hundred servings of sausage grilled by the Zimmerhanzels, and drained sixteen kegs of beer. At nine o'clock the skies opened and it "rained like mischief," in Pickle's wonderful phrase, giving partygoers a taste of the storms that could turn ugly on the Black Waxy. Afterwards, the *Granger News* shouted, "Holy Mudballs, Super Pickle it Rained!"[31] Hundreds of cars and trucks got stuck in the fabled black "gumbo"; it took a week to haul them out.[32]

"As to the liquid," Wilson Fox later wrote to Pickle, with uncharacteristic jocularity,

> [I] will say that the final consumption sounds somewhat like the national debt. . . . The beer boys tell me that there are about three and a half cups to each quart; and then by continuing the mathematics we learned in school, we find that there were some 10,528 cups of beer consumed. If we want to really get into the mathematics, we can say there were 10,528 trips to the kegs, varying anywhere from two feet to 100 feet of travel distance; and by a little more multiplication we can find out about how many miles people walk to get the beer. However, we might have to drop that somewhat as quite a few of them never moved out of their tracks which were imprinted very close to the keg. There was another type of liquid on the ground I believe known as soda water, but the amount consumed there pales into insignificance as compared to the more potent beverage. . . . All I can say is that you really do have some beer drinking friends.[33]

From then on, Pickle addressed Wilson as "My Beer Friend."[34]

---

[31] "Holy Mudballs, Super Pickle it Rained!" *Granger News,* Sept. 8, 1966. Pickle liked this piece so much he passed it around the House of Representatives; "holy mackerel it was good," he wrote author Don Scarbrough. See Pickle Papers (Center for American History, University of Texas at Austin).

[32] Interview, J. J. "Jake" Pickle, Apr. 18, 2000, Austin.

[33] Letter, Wilson Fox to J. J. "Jake" Pickle, Sept. 14, 1966, Taylor BBQ 1966, Box 95-112/107, Pickle Papers (Center for American History, University of Texas at Austin).

[34] Letter, J. J. "Jake" Pickle to Fox, Sept. 19, 1966, Pickle Papers, ibid.

The deal stayed "hitched." In October 1967, less than three years after he entered Congress, Pickle pushed a $1.8 million appropriation through the labyrinth of lawmakers to start construction on the Laneport and North Fork dams—the first big money the project had ever garnered. President Johnson signed it into law just after Thanksgiving.[35] Taylor and Georgetown never stopped haggling. "Both cities suspected each other," Pickle said. "We had to promise we'd start the two dams at the same *hour!*"[36] But except for the anger and sadness rife within the 152 farming families whose forced departure was necessary to make way for the Laneport dam, the opposition collapsed.[37] Pickle's compromise had split up the opposition to Laneport, forcing Georgetown to backpeddle away from its old support of the Czech farmers if it wanted a dam on the North Fork.

In the summer of 1970 the U.S. House of Representatives appropriated four million dollars—more than President Johnson had requested—for dam construction on the San Gabriel. It was a tough time for dam appropriations: Americans were deeply torn over the

---

[35] "Dam Funds Get 'Okay' from House," *Williamson County Sun* (Georgetown), July 27, 1967; letter, Walter J. Wells to J. J. "Jake" Pickle, July 28, 1967, Box 59 (Brazos River Authority, Waco); "Senate Restores 'Start Money' for Dams," *Williamson County Sun* (Georgetown), Oct. 5, 1967; "President Johnson Signs Bill," ibid., Nov. 30, 1967. The final appropriation contained $1.5 million for the San Gabriel projects, mostly for land acquisiton. President Johnson signed the Appropriation Bill for Public Works on November 27. See letter, J. J. "Jake" Pickle to Oxsheer Smith, Dec. 6, 1967, Box 59 (Brazos River Authority, Waco).

[36] Interview, J. J. "Jake" Pickle, July 3, 1998, Austin.

[37] There was a great deal of bitterness, especially among the Czech landowners near the Laneport Dam site, not only about being forced to abandon their family homes, but also at the "horsetrading" mentality they encountered in the Corps. The Laneport Basin Landowners Association formed to bolster landowners' spirits in the face of the overwhelming power of the United States government, promising to help defray court costs for anyone who believed the Corps had "lowballed" its first offer. "These landowners aren't just haggling over the price of a horse, they're dealing against their will over their life's work," read an undated article in the *Taylor Daily Press*. But this was not unusual for Corps dam projects, as has been well documented in Marc Reisner's *Cadillac Desert* and in a meticulous study by Gordon L. Bultena, "Dynamics of Agency–Public Relations in Water Resource Planning," in *Water and Community Development; Social and Economic Perspectives,* ed. Donald R. Field, James C. Barron, and Burl F. Long (Ann Arbor, Mich.: Ann Arbor Science Publishing, 1974).

Vietnam War, questioning the government in a way it had not been questioned since before World War II, and environmentalists were stopping some dam projects. Pickle's feat of securing critical start-up funds on dam projects that had been contentiously disputed for twenty years was practically miraculous. The BRA's Wells, an experienced Corps hand who knew the process well, was deeply impressed. He wrote Pickle,

> It is most unusual for the House to add anything to the President's budget. This is the first time I have ever had it happen on any project I've been associated with. It is rare indeed, and getting it done is a great tribute to your ability and effectiveness as a Congressman. I don't know how you did it, but you deserve all the credit.[38]

It was true. For all practical purposes, Pickle dammed the San Gabriel River. He willed it to happen, manipulating his constituents into compromises that made them squirm. Pickle closed in fast on the San Gabriel quagmire, shrugging off the doubts, snares, and entreaties of dam supporters and opponents alike, from Owen Sherrill to Wilson Fox to Henry Fox. Without Pickle's 1965 "oak tree" talk, it is hard to imagine Congress financing any dam on the San Gabriel River. In truth, local opposition to Laneport ran nearly as fervently as ever, despite Pickle's "compromise."

Across the nation, resistance to American military forays in Vietnam was stiffening, eroding support for President Johnson's domestic programs, while opposition to large federal dams was starting to swell. In early 1966 the Bureau of Reclamation started work on two Colorado River dams that would have partially flooded the Grand Canyon, triggering a wave of outrage that led to a paradigm shift in the politics of environmentalism. Almost immediately, Congress pulled the plug on the Colorado projects and passed the Wild and Scenic Rivers Act, ensuring that lawmakers would be forced to consider rivers as precious resources.[39] The American public was dis-

[38] Letter, Walter J. Wells to J. J. "Jake" Pickle, June 22, 1970, San Gabriel (Brazos River Authority, Waco).

[39] Reisner, *Cadillac Desert*, 243, 273–274, 285–288.

Three Dam Plans

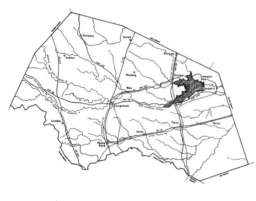

1948: Corps of Engineers'
Laneport Dam Plan

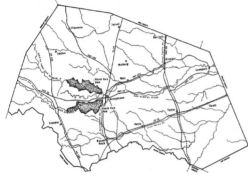

1948: Owen Sherrill's
Upper San Gabriel Dams Plan

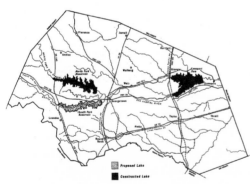

1965: Jake Pickle's "Compromi
Three Dam Package

Three versions of plans to dam the San Gabriel River: Top, the Corps of Engineers' 1948 Laneport Dam plan. Middle, the Brazos River Authority's proposed
dams on the North and South Forks of the San Gabriel, embraced and pushed in
1948 by Owen W. Sherrill. Bottom, the three-dam plan advocated by Congressman J. J. ("Jake") Pickle, requiring three dams to be built as a single package.
Drawn by Kristen Tucker Pierce.

gusted with projects like the proposed Marble and Bridge Canyon dams on the Colorado. In Texas the grandiose Texas Water Plan was defeated at the polls.

Had Pickle not broken the San Gabriel River logjam when he did, the dams almost certainly would not have been built. The San Gabriel River would still be flowing freely, drawing whitewater canoeists to its rapids during the spring and fall, enriching the Black Waxy's alluvial deposits in Williamson and Milam Counties, and periodically flooding out the bottomlands along the river valley. Lake Georgetown would be an engineer's schematic tucked away in the National Archives, instead of the reason Round Rock exploded in the early 1970s, growing into a mecca for the nation's burgeoning computer industry, and turning Williamson County, at least for a time, into the second-fastest-growing county in the United States.

DIASPORA

[T]hey thrill to the thought of the vine and fig tree, of the
family hearthstone that survives the mutability of the years.
The thing is in the blood of the race.

WILLIAM E. SMYTHE, *City Homes on Country Lanes*

Afer nearly three decades of deadlock, the Army engineers
were unleashed to dam the San Gabriel River.[1] The compromise agreement to build three dams (though one, South
Fork, was stillborn) on an inconsequential stream was far from the
Corps' ideal plan favoring one giant reservoir at Laneport, but from
the engineers' point of view it was better than nothing. All the Army
had to do now was acquire land from mostly unwilling sellers, with-
stand legal challenges, and maintain momentum in Congress's annual
appropriations grind. The San Gabriel compromise still might
founder on the shoals of an increasingly vexatious Vietnam conflict,
but if the Army engineers and the politicians stayed their course, their
power of eminent domain was nearly invincible. A government attorney underscored this point as he took the Corps' real estate operatives
to task over their negotiating tactics at the Laneport dam site: "from
the fact of the relative strength of the United States Government and
the individual property owners . . . it seems only fair that the property
owner be told as much as possible about the means employed to
arrive at value."[2]

---

[1] Surely this was one of the longest-delayed projects ever resurrected in Army Corps
of Engineers history.

[2] Letter, William C. Black to Woodrow Berge, May 23, 1969, W. R. Poage Papers,

In Williamson County 152 farm families, approximately five hundred Czech-Americans, faced a shattering diaspora.[3] As once they had sown seeds to the wind, they now were being blown from their lodgings. Fearing expensive legal battles which their congressmen advised them against making, most families whose farms lay in the way of the dams accepted settlements they considered insulting—a word that appeared frequently in protests.[4] Few of the Czechs had ever negotiated a major real estate deal, much less one that involved their entire life's work. Some spoke English haltingly; others were widows with little business experience; most had inherited or purchased their land from friends or relatives. The Army's land men were hardened professionals—they knew the "takings" game by heart. Usually the Army's advance land men would start with the least experienced—or most desperate—landowners controlling crucial segments of the project. If the landowner refused the Army's first offer, and negotiations failed to produce a settlement, the Army could sue and let the court decide. Or it could keep dickering, and settle with holdouts as needed. It was a practical approach, designed to save taxpayers' money and impress congressional penny-pinchers. But the fact was that many small landowners suffered greatly, particularly those who settled early with the government. Those who resisted the Army did far better, in some cases netting nearly four times as much per acre as landowners who agreed to the Corps' first offer.

Over the years, the Army had honed its response to landowners' pleas of inadequate compensation, tending to view property owners as being greedy whiners and hopelessly antiprogressive. It was a view the Army successfully proselytized in Congress, allowing the people's elected representatives to feel a little better about the forced evictions of their constituents. Just as Native Americans had been stripped of

Box 692, File 7 (Baylor Collections of Political Materials, Waco). At this stage, the politicians who had led efforts to build the dams sometimes interceded with the Corps when it appeared that a landowner might be getting a raw deal. And yet the crucial politicians, Poage and Pickle, continued to defend the overall acquisition process as fair.

[3] Interview, Loretta Mikulencak, July 6, 1998, Granger. Mikulencak was tax assessor collector for the Granger school district for many years, as well as a local historian.

[4] Letter, Poage to Stacy Labaj, Mar. 15, 1972, Box 692, File 7, W. R. Poage Papers (Baylor Collections of Political Materials, Waco).

their choice hunting and fishing grounds when the United States "opened" new territories to waves of Anglo-European settlers, the descendents of many of Williamson County's earliest entrepreneurs—the European immigrants who had possessed the vision and backbone to cultivate the rich Black Waxy—had to give up the land, homes, and even the communities they had developed to a government goal they barely comprehended. They did receive government compensation—"market price" the Army called it, though few believed in it.

No matter. They had to leave. A few fared well; many did not. Unable to purchase comparable land with their government settlements since little land comparable to the Blackland Prairie was available, many of the ousted Czech farmers "retired" to Granger and Taylor, where they eked out their days working at low-status jobs such as school custodians, mail carriers, or the like, stripped of the land that was their proudest possession. The things they cared most about—farming the Black Waxy, creating bounteous home gardens and intricately grafted orchards, and living in close-knit communities—became bittersweet memories.[5] Gone too were Friendship, Machu, and Moravia, where two and three generations of Bohemian, Moravian, and German children had become Americans.

Generally the ranchers who gave ground for the San Gabriel's North Fork dam fared better than their counterparts near Laneport. There were several reasons for this. First, only a handful of families actively ranched the San Gabriel's North Fork Valley, and most of them had been preparing themselves for the prospect of a dam for fifty years.[6] By contrast, more than 150 families had to leave their farms for the Laneport dam, and, unlike the ranchers, they had believed the project dead after the small dams project had seemed to

[5] Field, Barron, and Long (eds.), *Water and Community Development*.

[6] Depending on how tightly one defines the word "family," the count ranged from five active ranch clans to ten. See "Tract Register," R.E. Audit Files, Department of the Army, Fort Worth District, Corps of Engineers, North San Gabriel Dam, undated, Vol. I, undated, FTW-2-0010; and "N. San Gabriel Dam & Reservoir above Georgetown," *Williamson County Sun* (Georgetown), June 16, 1966, Section III.

prevail.[7] Second, the ranchers owned much larger spreads—a thousand-plus acres—compared to the typical hundred-acre Blackland Prairie farm. After the government took what it needed for the North Fork reservoir, the ranchers still retained large chunks of real estate to sell, develop, or run cattle on. If they wanted to start fresh on comparable land, they could. But for the Czech farmers there was nothing like the Black Waxy. If it could be found it was impossibly expensive and barren of the Czech culture that had grounded their lives. Third, the ranchers came from a trading culture. A lifetime of buying and selling livestock prepared them for dealing with the government's land men. Fourth, now that the ranchers' scenic, hilly "waste" land was destined to overlook a lake, it was becoming desirable residential real estate, especially in light of the new interstate highway nearby. Finally, while the affected ranchers were no keener to leave their homes than the farmers on the opposite end of the county, the ranchers did not cling to a common cultural heritage, as did the Czechs. Hence the ranchers, for whom movement had been a defining pattern, could easily imagine themselves in new places.

Williamson County business leaders thought the lake created by North Fork Dam would spark a real estate boom like the one created by the lakes along the Colorado River, stretching from Austin to Marble Falls, seventy-five miles away. Believing the dam would pay handsome dividends, bankers and real estate speculators immediately started sinking financial capital into the area, and prices for ranch land quickly rose.

At the Laneport dam site it was a different story. For farm families facing dispersal, the economic forecast was grim. The five or ten acres they might retain after the Army engineers finished assembling their

[7] It is not clear precisely how many families were dislodged from active farming by the Laneport dam. One hundred and fifty families was the number most often cited by forces opposed to the dam, but the estimates of families forced off their land ranged from 125 to 200. The Corps' final audit of the project named 180 separate individuals owning property large enough to support a farm, plus several corporations and cemeteries. Tenant farmers worked some of the 15,303 acres "taken" for the dam, but in this Czech stronghold, tenancy was rare compared to the rest of Texas's Blackland Prairie.

reservoir site would be useless for farming. Nor did the flat, treeless Blackland Prairie suggest vacation homes or horse farms, despite farm lobbyist Ralph Moore's dreams.

What the Czechs loved about the Black Waxy was the fact that they *owned* it. They had planted roots that had grown deep and strong, formed an intricate web of schools, church life, and social customs that harked back to Moravia and Bohemia, and evoked sentimental memories and intimate attachments between pioneer families. After choosing the uncertainties of frontier Texas over the Hapsburg Empire's caste system and climbing the ladder to social acceptance and political dominance in eastern Williamson County, these Czechs—there were almost two thousand of them in a triangle roughly bounded by Granger, Laneport, and Taylor—anguished at the breaking of their bonds with the soil.

———

Anastasia "Stacy" Mikulencak was born in 1903 on a farm founded by her mother's parents, the Zarskys, near the hamlets of Moravia and Machu, between the San Gabriel River and Willis Creek south of Granger. She was the fifth of ten children. When Stacy reached school age, her parents moved to Granger so she could attend "superior" schools. But Stacy loved visiting her Grandfather Zarsky's farm. It must have seemed a slice of heaven, like Willa Cather's fictional Nebraska farm of the young Bohemian Åntonia in *My Åntonia*—an oasis of fruit trees, serenity, and civilized husbandry.[8] At the Zarskys' farm, peach, apricot, and mulberry trees shaded the front yard, and beehives bulged with honey. A cottonwood towered over the front porch, making delicious shimmery sounds whenever a breeze stirred; a big garden produced sumptuous vegetables, including kohlrabi and poppy seed from the Old World.[9] Like Åntonia, Stacy explored the creeks and river bottoms, deciphering Nature's signs and the remnants of departed settlements. Of an abandoned school she wrote, "nothing remains . . . except a plum thicket which sprang up from the countless

[8] Willa Cather, *My Åntonia* (Boston: Houghton Mifflin, 1949), 339–342.
[9] Ben A. Merrick (Mikulencak), "Granger Farm" (1990, unpublished manuscript, lent by Loretta Mikulencak, Granger), chapt. 4, p. 4.

plum pits that were spit out; the thicket indicates the exact location of the once lively school. The bluebonnets were knee high and the native grasses were shoulder high to the school kid. . . . This was a lush, fertile land."[10]

Before they moved to town, the Mikulencak children shivered with fright as wolves howled at night from the wooded river bottoms. They quailed at Grandfather Zarsky's tales of "Old World demons, devil-like riders in the night, dark and sinister castles, bloodthirsty bandits, eerie lights dancing over shadowy marshes"—a vanished European mythological landscape.[11] Like other neighborhood families, the Mikulencak family practiced the Old Country custom of *Na Jozefka*, during which local "Josefs" were serenaded under glowing Japanese lanterns, while friends sang traditional songs to accordian music, drank beer, and ate steaming fresh *kolaches*. On special feast days the men and women dressed in Bohemian or Moravian costume, but otherwise they blended in with their German, Anglo-American, Swedish, and Wend neighbors, enthusiastically adopting American scientific-farming practices.

Like many of her friends, Stacy Mikulencak spoke only Czech until she started first grade, but she graduated from Southwest Texas State University with honors in voice and art.[12] She taught briefly at the Moravia School but quit after she married Henry Allen Labaj and started a family. During the Depression and World War II, the Labajs lived on Henry's parents' farm, scrimping to buy their own place. Stacy wrote a chatty column for the *Granger News* and collected histories of the Czech community that dominated her world. In 1946

[10] Stacy Labaj, "Friendship, Texas" (unpublished manuscript compiled between 1966 and 1972, owned by Tesse Mikulencak Knox, Glen Rose, Tex.). The school Labaj described was located at Enterprise, which existed from 1900 to 1922, when it consolidated with Friendship, according to Clara Scarbrough in *Land of Good Water*.

[11] Merrick, "Granger Farm," chapt. 4, p. 9.

[12] Much of this section is drawn from an interview with Stacy Labaj's daughter, Dorothy "Dot" Daniel, June 12, 2001, in Round Rock. Of Stacy Mikulencak's siblings, one became a physician who served in the 56th Evacuation Hospital in North Africa and Italy and later wrote a history of Baylor University; two were registered nurses, two were teachers, and two were businessmen. Two sisters married Moravian Brethern pastors, deeply involving the Mikulencak family with this branch of Czech Protestantism.

she and Henry purchased one hundred and fifty acres of what she called the "renown" Blackland Prairie between the old Moravia School, Machu Cemetery, and Friendship. Two years later they were stunned to learn that the Army Corps of Engineers planned to flood their land with an elephantine lake that for most of its expanse would be less than ten feet deep.

To Stacy Labaj, the fact of being Czech, the "fabulous" Blackland soil, and the San Gabriel River were inexorably linked. In 1963 she interviewed Frank Machu, whose father established the Moravia School. Machu recalled when a teacher encouraged a Czech-speaking sixth grader to read along with him: "Honey, this says 'Run, Rover, run,'" the teacher intoned. The student parroted his teacher exactly, but substituted "River" for "Rover": "Honey, this says 'Run, River, run.' "[13] For Labaj, that was a natural (and delightful) slip; the river held together everything important in her life—family, religion, language, customs, farm, landscape.

When the Army first announced, in November 1948, that it would dam the San Gabriel River at Laneport, Stacy and Henry Labaj— along with most of Williamson County—were incredulous. It had to be a mistake. They could not believe that the United States would eliminate a thriving farm community working the most productive nonirrigated land in Texas.[14] Initially they thought that if their elected representatives and the Army engineers were made aware of the facts, they would never allow such a calamity.

A month after the Army's announcement, Stacy Labaj waded into the fray. In the 1948 Christmas edition of the *Granger News* she reported, "Mr. Minzenmayer was railroaded into playing the Santa

[13] Frank Machu, interviewed by Stacy Labaj, Jan. 28, 1963, S.P.J.S.T. Rest Home, Granger, in an unpublished fragment of her history of Friendship (Family papers owned by Dorothy "Dot" Daniel, Murfreesboro, Tenn.).

[14] Throughout the 1950s, the anti-Laneport coalition repeatedly claimed that the land to be flooded by the government consistently produced between $1 million and $1.5 million a year. It is difficult to establish how accurate that claim was, but one credible source, Joe Zimmerhanzel, who owned the general store at Circleville, agreed that the land to be covered "is easily producing $1,000,000 a year in cotton, corn, maize, pecans and cattle . . . " See Zimmerhanzel to Pickle, Sept. 5, 1964, J. J. "Jake" Pickle Papers, BBQ (Center for American History, University of Texas at Austin).

Claus for the local small fry—it was so sudden and unexpected he hardly had time to adjust his personality to match the flowing beard and the bug eyes." Then she switched topics, arguing that productive agricultural soil, an increasingly scarce commodity, must not be destroyed. The U.S. government had preached this message and funded soil conservation programs as a result of the 1933–1937 Dust Bowl years, after the great "plow-up" of the Great Plains and the dust storms that followed.[15] Labaj's piece built on the lessons learned from that ecological disaster:

> That is why we feel so strongly against submerging our rich
> lands. . . . If there were no alternative, we would let our land
> go . . . however, there is an alternative . . . two smaller dams in
> the cedar brakes and waste hill country of Georgetown. . . .
> Since Georgetown is eager to have these dams, and since the
> harm it will do there is inconsequential as compared to the
> vast yearly loss it will create here, it is sheer short-sightedness
> to fold one's hand and let criminal damage be inflicted.[16]

Stacy Labaj was a traditional Czech-American wife who always signed her name "Mrs. Henry Labaj." She took pride in her ability as a clever seamstress and good cook. It took courage, she confessed to her family, to speak out at public meetings.[17] But she spoke out, repeatedly, throughout the 1950s and 1960s. Amid the thousands of letters politicians received protesting the Army's plan for the Laneport dam, hers shine with eloquence. Though she was not an "important" constituent, like Wilson Fox (or Wilson's brother, Henry), or Owen Sherrill, or Taylor's Virginia Forwood Lawrence, whose father built the Morning Glory Mattress empire, Labaj's congressional correspondents took pains answering her letters.[18] In 1954

---

[15] Donald Worster, *Dust Bowl: The Southern Plains in the 1930s* (New York: Oxford University Press, 1982), 30, 34–35, 42–43, 182–230.

[16] Mrs. Henry [Stacy] Labaj, "Friendship Facts," *Granger News*, Dec. 23, 1948.

[17] Letter fragment, Stacy Labaj to family, undated but undoubtedly Sept. 15, 1964, held by Dorothy "Dot" Daniel, Murfreesboro, Tenn.

[18] *Welcome to Taylor*, ed. Alma Lee Holman (Taylor: Taylor Daily Press, 1994), 71; Clara Scarbrough, *Land of Good Water*, 238; *Taylor and its Opportunities* (Taylor: Taylor Chamber of Commerce, 1940), 18–19, "Williamson County Scrapbook" (Center for

Left: Stacy and Henry Labaj on their land in April 1951. *Courtesy Dorothy "Dot" Labaj Daniel, Murfreesboro, Tennessee.*

Below: The Labaj home, summer 1955. *Courtesy Dorothy "Dot" Labaj Daniel, Murfreesboro, Tennessee.*

Bottom: Henry Labaj and his children harvest cotton with the help of hired "hands." *Courtesy Dorothy "Dot" Labaj Daniel, Murfreesboro, Tennessee.*

Labaj addressed Poage, champion of farmers, in whom she was deeply disappointed:

> I regret but feel urged to say you seem not to realize that the passage of the Laneport Dam project is like snuffing out our lives. Perhaps you have come easier by your livelihood, but to us our farm is the product of our labor of our entire married life *plus* the share received from both sides of our parents as inheritance. . . . [O]ur children grew up treasuring a nickle, we made clothes over from out-moded styles and fresh fruit was a delicacy; to hold on to insurance to insure an education for our children was a real struggle. . . . Acquiring our farm seemed almost too good to be a reality. . . . A wisely located dam protects rich lowlands, this dam would *destroy* the rich productive valley . . . it is against all laws of good judgment.[19]

As the Laneport controversy flared, family tragedy struck. In 1963 Stacy's daughter Joy died of a brain aneurysm.[20] Labaj pressed on, penning dozens of letters to politicians and organizing petition drives. In 1964 she recommended that Congressman Pickle read historian Walter Prescott Webb on the importance of preserving good soil. She stated publicly what Wilson Fox and his allies hated to see in print: Taylor "will make another slide toward a 'has been' if they succeed in their clamor to destroy these 16,000 acres."[21]

American History, University of Texas at Austin). In 1946 the Taylor Bedding Manufacturing Company, maker of the Morning Glory Mattress, claimed it was "the world's largest bedding plant." The U.S. military was one of its biggest clients. After a postwar disagreement over cotton-waste price controls, the conflict was resolved in the company's favor with the assistance of Congressman Lyndon Johnson. Johnson aide Walter Jenkins did the "grunt" work and received a Morning Glory mattress as a "token of our appreciation," a Taylor Bedding official wrote. L. D. Hammack to Jenkins, Oct. 29, 1946, House Papers 1937–49, Container 282, OPA Taylor Bedding (LBJ Library, Austin).

[19] Letter, Mr. and Mrs. Henry Labaj to Poage, Nov. 3, 1954, W. R. Poage Papers, Box 692, File 3 (Baylor Collections of Political Materials, Waco).

[20] Interview, Dorothy "Dot" Daniel, June 12, 2001, Round Rock.

[21] Letter, Mr. and Mrs. Henry Labaj to J. J. "Jake" Pickle, received Mar. 23, 1964, Pickle Papers (Center for American History, University of Texas at Austin).

After an unsettling public meeting in 1964, she fumed privately to her daughter,

> The Army Engineer talked long and holy and as tho' the thing is irrevocably settled—tho' several thin places shone thru; two of the greeds from Brazos River Authority gave their little smart alec speech. . . . I pointed out we in Friendship *are an industry*—the thing all small towns are eagerly trying to acquire—and its one and a half million dollar gross annual income helps Taylor-Granger business in way of tractors—cars—food—clothing—schools—churches etc. We need no artificial aid such as irrigation to produce fabulous yields. . . . Perhaps by their coolness we can judge we won an edge on them . . . tho' victory is far away and elusive.[22]

In another note to a family member she despaired, "Even Granger is determined to cut its own throat for a possible boat ride over our farm; some of our friends are quite cool to us because we resisted putting our heads on the block so they could have a little fun. Ignorant people. I told Raymond Holubec . . . [the dam] will be death blow to Granger."[23] By that time, in early 1965, Congressman Pickle was stitching together his "compromise solution," though neither Stacy Labaj nor anyone else in the county knew it.

By 1970 the local fight against the Laneport dam was over. The Labajs, like most of their neighbors, simply wanted what they considered a fair price for their precious farm. The Army wanted 119 of their 150 acres, including their rose-covered farm house, and offered the Labajs 440 dollars an acre—$52,360 for the whole works.[24] The Labajs declined. The Army condemned the Labajs's property but allowed Stacy and Henry to rent their own house. For a year, the

[22] Letter fragment, Stacy Labaj to unknown, not dated but probably written Sept. 15, 1964, held by Dorothy "Dot" Daniel, Murfreesboro, Tenn.

[23] Letter fragment, Stacy Labaj to unknown family member, not dated, but most likely Sept. 1964, ibid. Holubec was a director of the Brazos River Authority and owned land below the proposed Laneport dam.

[24] Letter, Col. Floyd H. Henk to Poage, Mar. 14, 1972, W. R. Poage Papers, Box 692, File 7 (Baylor Collections of Political Materials, Waco).

Labajs paid. Then they stopped paying rent. The Army threatened to evict them. The Labajs wanted six hundred dollars an acre—ten thousand dollars more than the government's top price.[25]

Again Stacy Labaj appealed to Congressman Poage, who had recently become her representative. "We are truly weary of this demeaning haggling," she wrote,

> and cannot comprehend how one arm of the law states a party must be re-imbursed at market value for any property thus taken while another arm of the government rides rough shod over these same parties. . . . This is truly unreal especially in view of the fact this dam will serve no good purpose, there is absolutely no need for it. . . . Mr. Poage, do be so kind and take time to correct this wrong to make a call in our behalf. My husband has developed an ulcerated stomach from all this, the neighbor to the east of us suffered a heart attack and the one to the west of us, a mild stroke which left after effects.[26]

Finally, the Labajs bowed to the inevitable. They accepted the Army's "final" offer of 521 dollars an acre, ten thousand dollars less than they considered fair, and moved to Granger.[27] Stacy Labaj compiled a history of Machu Cemetery, which the reservoir would soon cover, for the Corps of Engineers. In 1976 she was helping decorate floats for "Granger Days" when she felt ill and went home, suspecting the flu. Soon after, she died of pancreatic cancer.[28]

Some of Stacy's Labaj's closest friends blamed Laneport Dam. "We always thought she died because of it. She fought it for twenty years, and when she had to leave her home, she grieved and grieved. It tormented her," said Loretta Mikulencak, Granger's school tax assessor-collector. "The Corps offered certain people high dollar for their land, but everyone was at their mercy. The chief damage to Granger was getting those [Czech] families out of here; they were stable farm

[25] Ibid.

[26] Letter, Mr. and Mrs. Henry Labaj to Poage, Mar. 8, 1972, W. R. Poage Papers, Box 692, File 7 (Baylor Collections of Political Materials, Waco).

[27] "Granger Lake, Texas," Audit #Ft W-2-0008, Index, Tract Register, Aug. 14, 1978, Vol. 1 (Department of the Army, Fort Worth District, Corps of Engineers, Fort Worth).

[28] Interview, Dorothy "Dot" Daniel, June 12, 2001, Round Rock.

families who had inherited their land and they were not going to leave. They never recovered . . . they just died, one by one. And the worst of it was that it made us bitter; it made us what we weren't. It made us different people."[29]

The Labajs's 521 dollars per acre did not come close to the top price paid for Laneport Dam land. Among the more fortunate were Lee and Virginia Forwood Lawrence of the Taylor Bedding Company fortune, whose land sold for 750 dollars an acre. The Lawrences fought the Corps of Engineeers in court and walked away with a $268,553 settlement.[30] Ironically, one of the lowest prices paid per acre for Laneport land went to Texva Realty, Ralph Moore's company, which speculated on land in the area well before the Corps announced its plans to build the dam. In one of the earliest sales, Texva sold two tracts to the U.S. government for 392 dollars an acre, the lowest recorded price paid for a substantial amount of land at the Laneport site.[31] After Moore died in 1971 and his widow Lois liquidated Texva Realty, Gabriel Farms, which took over Texva's assets and

---

[29] Interview, Loretta Mikulencak, July 6, 1998, Granger. Mikulencak, an invaluable source for this book, passed away on Nov. 21, 2003, after a mercifully brief bout with cancer.

[30] "Granger Lake, Texas," Audit #FTW-2-0008, Index, Tract Register, Aug. 14, 1978, Vol. 1 (Department of the Army, Fort Worth District, Corps of Engineers, Fort Worth). The Corps "took" 358 acres of the Lawrences' land, a portion of which was planted in a pecan orchard, which generally was more highly valued than row cropland. The case concluded on July 9, 1973.

[31] It appears that Texva sold cheap because it had to—it was in dire financial straits. "Granger Lake, Texas," Audit #FTW-2-0008, Acquisition, Vol. 4 (Department of the Army, Fort Worth District, Corps of Engineers, Fort Worth). Moore's 1946–1948 land acquisition program resulted in Friendship Farms, Inc., which in 1957 could not pay its franchise taxes. The company evolved into Texva Realty, Inc., with Thomas Sully partnering with Lois and Ralph Moore. After Moore died, Texva again fell into arrears and was liquidated by Lois Moore and her board of directors on the advice of Frank Scofield, the powerful former director of the Internal Revenue Service in Austin and old friend of Lyndon Johnson's. On August 14, 1972, Texva conveyed title to all of its land to a new partnership named Gabriel Farms, formed by Granger banker and former Texva director G. Truett Beard and his wife Pauline Beard (trustee for George Beard); Glen D. and Martha Katherine Chilek of Austin; and Paul L., Frank E., and Alice K. Scofield of Austin. Scofield was well known in the area as a former director of the Inter-

debts, sold 166 acres to the Corps for 837 dollars an acre—the highest price on record paid at the Laneport site.[32]

The Fox brothers—Wilson, Henry, and Bryan—presented a thorny problem to Corps appraisers. Their properties were not comparable; nor were the brothers. If Wilson Fox did not create Laneport Dam, as some believed, he was the driving force behind it—not only its most influential local supporter but also the architect of several key maneuvers that kept the dam plan alive when it had seemed hopelessly moribund.[33] The Army took 313 acres of Fox's Riverside Ranch, including Hilda's Bottom, leaving him his handsome home and more than a hundred acres. Henry Fox had been Laneport Dam's most serious adversary, along with the Owen Sherrill–led Georgetown dam coalition. Henry Fox had led ongoing grassroots campaigns against the Laneport dam and for the small dams on the San Gabriel, successfully freezing Laneport's prospects for sixteen years. The Army condemned 182 acres of Henry's land, including a pecan orchard whose trees he had personally planted, but his house and another 190 acres lay safely out of the reservoir's reach. Bryan Fox was a likeable educator and fiery public speaker who sent his son Paul to Harvard and loved throwing parties at his beautiful home, which sat high on a bluff looking out over the San Gabriel River. He would lose his entire place—house, swimming pool, pecan orchard, and ninety-four acres. Like Henry, Bryan Fox had stoutly opposed Laneport.

How could the Army reward Wilson Fox for his longtime support without appearing churlish toward his recalcitrant brothers? Wilson's sudden death in February 1974 may have resolved the problem.

---

nal Revenue Service in Austin. Ralph W. Moore's widow, Lois, owned no portion of Gabriel Farms; she spent her last years in a Bartlett nursing home.

[32] Audit #FTW-2-0008, Acquisition, Vol 4 (Department of the Army, Fort Worth District, Corps of Engineers, Fort Worth). Gabriel Farms settled its business with the Corps in two segments, on May 1, 1974, and Oct. 1, 1975. The first group of sales averaged $473 an acre; the second $830 an acre, proving the folk wisdom that says it is better to hold out when facing government condemnation.

[33] Wilson Fox "worked" the Brazos River Authority so artfully that its managers often conferred with him alone in Williamson County on important policy matters, as if he presided over the BRA's board of directors as his brother, Howard Fox, had at one time.

Before Wilson's death the Army had agreed to pay Bryan Fox seven hundred dollars an acre for his property.[34] It had also concluded two civil actions against Henry Fox, resulting in an award of 636 dollars per acre for 69.56 acres, with the price of another 112.7 acres still to be determined in the courtroom.[35] After Wilson Fox's death, Hilda Fox settled with the Corps for 679 dollars an acre for 313.61 acres. Shortly thereafter, Henry Fox's last parcel was condemned for 761 dollars an acre, bringing his overall average per acre sale price exactly in line with Bryan's—700 dollars an acre.

In the end Wilson Fox's estate received less per acre for land crucial to the Laneport reservoir's development than the brothers who tried to stop the project. Had he lived, Wilson almost certainly would have negotiated a higher settlement. His legal and political value surely would have ceded him a higher price for Riverside Ranch. But it didn't turn out that way. Wilson Fox went to his grave bitterly resenting his brothers' resistance to Laneport Dam, and in the end their land fetched more than his.[36]

As a young man, Roy Gunn ranched on Hamilton Creek, a fingerling feeding the Colorado River in Burnet County. In those days, before the Highland Lakes replaced the Colorado through much of the Texas Hill Country, ranching was all one could do in the eroded

[34] Civil Action A-73-CA-51, "Granger Lake, Texas," Audit #FTW-2-0008, Acquisition, Vol 1 (Department of the Army, Fort Worth District, Corps of Engineers, Fort Worth).

[35] Civil Action A-73-CA-194, ibid.

[36] After Laneport Dam was a "done deal," Wilson Fox started setting up arguments for a high settlement in several letters to Jake Pickle, including one on Sept. 9, 1966: "There certainly is a peculiar situation existing when a man is forced to give up his property for the good of all concerned. I think we should not hesitate to make this sacrifice but at the same time should not be penalized for it. In other words, the public as a whole should bear the burden as money compensation is about the only compensation." On November 27, 1967, he noted "I know no one wants to hold up anyone, but at the same time land is selling at a high price, and where a person gives up his land for the good of the country, then it is only right that he be paid a fair price." His thoughts anticipated the property rights movement of the 1990s that was triggered by environmental regulations that lowered land values. See Pickle Papers, Box 95-112/107 (Center for American History, University of Texas at Austin).

hills west of the Balcones Escarpment. A successful rancher was one who controlled access to water, and Roy Gunn had accomplished that. He and his wife Maggie Lee ("Bob" to friends and family because she was the first among her set to "bob" her thick curly hair) ranched with passion and steered their children toward ranching.[37]

Then, starting with Buchanan Dam in the 1930s, a string of six dams was built on the Colorado, ramrodded by the brilliant Texas attorney-lobbyist Alvin J. Wirtz and his political protégé, Lyndon Baines Johnson.[38] In the 1940s the Lower Colorado River Authority condemned most of Roy Gunn's land for Wirtz Dam and the "greater good." Today, part of Gunn's first ranch lies under Lake LBJ; the rest became part of Horseshoe Bay, a luxury development of condos and second homes.[39]

Gunn did not relish leaving his friends in nearby Kingsland, but he found a new ranch in isolated western Williamson County—two thousand acres along the San Gabriel River's North Fork. The Gunn family moved there in 1946. Tate, the eldest son, studied agriculture at Texas A&M. Pat, Gunn's only daughter, loved books, horses, and the San Gabriel River. Sam, the charismatic younger son, spent his teenage years drinking, romancing, and riding rodeo broncos. All the Gunn children envisioned ranching as part of their future lives.[40]

On March 5, 1968, the Corps of Engineers announced its land acquisition program for the North Fork and Laneport dams. The decision seemed final.[41] That year, Tate Gunn visited Australia and liked what he saw. On his recommendation, Bob flew to Queensland to scout the territory after her husband got "cold feet."[42] On March 4, 1971, Roy Gunn sold the Army 593 acres of choice river bottom-

[37] Interview, Pat Gunn Spencer, Sept. 2, 2001, Georgetown.

[38] Bill McCann, *The State of the River* (Austin: Lower Colorado River Authority, 1993), 13, 31. For background on Wirtz, see Caro, *The Path to Power.*

[39] Interview, Pat Gunn Spencer, Sept. 2, 2001, Georgetown.

[40] Linda Scarbrough, "A Tale of the Gunns," *Sunday Sun* (Georgetown), Mar. 23, 1986.

[41] "Opposition to Laneport Dam Aired at Georgetown Hearing," *Temple Telegram,* Mar. 6, 1968, undated file, Master File (Brazos River Authority, Waco). Also in the file were articles from the *Austin Statesman* and the *Taylor Press.*

[42] Telephone interview, Tate Gunn, Sept. 4, 2001, Rockport.

The Gunn family in 1947—Tate, Roy, Maggie ("Bob"), Sam, and Patsy—astride their horses in front of their ranch home. *Edward A. Lane Photo, Courtesy Pat Gunn Spencer, Georgetown.*

In 1968 Roy and "Bob" Gunn face the government "taking" of all their land with access to the San Gabriel River. *Jim Smith Photo, Courtesy Pat Gunn Spencer, Georgetown.*

The Gunns, like thousands of yearly visitors to the "Booty's" Crossings, loved playing in the North Fork's shallow, clear rapids. In 1967 Roy Gunn, left, lassoes an inner tube ridden by grandson Jack Spencer. Also shown, left to right, are Tommy Gunn, Roy's sister Oleta, "Bob" Gunn, an unidentified friend, and grandson Mike Spencer. This spot is now buried underneath the City of Georgetown's water intake system. *Jim Smith Photo, Courtesy Pat Gunn Spencer, Georgetown.*

Sam Gunn at Kirrama, Australia, February 1986. *David Sprague Photo, Courtesy Williamson County Sun, Georgetown.*

land for 180,000 dollars—303 dollars per acre.[43] He still had fourteen hundred acres left, most of which he later sold to an Atlanta development firm and a Dallas businessman.[44] Gunn reinvested his government money in a "station"—an Australian ranch—fifty times larger than his old Williamson County place. The station, in a North Queensland rain forest, was called Kirrama after the nearby Kirrama mountain range.

By 1973 Roy and Bob Gunn had assembled their entire clan in Australia: Tate and his wife, Barbara; Sam and his wife, Christine; Pat and her husband, Jim Spencer; and all the children. The Gunn headquarters was an 1889 Kirrama homestead. In 1986 I visited the ranch. Sitting on the veranda of his family's home, Sam Gunn, Roy and Bob's youngest son, gestured toward the lush green paddocks falling off in successive waves below the house, and defined his world. "We've got 100,000 acres, or 166 square miles, and 5,000 to 6,000 cattle," he said.[45]

It looked like Shangri-La. At twenty-four-hundred feet, the air crackled with Alpine crispness; streams rippled through emerald paddocks and joined the Herbert River to feed Blencoe Falls, which plunges three hundred feet down a granite gorge. Blencoe Falls, part of Herbert River Falls National Park, bordered on the Gunn property.

Unfortunately Kirrama was not entirely a paradise. The Gunns had to work brutally hard to keep up with their cattle, hiring "Jackaroos" and "Jillaroos" during mustering, or round-up, season. There was no electricity. The closest schools were three and a half hours away; Pat and Barbara taught the children at home. But the family buckled when Bob, the family's steadying heart, died in 1974, barely six weeks after arriving at Kirrama. Roy and Sam were off on a cattle-buying trip. She was buried in a little cemetery at Cardwell, overlooking the Coral Sea.

The Gunns began to fall apart. Tempers frayed. The station could

[43] "North San Gabriel Dam-Lake Georgetown," Audit #FT W-2-0010, Real Property Title/Historical Files, May 4, 1990, Department of the Army, Fort Worth District, Corps of Engineers (National Archives, Fort Worth).

[44] Interview, Pat Gunn Spencer, Sept. 2, 2001, Georgetown.

[45] Linda Scarbrough, "A Tale of the Gunns," *Sunday Sun* (Georgetown), Mar. 23, 1986.

not support four families; Kirrama was draining the Gunns' bank account. Pat moved back to Texas. Tate and Barbara sold their share of Kirrama and purchased a station on the Coral Sea. Sam stayed on at Kirrama, with his father, hoping his family could somehow persist. Tate thought Kirrama impractical, but he acknowledged that it suited his dad. "He can ride on a motorcycle or a horse without using a gate until his butt gets sore," Tate said in 1986.

> He can carry a scope pistol—a .221 Remington Fireball with single shot bolt action—and shoot 'roos to feed to the stock dogs. He shot 60 once; they were a plague. He can shoot wild pigs, or dingos once in a while. He can . . . ride until he wants to come home, and he's never left Kirrama. How can a man like that be satisfied with 2,000 acres?[46]

But a year later the Gunns sold Kirrama. They made money, but it was a heartbreak. Sam and Christine bought a "small" fifteen-thousand-acre station, Minerva Hills, in Central Queensland. Tate and Barbara abandoned their "croc-infested" river by the Coral Sea and bought the Suntan Motel in Rockport, Texas, where Tate guided fishing trips on the Gulf of Mexico. Pat operated a canoe retail and rental store on the Blanco River near Wimberly. Still searching for another ranch, Roy Gunn "temporarily" moved in with Pat. "He wouldn't buy another ranch because my mother wasn't here," Pat said. "He could never get over the price of land, and he couldn't replace his carrying capacity anywhere in Central Texas." At ninety-one Roy Gunn died in a car accident while driving to an H.E.B. grocery store in San Marcos.[47]

Kirrama Station vanished. The original house was diassembled and moved to another property. The place is now a popular camping and picnic spot on the itinerary of numerous "survivor," birding and eco-tourism treks through North Queensland. During Mark Burnett's

---

[46] Ibid.

[47] Telephone interview, Sam Gunn, Sept. 5, 2001, Minerva Hills, Australia; telephone interview, Tate Gunn, Sept. 4, 2001, Rockport; interview, Pat Gunn Spencer, Sept. 2, 2001, Georgetown. "Carrying capacity" is a term used by ranchers to define how many "units" of livestock a piece of land can feed and water without danger of overgrazing and erosion.

1997 Eco Challenge Australia, contestants rappelled up Blencoe Falls. Sam and Christine Gunn live at Minerva Hills and say they will never leave. "I love this place; it's very quiet," Sam said. Since he emigrated to Australia thirty years ago, Sam Gunn has returned to the States just once, for his father's funeral. Pat settled in a Georgetown suburb and produces outdoor adventure videos. Tate and Barbara sold the Suntan Motel. "I can't decide what to do," said Tate. "It's either Homer, Alaska, or Belize. We've also thought about Port Mansfield—it's pretty much unchanged."[48]

A Georgetown banker and old family friend, Jay Sloan, believes Roy Gunn's forced exodus from Williamson County was a boon for his family. "Roy Gunn had two fortunate things that happened in his life: his land on what became Lake LBJ was condemned and he sold it for more than it was worth. Then, fifteen or twenty years later, the same thing happened again at Lake Georgetown."[49] Of course, Sloan's analysis focused on the financial picture.

Thirty years after the U.S. government changed the trajectory of their lives, Roy Gunn's children view the consequences of losing their Williamson County ranch differently. "I think it's kind of simple," Tate said. "I went to A&M and majored in animal husbandry but we would have starved to death if we had stayed [in Williamson County]. That was the life Dad and I wanted to live, but there was no money in it. And so we had a chance to ranch big over in Australia—*really* big— big enough for all the family." The experience cost the family, "but I wouldn't trade any of it for our trip. There was just one bad thing about going over there. When we came back, I haven't been satisfied, even though this is the best country in the world."[50]

Sam, now more Australian that American, is "as happy as a tick on a dog," said his sister. "I kind of lucked out," he agreed. "Minerva Hills is lovely." His and Christine's children are nearby. And yet, he said, "It split the family. Everybody had to survive out of one checkbook. My Dad, working with him . . . it was difficult. Sometimes I loved him and

[48] Port Mansfield is a fishing village near the tip of South Texas, facing the Laguna Madre from the mainland.

[49] Interview, Jay Sloan, Nov. 8, 1999, Georgetown.

[50] Telephone interview, Tate Gunn, Sept. 4, 2001, Rockport.

the next minute I hated him. He was a fine fellow but he wanted everybody under his thumb."[51]

Pat still saddens at the loss of the Gunn Ranch on the San Gabriel River. "We got a check from the Corps for $176,000," she remembers. "But Dad ended up in a tight cash bind and he *had* to sell the rest [of our land] to pay for Kirrama. We didn't choose to sell." With her finger, she traces the photographic images of a family frozen in time, cavorting in a river, framed by limestone boulders that rear up out of the water like totems. The photo shows First Booty's Crossing on the San Gabriel, part of the Gunn Ranch. First Booty's was the most popular of four low-water bridges that crossed the San Gabriel's North Fork, where thousands came to fish, swim, picnic, and enjoy outdoor parties. The family in the picture was hers. Pat's boys, Mike and Jack Spencer, splash water at Tommy Gunn, Sam's eldest son. Roy and Bob Gunn's inner tubes are stuffed with children. Today First Booty's Crossing lies invisible underneath Georgetown's municipal water treatment intake structure. "There is nothing I have that I wouldn't trade for that land," Pat says. "Money in the bank means nothing. I would rather have had my sons out there on that river for another twenty years. I'm the one in the family who needs roots, and I don't have roots."[52]

Indeed, the Gunn Ranch, once an integrated "place" that had meaning and, along with other ranches of the time, comprised a cultural geography of the American West, was piecemealed into unrecognizable bits—a change so drastic that it calls to mind the change that occurred when Anglo-American settlers displaced Comanches and Tonkawans. Today part of the Gunn place lies underneath Lake Georgetown. Part of it has become public parkland featuring a man-made beach, barbecue pits, boat ramps, and camper sites.[53] The rest is Fountainwood Estates, a prestigious Georgetown subdivision, where a small swatch of Roy Gunn's old ranch, with a custom-built house, fetches up to a million dollars.[54]

---

[51] Telephone interview, Sam Gunn, Sept. 5, 2001, Minerva Hills, Australia.

[52] Interview, Pat Gunn Spencer, Sept. 2, 2001, Georgetown.

[53] Jim Hogg and Russell Parks, Lake Georgetown's two most popular parks, attracted what the Corps tallied as 3,963,249 million visitor *hours* in 1993.

[54] Interview, Glenda DuBose, Sept. 19, 2001, Georgetown. Fountainwood Estates

Roy Gunn sold too fast. True to its general operating procedure, the Corps dealt with Gunn early, since he controlled a critical piece of land needed for North Fork Dam. And though the Army's price— 303 dollars an acre—added up to a fat check for Gunn, other ranchers parlayed their cedar-choked hills into considerably larger fortunes.[55] A year after Gunn sold, for example, Judge D. B. Wood, who forced the Army to take him to court, received 836 dollars an acre for 301 acres.[56] One more year passed, and as a result of another civil action, Carl E. Allen got 1,024 dollars an acre for his 359-acre tract. And three brothers—James, Jerry, and Rex Hawes—who owned land at the tail end of the reservoir, waited out the Corps with spectacular results. In 1974 the Hawes boys, as they are known locally, collected $1,124, $1,142 and $1,154 per acre, respectively, for the land the Corps took—nearly quadruple Roy Gunn's selling price.[57]

Without conducting an exhaustive geographical and financial analysis, it is impossible to conclude whether or not the Army equitably treated owners of the approximately twenty-two thousand acres of land the engineers needed for the Laneport and North Fork dams. From a close examination of the Corps' records, however, several signposts stand out. First, those who contested the taking of their land

---

itself is subdivided into four separate sections of homes, ranging from a fairly typical upper-middle-class neighborhood where homes start at $190,000 to a "gated" community where five-acre lots are the norm and homes are marketed at just under $1 million. Across the road from Fountainwood Estates is Sun City Texas, an age-restricted planned community projected to grow to a population of 19,000.

[55] "North San Gabriel Dam," R.E. Audit Files, Index, Tract Register, Vol. 1, FTW-2-00010 (Department of the Army, Fort Worth District, Corps of Engineers, Fort Worth). In this tract register, the breakdown of each tract number sold to the Corps shows the Army settling with Gunn on March 4, 1971, for a total of $180,680. This closely agrees with Pat Gunn Spencer's recollection that her father received a check from the Corps for $176,000. However, on the tract register's summary page, the "total value" of the Gunn property was listed at $108,680, an apparent typographical error on the part of the Corps.

[56] Letter, D. B. Wood to Poage, Aug. 24, 1971, W. R. Poage Papers (Baylor Collections of Political Materials, Waco). Judge Wood wrote Poage to congratulate him for taking Williamson County into his congressional district, and mentioned that he and the Corps were "squabbling" over the price for his land.

[57] "North San Gabriel Dam," R.E. Audit Files, Index, Tract Register, Vol. 1, FTW-2-00010 (Department of the Army, Fort Worth District, Corps of Engineers, Fort Worth).

or held out longest against settling with the Army received substantially higher prices than those who did not, despite advice to the contrary by anxious constituents' elected representatives. Second, politically connected landowners, or those with strong legal representation, garnered higher land prices than their neighbors. Whether this relatively powerful group owned superior land, or whether political clout made the difference, or whether some owners traded more sharply than others is impossible to say. But the available evidence suggests that the latter two factors—political-legal connections and horse-trading ability—boosted the Army's "best" offer more than the intrinsic value of condemned land.

If this is indeed the case, the Army's claim that its condemnation proceedings were based on the fairest possible formula—"market value"—was simply not true. If government "takings" of hundreds of tracts resulted in some property owners receiving two, three, or four times as much as others for similar land, the democratic system, as Stacy Hajda so eloquently suggested, was terribly out of kilter.

Eventually the U.S. government spent 8.1 million dollars purchasing 15,303 acres for the Laneport Dam site, an average 532 dollars an acre. For the North Fork Dam project, it paid 2.5 million dollars for 6,300 acres, an average of 390 dollars an acre. Obviously, the "fabulous" Black Waxy soil counted for more than the sorry "waste" land of the Balcones Escarpment, but not as much as the Blackland Prairie farmers had expected.[58] Riverfront property usually fetched a higher price than its off-river counterpart.[59] Beyond those two givens, though, the only conclusion one can safely reach is that some landowners deftly stretched the standard "takings" formula in their favor. Precisely how they managed this remains murky. One thing, however, is clear: in Williamson County, Texas, the Army Corps of Engineers did not condemn property on behalf of the U.S. government with an even hand—even if the land was taken for society's "greater good," as Wilson Fox always insisted.

[58] In the year 2001, Black Waxy landowners still referred, with contempt, to land west of the Balcones Escarpment—land that has boomed in value through the last three decades—as "that sorry old caliche stuff."

[59] This generalization works better when applied to the Laneport dam "takings" than at North Fork Dam, where Roy Gunn's land bordered the river but Judge D. B. Wood's land did not; the judge received $836 an acre compared to Gunn's $303.

T E N

SELLING THE WATER

Under the arid-region doctrine . . . not to use water . . . is to forfeit it.

WALTER PRESCOTT WEBB, *The Great Plains*

I n September 1967 the Brazos River Authority's directors bet the agency's economic life on a scenario sketched by their regional planning team.[1] Encouraged by cheap federal interest rates that encouraged new dams for future water supplies throughout the American Southwest, the directors guaranteed that the BRA would pay all construction costs related to water conservation, or storage—44 percent of a total projected cost then estimated at 46.5 million dollars—for the San Gabriel River dams.[2] There was a hitch. If the BRA

---

[1] These planning experts completely missed the possibility that Round Rock might grow, despite the fact that it lay close to Austin, on the path of Interstate 35, and in close proximity to potential water sources. It was not until Round Rock leaders invited themselves to meetings in 1964 and later requested information about water supplies that the BRA and state water-planners noticed the city's existence. Round Rock, of course, became one of the most dramatic growth stories in Texas during the 1970s, when the city's population surged from 2,800 to 15,000 in a decade. In 2004 it counted 78,000 residents. As late as 1967, BRA manager Walter Wells predicted that "most of the water" from the Laneport and North Fork dams would serve the "growing industrial area of Brazoria, Galveston and Fort Bend Counties"—an assertion that proved dead wrong. See letter, Walter J. Wells to Robert G. Fleming, Dec. 20, 1967, Box 257 (Brazos River Authority, Waco). Fleming was operations director for the Texas Water Quality Board.

[2] The U.S. government fronted the money for the water conservation, or storage,

failed to sell the water it "owned" in the new reservoirs, it would take a horrendous financial hit. As a BRA director put it back in 1964, well before the San Gabriel dams upped the ante, "We have run our necks into a $24 millions noose."[3] Since the BRA brass did not believe that Williamson County's water needs could make a dent in North Fork and Laneport's supplies, the directors' gamble was a gutsy—some might say foolhardy—move.

In such a manner, on dams much like those planned for the San Gabriel, the Brazos River Authority had accrued a colossal debt in exchange for congressional financing of the flood-control portions of dams throughout the Brazos watershed, bestowing massive plumbing projects on the San Gabriel, Lampasas, Leon, Bosque, Aquilla, and Yegua tributaries—nine huge public works projects in all. These nine dams collectively cost 333 million dollars; the BRA's water conservation share varied from project to project, but ultimately reached one hundred million dollars. It was the most extensive joint local and federal dam-building effort ever mounted in Texas, save for the Trinity River watershed, where in cooperation with the Trinity River Authority the Corps built a system of dams and channels designed to protect Dallas and Fort Worth from floods, leaving the Metroplex pleasantly awash in surface water supplies and electrical power.[4] The

---

portion of flood-control dams built by the Corps of Engineers at low interest rates— between 2.5 and 3 percent in the 1950s and 1960s. Local governments, such as the BRA, borrowing under these terms had ten years from the time a dam was declared "finished" for water conservation before it had to start making payments, which were not supposed to exceed 30 percent of the total cost of the project. The cost of Laneport and North Fork Dams eventually reached $100 million. See memorandum Aug. 13, 1964, Chief Engineer Burke G. Bryan, Box 59 (Brazos River Authority, Waco); Water Resources Development in Texas 1995 (Dallas: U.S. Army Corps of Engineers, Southwestern Division, 1995), 15–16; "Engineers to Acquire Land for Williamson County Dams," Austin American-Statesman, Mar. 6, 1968.

[3] Harry Provence to J. J. "Jake" Pickle, Feb. 29, 1964, Pickle Papers (Center for American History, University of Texas at Austin). Provence, publisher of the *Waco Tribune*, was an old friend of Pickle's.

[4] "Water Resource Development in Texas 1995," 11–19, 77–92. The Colorado River basin also boasts an extensive system of dams, but most of these were built with Federal funds funneled through agencies other than the Corps of Engineers. The Trinity River Authority–Corps projects were extensive, but two of them, built in the 1980s to satisfy the water and power needs of Dallas, drove costs to staggering heights—$618 million for Joe Pool Dam, completed in 1986, and Ray Roberts Dam, completed in 1991.

Corps' final Brazos dams—North Fork, Laneport, and Aquilla Lake north of Waco—represented the last flexing of political muscle that drove the United States' great dam-building era of the fifties and sixties, characterized by critics as "log rolling" and "pork barrel." This in turn helped spark the rise of an anti-dam environmental movement, at almost the same time that the Laneport and North Fork dams were completed.[5]

To a great degree, what followed in Williamson County regarding the San Gabriel River dams, both politically and bureaucratically, was fired by that 1967 bet on the "come"—the BRA directors' optimism that by the time they had to start making payments to "Uncle," their stored waters would be turning a tidy profit. Thus was the BRA committed. It could not turn back. It *had* to make the North Fork and Laneport dams work, whether that meant convincing Corps engineers to alter the architecture of their dams or "playing God" with the future of cities by withholding or selling them water. To succeed, the BRA had to adapt to a rapidly evolving demographic landscape in Central Texas, making sure that San Gabriel reservoir water was priced low enough to sell, and inducing or discouraging potential customers to buy it based on what the BRA thought their ability to pay would be.

This was not simple math. In 1964 BRA engineers had realized with horror that the Corps' proposed San Gabriel dams would result in outrageously expensive water—water costing about twenty dollars an acre-foot from the North Fork reservoir, for example. At that time, water from other Brazos basin reservoirs was selling for one dollar to six dollars an acre-foot. Throughout the process of finding a political solution that would allow the dams to go forward, no one had noticed.[6]

From 1967 on, most Williamson County people saw the continuing dam saga as a purely local cluster of issues: individual landowners' struggles to maximize compensation for their property lost to the government's power of eminent domain, presumed loss of a significant agricultural economy and what that might mean to Taylor and

[5] Reisner, *Cadillac Desert*, 307–331.

[6] Memorandum by T. B. Hunter, assistant general manager, BRA, Aug. 12, 1964, Box 59 (Brazos River Authority, Waco).

Granger versus the promised economic miracle that the lakes were supposed to bring, the surprising aggressiveness of tiny Round Rock's battle to acquire reservoir water, and a last-ditch environmental battle against the North Fork dam.

Crucial to the eventual outcome was an invisible struggle between the two cooperating water bureaucracies, the BRA in Waco and the Corps in Fort Worth, which played out in a series of memoranda, letters, and computer printouts. Once the BRA's Wells realized that the surface water stored by the San Gabriel dams would be too expensive to sell, he asked the Army engineers to reduce the Laneport reservoir's available "yield," roughly analogous to marketable water. Logically one might assume that the bigger the reservoir, the cheaper its water would be. The truth is, when a reservoir exceeds its optimum size (and this changes when other dams enter the picture), its water becomes *more* expensive. North Fork combined with Laneport had produced this result.[7] Laneport's size had to be reduced or its water could not be sold. But cutting its size (and hence its water cost) decreased its effectiveness as a flood-control dam. Thus, the water-selling agency (the BRA) and the flood-control agency (the Corps) found themselves at odds over their central missions. In an effort to work out the problem, over a six-month period BRA and Corps engineers quietly exchanged combinations of numbers in the form of "yield studies."

When they finally agreed on a number, Laneport Dam and its reservoir shrunk. Its water could be sold more cheaply, and its flood control capacity was reduced. Both agencies seemed satisfied. Outside the two agencies, no one seems to have been aware of this delicate negotiation.[8] The dam projects went forward. Had the two water

[7] South Fork Dam had compounded the problem; that dam project was killed partly because its cost could not be justified, and partly because it would increase the price of water from the other San Gabriel reservoirs.

[8] Memorandum, Burke G. Bryan, BRA directors meeting, Apr. 18, 1966, pp. 3831–3832; with Exhibit E, pp. 3842–3843, Box 59 (Brazos River Authority, Waco); memo, Walter J. Wells, Sept. 20, 1967, Box 257, ibid.; letter, Walter J. Wells to Col. Jack W. Fickessen, Sept. 29, 1967, Box 257, ibid.; letter, Carson Hoge to W. H. Sims, Forrest & Cotton, Inc., Nov. 6, 1967, Box 59, ibid.; letter, Hoge to District Engineer, Nov. 6, 1967, Box 59, ibid.; letter, Hoge to Sims, Nov. 8, 1967, Box 59, ibid.; memo from Hoge, Dec. 28, 1967, Box 59, ibid.; letter, Sims to Wells with memo on Laneport Reservoir and

bureaucracies not worked out their differences, the Brazos River Authority might have suffered acute financial distress, which might have rippled through the entire Brazos Valley. But they did work things out, and the two big bureaucracies soldiered on.

_____

In 1948, when Georgetown started campaigning for dams on the San Gabriel's upper forks, its four thousand citizens assumed that when the dams were built, the stored water would be theirs. If Taylor wanted to buy water, the thinking went, that city of eight thousand could tap the river at Circleville and pipe it south. Besides Taylor, no one else seemed vaguely interested. About a thousand people lived in Round Rock; Granger and Bartlett combined had about three thousand. All of Williamson County's little towns had bored wells into the Edwards Aquifer, on the west side, or the Upper Taylor Marl Formation to the east. There was plenty of water for everyone except in drought years, when farmers and ranchers suffered.

When the Corps picked Laneport as the dam site, Taylor's ruling elite assumed that its water would be theirs. All through the fifties, when drought reigned supreme, lakes and water became magical elixirs in people's minds. It became an article of Texas political faith that every trickle of water should be caught, whatever the cost, before it flowed "wastefully" into the Gulf. Wilson Fox fervently believed that a lake at Laneport would guarantee Taylor's future, in much the same way that the Black Waxy had put money in Taylor's banks.[9] No one in Williamson County seems to have considered that the new water supply might cost money. And few seemed aware that the BRA, the "local" public agency that would control any water dammed on the San Gabriel, was legally obligated to provide downstream rice farmers and Dow Chemical with a contractually regulated amount of Brazos watershed flow, limiting how much of the San Gabriel could be held back for municipal use. In years of ample rain

_____

North Fork Reservoir Yield Studies, Mar. 20, 1968, Box 257, ibid.; memo, Bob Steele to Sims, Apr. 5, 1968, Box 257, ibid.; letter, Forrest & Cotton, Inc. to Hoge, Apr. 10, 1968, Box 257, ibid. Finally, in May 1968 the Brazos River Authority released a "Report on the San Gabriel Watershed," the result of months of Corps' tinkering with yields figures, which satisfied the BRA's need for salable water.

[9] Interview, Tom Bullion, May 12, 2000, Taylor.

this presented no problem, but during droughts the superior water rights of downstream interests would stymie the BRA from committing upstream reservoir water for local use when it was most needed.

It is not clear that anyone in Williamson County had thought this through even by 1967, when the dams were about to be built. Several realities were at war. The BRA needed "transparent" dams that would allow the waters of the Brazos to reach the Authority's primary downstream markets. At the same time, the BRA had discovered that small upstream dams, which utilized hardly any water for local needs, could be quite advantageous. Too, the BRA desperately needed income. Cajoling the Army engineers and their congressional backers to build dams near potential water markets became a major BRA strategy to buttress its bottom line. But sometimes, as in Williamson County, these needs conflicted.

After its 1956 experience with Belton Dam, when the unsold Belton reservoir kept Dow Chemical operating, the BRA had warmed to the concept of a North Fork dam. In 1967 North Fork seemed a potential clone of Belton Dam's reservoir. In the words of an old-timer BRA executive, North Fork Dam would provide "another bucket that's full . . . or fuller."[10] Williamson County looked hopelessly stagnant. Despite the arrival of Interstate 35, the county's population had barely grown, from 35,044 in 1960 to 37,305 in 1970. Georgetown gained nearly a thousand new citizens. Taylor, still considered the county's leading city, garnered just 116 new residents during the decade. Round Rock picked up 353 residents, pushing its 1970 population total to 2,811 citizens, for the county's highest growth rate, 12.5 percent. East Williamson County was now an economic disaster zone, especially around Granger, where Laneport Dam was emptying the once thickly farmed Black Waxy. Granger lost 10 percent of its population.[11]

By 1970, despite the new interregional highway and visions of lake resorts dancing in a few entrepreneurial heads, most of Williamson County's residents lived quiet, small-town lives, still dependent on money from farming, ranching, education, government, and lime-

[10] Interview, Mike Bukala, Oct. 31, 2000, Waco.
[11] Clara Scarbrough, *Land of Good Water*, 345–346.

stone strip-mining. Agriculture was the number one industry. Two-thirds of Williamson County's 722,560 acres were still in cultivation. It was pleasant living on a small scale. A few high school graduates went to Texas A&M to study agriculture or veterinary science and returned to build their parents' agricultural businesses, but most migrated to the big cities, along with most other children of Texas's vast rural hinterland after World War II, to make their way onto more exhilarating urban stages.

Williamson County was sleepwalking into a new age. There was some talk about growth and industrial recruiting and quick drive times to Austin, but for the most part, even the most sophisticated local businessmen and politicians saw only faint glimmers of what was to come. "What limited vision we had then," exclaimed former Georgetown mayor and longtime banker Jay Sloan from a distance of thirty years. "We never in our wildest imagination would have imagined this"—a millennial population of 250,000 and bumper-to-bumper traffic crawling along Interstate 35.[12]

That included the Brazos River Authority's top 1970 planners. The exceptions could be counted on one hand, and they were all locals: Georgetown's Owen Sherrill, who had long preached that the new interstate and lakes on the upper San Gabriel would produce a population explosion; Taylor's Wilson Fox, who deeply feared his city might founder with the double loss of its old agricultural economy and the new interstate highway seventeen miles away, pinned his hopes on Laneport Dam; Round Rock banker-developer Tom Nelson, who helped put Round Rock on the map by recruiting a Westinghouse plant; land investor Tom Kouri of Austin, who in 1964 purchased the first of two perfectly placed Williamson County ranches on two important interstate highway intersections (one of which he helped create); a Burnet investment group that developed a commercial cave hard by Interstate 35; and a young Austin flying instructor named Bobby Stanton, who thought people might like to work in Austin and live "in the country."

At the time, these men were considered a bit loony on the subject of growth. The general view was that growth would occur, but noth-

---

[12] U.S. Census, 2000. A few years later the population crossed the 300,000 mark.

Road, River, and Ol' Boy Politics

ing spectacular. Many Williamson County businessmen thought the interstate would kill Georgetown and Round Rock. When the lakes filled, everyone was certain, there would be plenty of cheap water for everyone. A former Round Rock city manager recalls, "Round Rock and Georgetown thought when the government built the reservoir, they could just stick straws in the water and suck it out."[13] In 1968 Georgetown asked the BRA how much municipal water supply the new reservoir could be expected to yield.[14] The BRA replied that the agency's "priority" was to satisfy water needs in the "immediate vicinity of the reservoirs," which was not really the case. Water from either North Fork or Laneport, the BRA's chief planner said, would easily "meet the needs of Georgetown."[15] Six years passed before anyone in Williamson County again bothered to raise the question of water supply with the BRA.

In that interval an extraordinary thing happened. Westinghouse Corporation moved to Round Rock. In 1971–1972 the company built a giant turbine plant halfway between Round Rock and Georgetown. The plant lay in Round Rock's extraterritorial jurisdiction and initially employed 750 workers, but that number was expected to double in ten years. Westinghouse had purchased thirty-three hundred acres of ranch land that a company subsidiary planned to develop as a "new town."[16] Westinghouse embraced Williamson County and Texas mythology, purchasing a pair of Texas Longhorns to graze in front of the plant's I-35 entrance. Round Rock cut the company a sweet deal: no annexation for seven years (hence no taxes), new water and sewer works provided, free of cost, six miles north of town. All Westinghouse had to do was buy Round Rock's water.[17]

---

[13] Interview, Jim Hislop, Oct. 21, 2001, Austin.

[14] Letter, L. R. Hudson to Walter J. Wells, Dec. 23, 1968, Box 257 (Brazos River Authority, Waco). Hudson was a consulting engineer for W. H. Mullins, Inc., which had been hired by Georgetown, under the direction of Mayor Jay C. Sloan, to do a projection of "growth, water demands and possible treatment facilities" for the city.

[15] Letter, Carson H. Hoge to L. R. Hudson, Jan. 3, 1969, Box 257 (Brazos River Authority, Waco).

[16] Interview, Jim Hislop, Oct. 21, 2001, Austin.

[17] Interview, N. G. "Bunky" Whitlow, July 1, 1998, Round Rock. In 1970 Whitlow was vice president of Farmers State Bank (owned by Tom E. Nelson Jr.) and a rising star in Round Rock. Interview, Tom E. Nelson Jr., Oct. 23, 2001, Austin; interview, Jim His-

The deal was dicey for Round Rock. It depended on successful industrial, commercial, and/or multi-family development between Westinghouse and the city's core, none of which existed. "The way things worked in those days, we tried to get a deal going, and then we tried to play catch-up to make it work," Jim Hislop, the former city manager, said.[18]

Austin and Williamson County leaders were thunderstruck at Round Rock's success. Westinghouse's decision to move to Round Rock softened the county's "hick" reputation. Though Westinghouse's turbine market collapsed when the 1974 oil crisis shocked the United States, Round Rock kept seeking Blue Chip industries and getting them. In 1976 it lassoed McNeil Consumer Products, a Johnson and Johnson subsidiary, which started making Tylenol tablets on a handsome new campus north of town (between the old business district and Westinghouse). Industrial recruiters eyed Round Rock with new respect.

Across the county there was envy. Clearly the county's old dependence on well water drawn from aquifers would not satisfy the population explosion that was gathering steam along Interstate 35. In the summer of 1974 Georgetown's leadership decided it needed to secure rights to North Fork Dam's water. After informal conversations with Brazos River Authority officials, Mayor Joe E. Crawford announced that when the dam was completed, Georgetown would build a treatment plant and sell water from North Fork Dam to neighboring towns. The purchase price would be ten cents per thousand gallons, he said, "plus ten percent or some reasonable return on our investment."[19] Round Rock, Florence, Westinghouse, and the

lop, Oct. 21, 2001, Austin; letter, Mayor Dale Hester to George Chapman, president of Westinghouse, Apr. 23, 1971, Administration Department, City of Round Rock, City Hall, Round Rock. Westinghouse bought ranches from Leon E. Behrens, Raymond Pearson, and the Lomac Corporation. It also optioned James Garland Walsh's 1,400 acres for a million dollars, but backed out after the gas-turbine market collapsed. Later a group venture headed by Nelson purchased Walsh's ranch for $900 an acre—"a good investment," he says—which became Brushy Creek North and Stony Creek. Tobin Surveys Inc., San Antonio, "Williamson County, Texas," 1969. Author purchased Ownership Base Maps, Williamson Co. Tx., N/2 1–1228, S/2 1–1229.

[18] Interview, Jim Hislop, Oct. 21, 2001, Austin.

[19] "Georgetown Offers Water to Neighboring Cities," *Williamson County Sun* (Georgetown), July 25, 1974.

Jonah Water Supply Corporation expressed various degrees of interest. That fall Georgetown officially requested thirteen million gallons per day—more than the North Fork reservoir could supply—from the Brazos River Authority.[20]

The BRA's Wells reminded the mayor that the reservoir's maximum "dependable yield" would be only twelve thousand acre-feet a year, or twelve million gallons per day. When might Georgetown and its customers actually start using water from North Fork? Wells queried Crawford.[21] The mayor wrote back, "Our minimum needs will be eight million gallons per day for the first two years and eleven million thereafter for the next ten years."[22] This was a figure that seems to have had no basis in reality, even assuming that Round Rock and Westinghouse used copious amounts of North Fork water.

Wells clarified the situation for Georgetown. The water's cost, he wrote Mayor Crawford, would be on the order of forty to fifty dollars per acre-foot or about 125 to 155 dollars per million gallons—12.5 to 15.5 cents for one thousand gallons, not ten cents per thousand as Crawford seemed to think. He continued his analysis,

> At this rate, the cost for 11 mgd of raw water would be
> $500,000 to $620,000 per year. The cost for 8 mgd would be
> $365,000 to $450,000 per year. These represent very sizeable
> commitments, and before preparing a contract for considera-
> tion by the City of Georgetown calling for annual payments
> of these magnitudes, I thought I should call these anticipated
> costs to your attention.[23]

Crawford was staggered. "We were quite startled by your letter of October 29, 1974, in which you quoted prices of raw water to the City of Georgetown," he wrote back. Previous conversations, he reminded Wells, had alluded to a figure of ten cents per thousand, which would result in having to charge thirty cents per thousand "in the mains . . . not including the amortization of the plant and appur-

[20] Letter, Mayor Joe E. Crawford to Walter J. Wells, Oct. 1, 1974, Box 257 (Brazos River Authority, Waco).

[21] Letter, Walter J. Wells to Mayor Joe E. Crawford, Oct. 4, 1974, ibid.

[22] Letter, Mayor Joe E. Crawford to Walter J. Wells, Oct. 22, 1974, ibid.

[23] Letter, Walter J. Wells to Mayor Joe E. Crawford, Oct. 29, 1974, ibid.

tenances." That would represent an increase of seven to ten cents over the present cost of water in town. "In your letter," the mayor continued, "it would appear that Georgetown would be obligated to pay for the total amount of water contracted even though it was not used. This simply is a financial burden that cannot be assumed." In good faith, Crawford wrote, the city had gone public with its plan, pressed forward with land acquisiton near the dam, and had hired consultants to pursue the matter. Now the city had "much misgivings about our course of action."[24]

In reply, Wells softened his tone but maintained his message.[25] A few weeks later a chagrined Mayor Crawford revised Georgetown's request to the BRA for North Fork water down to one million gallons per day starting in 1979 or 1980, when the reservoir was expected to open for business. After five years Georgetown would probably need another one million gallons per day. "Beyond this time it becomes anybody's guess," Crawford wrote glumly.[26] Shortly thereafter the mayor died of a heart attack.

Part of what happened behind the scenes to change the complexion of the BRA's informal understanding with Georgetown was that after Georgetown went public with its grand plan, Round Rock city officials complained to the BRA about Georgetown's proposal to control all of North Fork Dam's water. "We said, 'We've got as much right to that water as Georgetown does, and we *need* it,'" former city manager Hislop would remember long afterwards.

> Then the BRA came back and said, "You know what? We're going to write a take-and-pay contract." What that meant was that rather than sticking your straws in and paying the BRA meter for just what you took out, like everybody had thought it would be, with a take-and-pay contract, if you signed for twelve million gallons, the day you stick your straw in you start paying for twelve million gallons right then.[27]

Indeed, Georgetown's mayor was right to be startled. The Brazos

[24] Letter, Mayor Joe E. Crawford to Walter J. Wells, Nov. 4, 1974, ibid.
[25] Letter, Walter J. Wells to Mayor Joe E. Crawford, Nov. 6, 1974, ibid.
[26] Letter, Mayor Joe E. Crawford to Walter J. Wells, Nov. 27, 1974, ibid.
[27] Interview, Jim Hislop, Oct. 21, 2001, Austin.

River Authority did sharply reverse gears on Georgetown during the summer of 1974. In July the *Williamson County Sun* quoted Wells as saying that ten cents per thousand gallons was "an outside figure. . . . We hope the actual price will be less than that."[28] Three months later Wells told Georgetown the price would be between 12.5 and 15.5 cents per thousand, and that a new take-and-pay contract would be required. After several years of jockeying, Georgetown and Round Rock split the North Fork reservoir's water. In the spring of 1978, each city agreed to buy 6,700 acre-feet at thirteen cents per thousand gallons.[29] Each city ended up building its own water treatment plant, using federal grant money to cover most of the expense. By the time the deal was concluded, each city considered itself lucky to have locked up that much water.

There was a final fallout—enmity between Georgetown and Round Rock that continues to this day. A former Georgetown mayor admits, "We didn't realize the need to reserve that water out there. Our growth had been two percent a year and the BRA was telling us we would have to pay a standby price. 'Who's going to use it?' we asked. Then, when we found out that Round Rock was reserving it, it was, 'How come we let Round Rock have all that water?' "[30] For its part, Round Rock delighted in one-upping Georgetown. "I'll never forget," Round Rock banker "Bunky" Whitlow later recalled with an impish grin, "ol' Thatcher Atkin standing up in a meeting and shouting 'til he was red in the face, 'The people of Round Rock came over here under the cover of night and stole our water!' He actually *said* that!"[31]

[28] "Georgetown Offers Water to Neighboring Cities," *Williamson County Sun* (Georgetown), July 25, 1974.

[29] "Dow Chemical Company's First Request for Production of Documents," Request RR-20, Fiscal Year 1989, Brazos River Authority, Box 362, Georgetown-RR-24 (Brazos River Authority, Waco).

[30] Interview, Jay Sloan, Sept. 11, 2000, Georgetown.

[31] Interview, N. G. "Bunky" Whitlow, Sept. 14, 2000, Round Rock.

ELEVEN

GOODBYE, BOOTY'S CROSSING

The most beautiful and most profound emotion we can experience is the sensation of the mystical. It is the source of all
true science.

ALBERT EINSTEIN, *The Universe and Dr. Einstein*

I
t is May in Central Texas. I am inner-tubing the rapids of the San
Gabriel in western Williamson County. There is nothing to do but
melt into the scenery as I tumble over small waterfalls, dawdle in
still emerald pools. Below me lurk three primeval-looking gar, magnified by the water into fearsome beasts. Wading sandpipers cool their
stilt legs in the water near the shore. Above me looms a forty-foot
limestone cliff, pocked with clumps of yellow wildflowers and caves.
Behind the sheer bluffs, hidden in thick stands of cedar, several pairs
of rare golden-cheeked warblers tend their young.[1]

In the spring of 1974 I abandoned my New York City home for
the San Gabriel River. Work had commenced on both dams. If one
avoided the dam sites themselves, where gigantic, ear-splitting earth-
moving machines were rearranging the shape of the land, one could
imagine that nothing had changed in the San Gabriel Valley since a
vigorous Paleo-Indian people first settled there about 9500 B.C.[2] I
floated the river with Linda Graves, a Georgetown minister's wife

---

[1] Field trip, author and Linda Graves, May 1974, Georgetown.

[2] *Final Environmental Impact Statement: Laneport, North Fork and South Fork Lakes, San
Gabriel River, Texas,* Feb. 24, 1972, U.S. Army Engineer District, Fort Worth (Department
of the Army, Corps of Engineers, Southwest Region, Fort Worth), Section II, pp. 22, 23.

with a sunny personality and a bent for archeology, the outdoors, and politics. Graves so loved the San Gabriel's North Fork that when it became clear that the river valley would be buried under a lake, she commissioned a movie about the North Fork's low water crossings, famous in Central Texas for their serene beauty.[3]

Dropping rapidly through steep limestone canyons at an average grade of seventeen feet per mile, the North Fork was indeed lovely.[4] Even the Corps acknowledged that it qualified as a United States "scenic river area," but only the Texas Legislature could give it that designation. It declined to do so in 1969. The Texas Senate did include the entire San Gabriel River in a proposed Texas Natural Streams System, but the House killed the bill.[5] The potential for protected status of the North Fork had a weird, Catch-22 quality: it handily met requirements for protection as a U.S. "wild river," except that human access to the river was not restrictive enough. On the other hand, it would have qualified as a U.S. "recreational river area"—except that it was too difficult to reach.[6]

White-water canoeists were captivated by the North Fork's winding, heady chutes through "steeply eroded hills, tall rocky bluffs, spurs, knobs and escarpments" rising between 100 and 250 feet over the river bed.[7] Bob Burleson practiced law in Temple, canoed avidly, and sat on the Texas Parks and Wildlife Commission. In 1974 his views of the San Gabriel's North Fork were rhapsodic, his attitude bitter toward the plan to dam it.

Ten miles to the northwest, near the Williamson County town of Florence, a team of archeologists from the University of Texas at Austin in 2000–2001 turned up a rich early Clovis site which threatened to rewrite archeological textbooks on the human species in North America. Whether similar evidence was buried deep in the layers of the Balcones Escarpment under North Fork Lake is unknown.

[3] Diane Koenig, *The San Gabriel River Crossings*, movie, circa 1976, Edward A. Clark Texana Collection (Southwestern University, Georgetown). Sadly, I was not able to schedule a "re-run" of my float trip with Graves. Shortly after I spoke with her about this project, she was diagnosed with an aggressive form of cancer. She died November 5, 2001.

[4] *Final Environmental Impact Statement,* 1972, Section II, p. 10

[5] Ibid., II, p. 29.

[6] Ibid., II, p. 28.

[7] Ibid., II, p. 20.

Goodbye, Booty's Crossing

The San Gabriel is as pretty a stream as you have in Texas. By any measure you have—scenic bluffs, beautiful clear water—it is the best canoeing around. The upper reaches are just super; as you go down, you see beaver, deer, moss-covered springs, little graveyards, old Indian camps. The canoeing is seasonal, there's no question about that. It's Spring and Fall, and any time there's a good thundershower. I've got a fellow in Liberty Hill who calls me any time there's a really good rain storm. If the phone rang right now, I'd say adios. I can name fifty people who canoe that stream regularly, but the numbers certainly go into the thousands.[8]

The North Fork's recreational use was indeed remarkable, considering that the place was never publicized.[9] A caliche-topped road followed the river's bends through long, green corridors of pecan, live oak, cottonwood, and cedar elm that met overhead. Booty's Road crossed the river four times, creating, in essence, four free "public" water parks.[10] Each crossing was named for an early settler—Booty, Jenkins, Box, Hunt—but because Booty's Crossing lay closest to Georgetown, all the crossings became incorrectly and collectively known as "the Booty's" or worse, "the Booties."[11] They all shared common characteristics—high limestone cliffs draped with mosses and wildflowers, stair-stepped shelves of hard limestone or river-smoothed pebbled beaches, sapphire swimming holes. Sandwiched between the crossings was the river itself, which, the Corps reported, "consisted of long stretches of barren bedrock riffles, short stretches of gravel riffles, with intermittent pools. . . . These streams, being primarily spring fed, are typically clear and rapid."[12] The cross-

[8] Interview, Bob Burleson, May 1974, Austin.

[9] No one ever studied how many people used the San Gabriel's North Fork for recreation, since it was not part of any public park system. Though the land in the valley was held entirely in private hands, its four low water crossings gave the public legal access to the river, and the public used it extensively.

[10] *Final Environmental Impact Statement*, Section II, p. 20, and Appendix A.

[11] This sort of thing would drive my mother crazy; for years she crusaded against the misuse of "Booty" and its variants. To her sorrow, she never completely succeeded in her mission.

[12] *Final Environmental Impact Statement*, 1972, Section II, p. 21.

ings presented a smorgasbord of topographical features that allowed Sunday picnickers, weekend smoochers, Boy Scout troops, fishermen, swimmers, arrowhead hunters, rock hounds, spelunkers, inner-tubers, and birdwatchers to simultaneously enjoy the ten-mile stretch of the San Gabriel North Fork without troubling each other or the river.[13]

It was clear why the Tonkawas called this country *takatchue pouetsu,* land of good water.[14] During their century of dominance, ranchers had marked the land lightly. Hundreds of springs still bubbled and seeped out of the ground. Tall grasses—bluestem, buffalo, grama, and others less common—still thrived. Season after season, all manner of wild creatures wintered, nested, fattened, and reproduced in the San Gabriel Valley.

It was a magnificent habitat for birds.[15] Five rare and endangered species visited frequently: Southern bald eagle, American peregrine falcon, whooping crane, greater sandhill crane, and green kingfisher. A sixth endangered species, the timid golden-cheeked warbler, soft and small as a mouse, nested in Williamson County's canyon country, its dwindling habitat composed of virgin stands of mature Ashe Juniper found in a handful of Edwards Plateau counties.[16] American osprey and the black-capped vireo, while not listed as endangered species, were quite rare, and often sighted along the North Fork. The road-runner, wild turkey, and anhinga made the valley their home, along with 272 other avian species. The county's mammalian life was less spectacular; still, nutria, long-tailed weasel, mink, beaver, white-tailed deer, cougar, bobcat, ringtailed cat, grey fox, raccoon, striped skunk, and six bat species were common.[17] (In underground caves west of the Balcones Escarpment five species of blind, albino invertebrates would be discovered and declared endangered in the 1980s, too late to affect environmental challenges to the dam, but posing significant

---

[13] "There is Just One Booty's Crossing," *Williamson County Sun* (Georgetown), Feb. 7, 1974.

[14] Clara Scarbrough, *Land of Good Water*, 25.

[15] *Final Environmental Impact Statement*, 1972, Appendix E.

[16] Interview, Eleanor Brogren, May 1974, Georgetown; *Final Environmental Impact Statement*, Section II, pp. 21–22; Clara Scarbrough, *Land of Good Water*, 33.

[17] *Final Environmental Impact Statement*, 1972, Appendix D.

challenges for west Williamson County developers.)[18] But with a dam rising on the North Fork and subdivisions multiplying nearby, the river and the animals were living on borrowed time.

The North Fork of the San Gabriel was a treasure trove of Hill Country ecological habitat, especially in light of its close proximity to Austin and the new interstate highway that was funneling thousands more people into the county every year. Few Georgetown voices protested the dam; the few exceptions were an odd mixture of ranch women, archeologists, and historians concerned about the loss of unexamined archeological sites.[19] Most of Georgetown, after all, had fought for decades to secure North Fork Dam and was not terribly bothered.

After Congress passed the National Environmental Policy Act of 1969 (which went into effect January 1, 1970), the Corps started working up an environmental impact statement on the San Gabriel dams, as required by the new national law.[20] On April 21, 1971, the Corps of Engineers filed what it termed the "final draft" of its *Final Environmental Statement* on Laneport, North Fork, and South Fork lakes and asked for comments from government agencies affected by the projects.[21] It was not the thorough and detailed examination

---

[18] Kim Tyson, "Cave Dwellers," *Austin American-Statesman*, Dec. 19, 1993, pp. G-1, G-8; J. B. Smith, "Endangered Cave Creature Could be Delisted," *Sunday Sun* (Georgetown), Apr. 16, 1995, pp. A-1, A-7; "Endangered Species Habitat Is To Be Carefully Safeguarded," undated press release, Dell Webb Corp., Sun City, Georgetown.

[19] "Hunting at North Fork Draws Landowner Gripes," *Williamson County Sun* (Georgetown), Feb. 14, 1974; interview, Linda Graves, May 1974, Georgetown. The local critics included Graves, "Bob" Gunn, Vera Allen, Agnes Wade, Judy Shepherd, Dr. Jud Custer, and Dr. Ed H. Steelman. Their opposition focused on limited targets, such as the Corps' poor control of hunters on what had become "public" lands, or on the inadequate period of time allocated to archeological and historical studies of the San Gabriel Valley before it became the bottom of a lake.

[20] *C. C. Allison v. Stanley R. Resor*, Civil Action No. A-71-CA-84, U.S. District Court, Western District of Texas, Austin Division, Aug. 13, 1971, included as part of the Corps' *Final Environmental Statement, 1972*, Section VIII, p. 107.

[21] Final Draft, *Final Environmental Statement Laneport, North Fork and South Fork Lakes San Gabriel River, Texas*, Apr. 21, 1971, San Gabriel General File (Brazos River Authority, Waco). Though South Fork Dam had been indefinitely deferred, the three dams were authorized as a unit and so were treated as a unit in the Corps' document.

Congress appears to have envisioned for the government's public works projects. The draft *Final Environmental Statement* was eleven pages long. One sketchy map of Williamson County and letters from three government agencies were attached. "Cursory" is a kind description.

Four months later, on August 13, 1971, an Austin veterinarian and Civil War buff, Dr. Charles Curtis Allison, who owned 366 acres of San Gabriel riverfront property, 326 of which the Army had condemned for Laneport Dam, filed a complaint for injunctive and declaratory relief with the U.S. District Court in Austin.[22] Perhaps Allison imagined himself emulating one of the Confederate generals he so admired for dash and daring; he was standing against the U.S. Army on behalf of Williamson County farmers.[23] Allison's land, sandwiched between Wilson Fox's and Virginia Forwood Lawrence's "ranches," reminded travelers of Virginia's rolling hunt country. Allison's attorney was a young environmental activist named Richard A. Shannon, who sat on the newly formed board of the Lone Star Chapter of the Sierra Club. Allison's complaint was joined by Austin's Sierra Club chapter, the Travis County chapter of the Audubon Society, Save Our Springs, the Texas Explorers' Club, and Bob Burleson, the aforementioned Texas Parks and Wildlife commissioner. The complaint attacked the Army Corps of Engineers projects on a number of grounds, chief among them that the engineers had so drastically altered the San Gabriel River project since Congress authorized the three dams eleven years earlier that the building of the Laneport and North Fork dams would constitute an illegal act.[24]

---

[22] *C. C. Allison v. Stanley R. Resor*, Civil Action A-71-CA-84, U.S. District Court, Western District of Texas, Austin Division, Aug. 13, 1971; *Final Environmental Statement*, 1972, Section VIII, pp. 107–129.

[23] Interview, Richard A. Shannon, Nov. 3, 2001, Austin. Allison was introduced to Shannon by Bob Clark, a young biologist from Taylor who worked for Senator A. R. "Babe" Schwartz on water pollution legislation. The lawsuit was later amended to read *Allison v. Froehlke*, the correct name of the secretary of the army, on May 7, 1972. The amended lawsuit asked for a preliminary injunction to stop all activity on the San Gabriel River projects.

[24] In the nine years since the dams were authorized, the lawsuit charged, their estimated cost had risen from $45 million to $72.3 million; their order of construction had changed from a staged plan to one of "simultaneous" construction of Laneport and

The Allison lawsuit also argued that the Corps' *Final Environmental Statement* on the San Gabriel Valley failed to meet Congress's mandate for an in-depth examination of the environmental impacts of proposed dams, as well as evaluating alternatives to building the dams, such as the Soil Conservation Service's small dams program that Williamson County had backed so overwhelmingly at the polls. The lawsuit also argued that: (1) Construction of the Laneport, North Fork, and South Fork dams would tend to "encourage the erection of structures in the flood prone area . . . because downstream property owners would be lulled into a sense of security by the presence of large dams upstream." Since the dams were designed to protect only against a fifty-year flood—not the hundred-year flood that the 1921 flood surely had been—the Corps could not "assure . . . safe surroundings."[25] (2) The value of the Upper San Gabriel River as a "free-flowing stream," as defined by the Wild and Scenic Rivers Act, had not been given "adequate consideration" by the Corps of Engineers.[26] (3) If all three dams were constructed as authorized, "one-fourth to one-third" of the San Gabriel River—"one of the most archeologically unstudied river drainage systems" in Texas—would be inundated. The acting director of the Texas Archeological Salvage Project assessed the scientific damage:

> The three reservoir area, considered as a unit, offers a highly unusual situation in that they bracket two distinctly different physiographic and biotic areas within a short river drainage segment. This fact alone is significant in evaluating their potential for ecologically oriented research. Without further and substantial investigation . . . another "laboratory" will be removed from the continually diminishing sampling universe available for study.[27]

North Fork; their storage capacity and effective yields had been reduced about 25 percent; their locations and designs had been altered; and their cost-benefits ratios had dropped from 2.8 to 1 to 1.7 to 1. See Judge Jack Roberts' "Findings of Fact," June 29, 1972, CA A-71-CA-84, U.S. District Court, Western District of Texas, Austin; *Final Environmental Statement*, 1972, Section VIII, pp. 111–112.

[25] Ibid., VIII, p. 116.

[26] Ibid., VIII, p. 117.

[27] Letter, David S. Dibble to Douglas Scovill, Jan. 8, 1971, "Comment on the USCE

(4) Though the Corps of Engineers had the authority to build multipurpose dams, federal funding for such projects was limited to three specific goals: flood control, wildlife enhancement, and recreation. A Corps-built dam was allowed to store, or "conserve," water for municipal or industrial use, but that portion of the cost of the dam had to be borne by a local government such as the Brazos River Authority. There was a cap on the water storage portion of any Corps project—thirty percent. Originally, the cost of water storage in the San Gabriel project would have greatly exceeded that limit: 40.6 percent for Laneport and 44 percent for the three-reservoir package had been allocated for water conservation.[28] (5) The Army's plans for the Laneport and North Fork dams would damage native wildlife, contradicting the Environmental Quality Protection Act of 1970. Both dams would increase the population of "rough fish" in the watershed, virtually eliminating a "good quality" natural fishery. A North Fork dam would also eliminate substantial habitat areas of the rare and endangered golden-cheeked warbler and wild turkey.[29] (6) Cameron Reservoir, a proposed dam in Milam County, would, if built, "permanently inundate about 45,750 acres of the lands of the Little River Basin, which includes most of the 33,600 acres which defendents purport to protect from flood by the proposed construction of Laneport Dam."[30] In other words, the chief original rationale for the Laneport dam—protection of Milam County from flooding—would disappear under another man-made lake if the Cameron dam were ever built.

Wilson Fox was irked. During the previous year, he had guided the Brazos River Authority's response to the Army's requests for comments for its environmental impact study of the San Gabriel Valley. At Walter Wells's request, he had edited the BRA's official text and

---

Environmental Statement: Laneport, North Fork, and South Fork Lakes, San Gabriel River, Texas," ibid, Section VIII, p. 21 included in the "Final Draft" of the "Final Environmental Statement," Apr. 21, 1971. Environmental Assurances (Brazos River Authority, Waco). Dibble ran the Texas Archeological Salvage Project at the University of Texas Balcones Research Center, Austin.

[28] *Final Environmental Impact Statement*, Feb. 24, 1972, Section VIII, pp. 120–123. Those high figures were ultimately reduced to roughly 25 percent of the cost of the overall project.

[29] Ibid., II, p. 21, VIII, p. 125.

[30] Ibid., VIII, pp. 125, 126.

become the essential, pivotal figure keeping Wells and Congressman Pickle on track with regard to Laneport's progress.[31] Now this lawsuit—and from a landowner who lived in Austin and had not been active in anti–Laneport activity until now. Fox sent a copy of the petition to Pickle, noting, "whoever worked this out really put forth a great deal of study. I am not personally acquainted with Richard A. Shannon and have asked a number of lawyers about him, but have not received any satisfactory information. No doubt someone is paying his bill, and possibly some of these outdoor-indoor organizations may be contributing something to the cause."[32]

By the time the case came to the courtroom of Judge Jack Robert nine months later, the Army had produced a brand new, nearly two-inch-thick *Final Environmental Statement* for the Laneport, North Fork, and South Fork lakes. At 235 pages, plus maps, tables, plates, and an extensive bibliography, it was a bloated green document that could not possibly be called cursory.[33] President Nixon had asked 7.5 million dollars for the San Gabriel project; then had frozen funds after Allison filed his lawsuit.[34] Political change had blown through Williamson County as the redistricting process replaced Congressman Pickle with Congressman Poage—an ironic development, since Poage had so often displayed displeasure at the county's intransigence regarding the Laneport dam.[35] The National Audubon Society had

[31] Letter, Walter J. Wells to Wilson Fox, Oct. 7, 1970; letter, Wilson Fox to Wells, Aug. 12, 1971; letter, Wells to Wilson Fox, Aug. 23, 1971; letter, J. J. "Jake" Pickle to Wells, Aug. 23, 1971; letter, D. L. Orendorff to Wells, Sept. 16, 1971; letter, Wells to Orendorff, Sept. 27, 1971, all in San Gabriel File, "Environmental Assurances" (Brazos River Authority, Waco). By now, Wells, Pickle, and Wilson Fox were quite chummy; Wells always addressed his letters to Fox to "Pillbox 192," an allusion whose meaning has been lost over time; Pickle saluted Fox as "My Beer Friend" after a Hilda's Bottom barbecue.

[32] Letter, Wilson Fox to J. J. "Jake" Pickle, Aug. 30, 1971, ibid.

[33] The final *Final Environmental Statement* was dated Feb. 24, 1972.

[34] Memo, J. J. "Jake" Pickle to Walter J. Wells, Wilson Fox, etc., Jan. 25, 1972, San Gabriel "General" File (Brazos River Authority, Waco); Connie Sherley, "San Gabriel Dam Suit Filed," *Austin American*, Aug. 19, 1971, p. A-17.

[35] Memo, J. J. "Jake" Pickle to Walter J. Wells, Jan. 25, 1972, San Gabriel General File (Brazos River Authority, Waco); letter, Richard A. Shannon to Robert Poage, Jan. 28, 1972, Box 692, file 7, W. R. Poage Papers (Baylor Collections of Political Materials); letter, Poage to Gene. N. Fondren, June 21, 1971, ibid.

weighed in with an opinion that Laneport would be acceptable from an environmental point of view, but not the North or South Fork dams.[36] The Texas Parks and Wildlife Department had withdrawn its previous approval of the San Gabriel projects, harshly criticizing the Corps of Engineers:

> When [we] reviewed the draft environmental statement . . . it was our impression that because of adverse impacts which the statement reflected the project would be abandoned and further alternative means sought. This has not been the case. We, therefore, are forced into withdrawing our concurrence of January 7, 1971. . . . If the only purpose of an environmental impact statement is to fulfill the physical requirements of NEPA and not to allow us to seriously consider the consequences of our actions upon the environment, then there seems to be little need for the continuation of the review process.[37]

This stance surprised nearly everyone. After the Parks and Wildlife letter had circulated through state water development circles, the governor's chief planner brushed off the protest. "The State's position remains unchanged," he reassured the Corps. "The actions of a single agency cannot alter the official position adopted."[38] For its part, the Corps carried on as if nothing untoward was happening. It pressed forward to advertise and award bids to start construction on the North Fork and Laneport dams.[39]

---

[36] Memo, John L. Spinks Jr. to J. J. "Jake" Pickle, May 1, 1972, with attached letter from Spinks to the Corps regarding the Audubon Society's comments on the Environmental Impact Statement on the San Gabriel River dams. Pickle sent the packet to Walter J. Wells on May 23, 1972 (Brazos River Authority, Waco).

[37] Letter, James U. Cross to Col. Floyd Henk, May 19, 1972 (Brazos River Authority, Waco). Cross was executive director of the Texas Department of Parks and Wildlife; Henk headed up the Fort Worth District of the Corps of Engineers.

[38] Letter, Col. Floyd Henk to Ed Grisham, May 30, 1972; letter, Grisham to Henk, June 1, 1972. Grisham was director, Division of Planning Coordination, Office of the Governor, Box 692, File 7, W. R. Poage Papers (Baylor Collections of Political Materials, Waco).

[39] Letter, Col. Floyd Henk to J. J. "Jake" Pickle, May 23, 1972 (Brazos River Authority, Waco).

On June 2, 1972, Judge Roberts eyed a packed Austin courtroom while Allison's two attorneys, Shannon and Lloyd Doggett, laid out the plaintiffs' case for temporarily halting the Corps' activities on the San Gabriel River.[40] The Army's top engineers looked on, along with Walter Wells and David Kultgen, manager and counsel for the Brazos River Authority.[41] Wilson Fox and a Pickle aide attended the proceedings, as did a slew of Austin environmentalists and a Georgetown Boy Scout troop.[42]

Shannon and Doggett called two key witnesses. Dennis Neal Russell, chief of the environmental branch of the Texas Parks and Wildlife Department, painted a dire picture of the proposed lakes' consequences. Their recreational benefits had been "over-inflated" by the Corps, he said, considering there were several other large recreational lakes within half an hour's drive from both Laneport and North Fork. In other words, there would be too much competition for the new lakes to attract much of a following, despite the Corps' claims to the contrary.

In addition, Russell testified, since the reservoirs were primarily flood-control structures, their levels would fluctuate "a great deal." North Fork Lake would fluctuate "inside of the canyon walls, from 20 to 40 feet" over a ten-year period, leaving ugly stains. The Laneport reservoir, sprawling shallowly over relatively flat land, would "move laterally in as flood waters are released, leaving large exposed mud banks." His assessment sounded a lot like Henry Fox's: "It is going to be awfully difficult to get a boat in the water to enjoy this boating recreation unless we use a chain hoist to lower it down the side of the cliff in two of the cases, or are willing to drive your car across an exposed mud flat in the other case," Russell said.[43]

---

[40] Eventually Doggett would become a justice on the Texas Supreme Court; still later he succeeded Congressman Pickle as congressman for the Tenth District.

[41] Memo for the record, Walter J. Wells, June 3, 1972 (Brazos River Authority, Waco).

[42] Letter, Walter J. Wells to J. J. "Jake" Pickle, June 5, 1972, ibid.; letter, Wilson Fox to Pearce Johnson, Parks and Wildlife Department, June 5, 1972, ibid.; "Opponents Appealing Dam Decision," *Austin Statesman,* June 8, 1972; Betty MacNabb, "Injunction Halts Laneport Dam Work," *Austin American,* June 9, 1972.

[43] *C. C. Allison vs. Robert Froehlke, Secretary of the Army, et al.,* No A-71-CA-84, June

Road, River, and Ol' Boy Politics

The Parks and Wildlife official also predicted that game-fishing would suffer when the lakes shut down the natural river system. When the Laneport reservoir is complete, he testified, "we can expect a buildup in game fish population in the first three or four years . . . but this will rapidly decline, leaving high rough fish population and very poor fishing."

Couldn't the lake be restocked? Doggett queried. Yes, Russell testified, but there were drawbacks: "We can go in and wait until the lake gets low . . . and come in with chemical toxicants, Rotenone, and poison out the lake and restock. . . . If we just restock the lake without removing the high rough fish population that are in them, then, all we are doing is feeding the existing fish already in the lake. When we throw our little fish in, the fish already there just eat them up."[44]

He added that the dams would change the very *nature* of the river's flow. He was referring to the impact of the changed flow on downstream fisheries, but his comments applied equally forcefully to the question of severe erosion along the river's banks—a point no one brought up. "When these reservoirs are constructed," Russell told the court, "there are very seldom ever guaranteed releases from them, so your entire stream either has periods of extremely high flow that will release a great deal of water, which will come rushing down, silting in some of the holes that the fish depend on, and then they [the BRA] will cut the water out, and the stream channel will completely dry up."[45]

A second critical witness, geologist Riser Everett, raised eyebrows when he testified that erecting dams on the San Gabriel near fault lines such as the Balcones Escarpment, where dams on the North and South forks were to be constructed, was a "very hazardous undertaking." He cited two reasons: One could never be sure when or whether

---

2, 1972, U.S. District Court for the Western District of Texas, Austin (U.S. District Court, Austin), 11–12, See also *Final Environmental Impact Statement*, 1972, Section III, p. 9. The two cases cited refer to the North Fork and South Fork dams.

[44] *C. C. Allison vs. Robert Froehlke, Secretary of the Army, et al.*, No. A-71-CA-84, June 2, 1972, U.S. District Court for the Western District of Texas, Austin (U.S. District Court, Austin), 13.

[45] Ibid., 14. Severe bank erosion has caused extensive loss of trees along the river since North Fork Dam was built.

seismic movement might start again; such movement could "rupture" the dams. But what was more likely was that the lakes might not hold water. Because of the extreme porousness of the Edwards Formation west of the Balcones Escarpment, the lake probably would leak, Everett testified. Worse, he said, it was likely to leak into Inner Space Caverns, the geologic find just south of Georgetown which had recently opened as a popular tourist attraction.[46]

The Army's witnesses rebutted with the argument that Congress had authorized the dams, so there was no question of legality. They maintained that their revised environmental impact statement on the San Gabriel dams was entirely sufficient. Certainly it was fat. Delaying the project, they testified, would cost taxpayers at least 49,000 dollars a month—145,000 dollars annually if one counted the loss of anticipated annual benefits from Laneport and North Fork.[47] They argued that their projections for lake users at Laneport and North Fork was not "over-inflated" at all, as the Texas Parks and Wildlife Department was saying, but "on the low side." Lewill "Bud" Horseman, chief planner for the Corps' Environmental Resource Section, testified that the immediate "market area" of a lake provides 80 percent of its "day users"—the overwhelming majority of any lake's recreational usage. The San Gabriel market, he said, contained "about fifty-five percent of your population right up through the center of the state, and, of course, the interstate highways make easy access for them."[48]

Col. William E. Wood, assistant chief of the engineering division, was drawn into making several embarassing confessions on the stand. While testifying that the Corps had carefully checked out the suitability of each dam's foundation through "extensive core boring"— two hundred at Laneport and fifty at North Fork—he added that it had been discovered that North Fork Reservoir "possibly" would not

---

[46] Ibid., 36, 41. Everett testified that he had worked for Standard Oil of New Jersey for thirty years as an oil and petroleum geologist. Oil and petroleum geologists are often considered experts in underground hydrology. But under cross-examination, Everett admitted that he had no prior experience with dam feasibility studies.

[47] *C. C. Allison vs. Robert Froehlke, Secretary of the Army, et al.,* No. A-71-CA-84, June 2, 1972, U.S. District Court for the Western District of Texas, Austin (U.S. District Court, Austin), 129.

[48] Ibid., 100, 107–110.

Road, River, and Ol' Boy Politics

hold water above 791 feet above sea level, so the dam had been redesigned.[49] The problem lay not in the Balcones fault, he said, but in the Edwards, "a very porous formation, and it is not—will not usually hold water."[50] Stillhouse Hollow Dam, Waco Dam, Canyon Dam—all like North Fork, very near the Balcones fault line and all on the Edwards—all had leaked, and all had required expensive reconstruction. At Waco Dam, Army engineers had experienced "a major slide" during construction, requiring an extra "four or five million dollars" to correct.

Shannon walked Wood through the story of how the Corps had altered the all-important conservation yields at the behest of the BRA, noting that projected water-storage yields in the reservoirs had suddenly dropped a remarkable 25 percent in 1967, which magically brought the San Gabriel system into compliance with the Corps' legal authority. Shannon asked Wood to read from the minutes of a January 7, 1965, conference between the Texas Water Commission, the Brazos River Authority, and the Corps of Engineers.

WOOD: "Colonel Wells stated that the cost of the water from these projects was considerably higher than that from any other reservoir in the Basin, and that they had gone to considerable length to negotiate water sales at these prices to the principal users in the Brazos Valley, with no success."

SHANNON: Now, isn't it a fact that due to the redesign ... that the amount of cost for the project allocated for water supply was reduced from approximately 44 percent to approximately 24 percent?

WOOD: There was some reduction, probably in that neighborhood. I don't know.

SHANNON: In that neighborhood. Was there also an increase in the portion of the cost allocated to recreation at the same time?

WOOD: There probably was, because at that time we had additional policy on recreation, and we were looked upon by

[49] Ibid., 86–87.
[50] Ibid., 90–91.

Congress to put in additional recreation facilities, which we hadn't done in the earlier project.

SHANNON: Was there also an increase in the percentage of cost allocated to flood control benefits at the same time?

WOOD: There could be.

SHANNON: Is your answer—can you make the answer more definite?

WOOD: I would say yes.[51]

The young attorney did not say so outright, but his drift was clear: he was suggesting that the Corps of Engineers had cooked the books to meet Congress's requirements for the San Gabriel dams.

Judge Roberts denied Allison and the environmental groups their request for a temporary injunction to halt the San Gabriel projects, disappointing those trying to block the dams.[52] On the other hand, the judge was quoted as saying he refused to impose an injunction "'solely on a time basis' because of the voluminous report and evidence requiring exhaustive study before a final judgment on the merits of the dam . . . can be given."[53] Several days later, Shannon and Doggett appealed Judge Roberts's decision to the U.S. Fifth District Court of Appeals, asking the higher court to stay dam work pending a hearing on appeal.[54] Judge John Minor Wisdom signed the temporary order, freezing the projects at the bid-letting stage.[55]

Back in Taylor, Wilson Fox, an astute attorney himself, fumed in a letter to Congressman Poage about the "so-called" Allison case: "With

[51] Ibid., 91–93.

[52] "Judge Refused to Block Laneport Dam Project," *Temple Daily Telegram*, June 4, 1972; letter, Walter J. Wells to J. J. "Jake" Pickle, June 5, 1972 (Brazos River Authority, Waco).

[53] "Opponents Appealing Dam Decision," *Austin Statesman*, June 8, 1972, *Austin Statesman* morgue.

[54] Ibid.; letter, Walter J. Wells to Otha F. Dent, chairman of the Texas Water Rights Commission, June 8, 1972 (Brazos River Authority, Waco).

[55] Betty MacNabb, "Injunction Halts Laneport Dam Work," *Austin American*, June 9, 1972; memo, Col. Floyd H. Henk to Prospective Bidders and Others Concerned, June 9, 1972 (Brazos River Authority, Waco); letter, J. J. "Jake" Pickle to Walter J. Wells, June 9, 1972, ibid.; "Nature Lovers Block Construction of $25-Million Dam at Laneport," *Waco News-Tribune*, June 9, 1972; memo, Wells to BRA Board Members, June 13, 1972 (Brazos River Authority, Waco).

the unpredictable attitude of our Federal Courts I am fearful of the outcome. I can anticipate that they will stay the proceedings and return the case for trial on its merits," he wrote. "If we have arrived at the ridiculous situation whereby a group of environmentalists can stop progress, then I am about ready to throw in the sponge. The San Gabriel River above Georgetown flows very little, and anybody that can canoe down that River should be a very strong character with strong legs and a strong back as he would be carrying his canoe most of the way. . . . So far as the golden-cheek warblers are concerned I would not know one when I met it in the road."[56]

A month later, in July 1972, a Granger farmer complained to Poage that the Army was withholding payment for land they had bought which he was supposed to vacate by January 1, 1973. Now the Corps had written that they were not sure whether "they would be permitted to complete the purchase of our land." Laneport Dam, Henry C. Rozacky Jr. wrote, "has uprooted so many people from their homes, torn up a community that cannot be rebuilt, and confused so many people to the point that they don't know what to expect next."[57]

Poage immediately turned for help to his former congressional colleague, Federal Judge Thornberry of the Fifth Circuit Court of Appeals. "This is the first time in my 36 years in Congress that I have written to a United States Judge regarding matters pending in his Court," Poage wrote. "I don't believe it is a good practice and yet, as one who knows something of legislative matters, you will realize that there are times when members of the Legislative Branch desperately need to know how the Courts plan to proceed. That is my problem now. I am not trying to suggest what should be done other than to point out the need for a decision."[58] Poage described "hundreds of land owners [who] had their property taken or contracted for by the United States." Many had "relied on commitments of the government, made payments and bought other property which they cannot pay for as long as the government does not pay." Poage ended his

[56] Letter, Wilson Fox to Poage, June 12, 1972, Box 692, File 7, W. R. Poage Papers (Baylor Collections of Political Materials, Waco).

[57] Letter, Henry C. Rozacky Jr. to Poage, July 16, 1972, ibid.

[58] Letter, Poage to Homer Thornberry, Aug. 4, 1972, ibid.

August 2 letter to Thornberry by requesting a "probable timetable" regarding the *Allison* lawsuit.

A week later Thornberry replied. After checking with other judges of the Fifth Circuit, he wrote, "I entered a stay pending consideration of the motion for injunction by a panel of this Court. We were in agreement that the case was too important to be decided without briefs." The environmentalists' brief had already been filed. The Army's brief was not yet in, but when it was received, Thornberry assured Poage, "we will decide whether the case requires oral argument. If it does, we shall set it for a prompt hearing."[59]

In mid-September, the Fifth Circuit Court lifted Judge Thornberry's stay against construction of the San Gabriel projects and set a hearing date.[60] On November 13, 1972, Chief Judge Brown, Judge Roney, and Judge Moore, on loan from the Second Circuit, heard Allison's case in New Orleans. Judge Thornberry did not participate. Appellates' attorney Shannon recapitulated much of what Judge Roberts had heard in Austin, focusing on the legality of the project and the sufficiency of the environmental impact statement. The judges seemed nettled that no trial had taken place, just a hearing. Judge Brown asked Shannon, "Why not have a full trial, then let Judge Roberts decide?" On December 27 the Fifth Circuit ordered that the case be sent back to District Court for a full trial, but refused to delay the project. There matters rested. Allison and his environmental posse let the matter drop, having run out of money and convinced they could not win.[61]

In many respects the environmentalists' fears about the dams' negative consequences were well founded. Some of their worst-case sce-

[59] Letter, Homer Thornberry to Poage, Aug. 11, 1972, ibid.

[60] Memo, Walter J. Wells to General Counsel, Sept. 15, 1972, San Gabriel "General" File (Brazos River Authority, Waco).

[61] Interview, Richard A. Shannon, Nov. 3, 2001, Austin. One financial contributor to *Allison vs. Froehlke*, according to Shannon, was the National Resource Defense Council, which at about that time realized its attorneys were having more success stopping highways than dams. Courts, Shannon said, were "reluctant to stop projects where congressional funding had taken place," whereas highways were funded largely through the Executive Branch, and proved easier targets.

narios did not pan out. No one seems to have anticipated one serious environmental consequence of the North Fork dam—losing Crockett Gardens. In a canyon snaking south from the North Fork of the San Gabriel River, near the projected shoreline of the new lake, Crockett Gardens stood as an exquisite example of Balcones Escarpment and Edwards Plateau ecosystem, characterized by honeycombed limestone and underground springs. Since its development by two Anglo-American settlers in 1855 as a mill and, later, a truck garden producing strawberries and other exotic fruits and vegetables, Crockett Gardens (originally known as Knight's Springs) had been visited, painted, and photographed by thousands of nature lovers.

At Crockett Gardens the Edwards Aquifer sprung from the rock and plunged down steep cliffs, creating a spectacular waterfall flanked by smaller falls. Over the eons, the water had shaped the gigantic rock shoulder over which it fell, carving caves into the stepped bluffs that sheltered a vast cape of maidenhair, fiddlehead fern, velvety moss, and watercress—an artist's palette of greens. In the 1950s a rancher had built a rock ranch house and swimming pool above the falls, where the springs bubbled out of the ground. Water from the springs filled his pool and danced through dozens of rivulets that switchbacked through the property. Coming upon this scene after a short climb from Knight-Jenkins Crossing at Second Booty's, Crockett Gardens appeared like a vision of the Garden of Eden, transported to Texas.[62]

Initially the Army engineers claimed Crockett Gardens would survive the North Fork lake. It would be the Corps' most prized possession. There was nothing official about this commitment, but that was the impression left by the Army's minions. The Army didn't mention Crockett Gardens in its environmental impact statement, nor was Crockett Gardens mentioned in *Allison v. Froehlke*. The place, however special, could not command the legal protection offered under the Endangered Species Act to a rarity like the golden-cheeked warbler. The Austin environmentalists may not even have known about Crockett Gardens. If they had, they might have considered calling in Eliot Porter, the brilliant Santa Fe nature photographer who immortalized Glen Canyon before its flooding by the U.S. Bureau of Recla-

[62] Clara Scarbrough, *Land of Good Water*, 434–435.

Above: Knight-Jenkins Crossing (second Booty's), a short hike from Crockett Gardens. The low water crossing was one of four along the North Fork of the San Gabriel River obliterated by Lake Georgetown. The limestone cliffs dripping with springs and sandbars typified the river landscape. *Courtesy* Williamson County Sun, *Georgetown.*

Below: A still life of fern-laced caves, wildflowers, and waterfalls at Crockett Gardens, circa 1978. *Courtesy* Williamson County Sun, *Georgetown.*

Author Linda Scarbrough stands in front of a Crockett Gardens waterfall, circa 1976. *Donna Scarbrough Photo, Courtesy Donna Scarbrough Josey, Georgetown.*

mation, inspiring nature lovers to rally and save the Grand Canyon from a similar fate.[63] Depending on one's point of view, the Army engineers either meant well and failed, or anticipated trouble and launched a brilliant public relations campaign to convince Crockett Gardens lovers that the San Gabriel's ecological jewel would remain safe in their hands.

While North Fork Dam was under construction, the Corps put out a slick four-color brochure. Crockett Gardens was its cover attraction. "Are You Ready?" the Corps' publication asked readers. "Picture

[63] Eliot Porter, *The Place No One Knew: Glen Canyon on the Colorado* (San Francisco: Sierra Club, 1963). Or, if Porter had not been available, the Lone Star Sierra Club probably could have gotten Texan Jim Bones, who had studied under Porter and shot many stories for *Audubon* magazine, to chronicle Crockett Gardens' beauty.

a lake surrounded by high bluffs, cool woodlands, and open meadows. Imagine walking along a trail and finding spring water cascading over limestone ledges bordered with mosses and ferns."[64] It was a specific description of Crockett Gardens.

Thus the Corps sold North Fork Lake, soon to be renamed Lake Georgetown. But North Fork was, first and foremost, a flood-control dam. If Corps engineers needed to raise or lower the height of the reservoir to control flooding, they would do so. That was why the Corps' environmental impact statement made the point that the lake level would fluctuate so wildly. It could have been worse. Periodically the Corps fended off pressure from the Brazos River Authority to *eliminate* recreational use of the North Fork reservoir entirely so that the BRA could "make full use" of the stored conservation pool. That would have eliminated the beach, boat docks, camping grounds, trails, *and* Crockett Gardens.[65] The truth was, North Fork Dam threatened Crockett Gardens by its very existence. After the lake opened to hikers and boaters, Crockett Gardens immediately suffered from overuse, "people languishing in the springs," as one lake manager described it. The delicate ferns and mosses were trampled; the springs started drying up. In 1992 heavy rains swelled the San Gabriel River and the Corps raised the lake level, flooding Crockett Gardens to contain downstream flooding. For four months most of Crockett Gardens was submerged. When the Corps lowered the lake to its "normal" level, Crockett Gardens was all but dead.[66]

---

[64] Brochure, "North Fork Lake," undated, U.S. Army Corps of Engineers, Fort Worth District.

[65] Letter, William T. Moore to Department of the Army, Nov. 27, 1973, Box 59 (Brazos River Authority, Waco). The BRA would have preferred that the Army not hold as much water in the lake as it planned to do during the dry summer months, when downstream rice farmers needed irrigation water. The Authority was not concerned about creating a "nice" lake for boaters and swimmers; it wanted a free hand to manipulate the lake's conservation storage waters. "From our point of view," Moore wrote, "it would be better to delete the recreation pool elevation and area; however, if it cannot be deleted, the footnote should be changed to read: 'Elevation and area will vary between the top and bottom of the conservation pool depending upon hydrological factors and consumers' needs.'"

[66] Brochure, "Good Water Trail," Aug. 1994, U.S. Army Corps of Engineers, Fort Worth District.

Road, River, and Ol' Boy Politics

In the 1970s Linda Crawford Graves was Georgetown's apple-cheeked dynamo, the chamber's "Woman of the Year," a tenacious and widely loved do-gooder. Town wags said the First United Methodist Church had *two* pastors—Linda and her husband, the Reverend Tom Graves.[67] The Graves had encountered the San Gabriel's North Fork on their first Sunday in Georgetown. A church member had invited them out to "the Booty's," and Linda's love affair with the river began. "Every free Sunday afternoon we spent out at the Booty's with our children," remembers Tom Graves. "We swam, picnicked, canoed, floated the river—all those things you did." The Graves had been enthusiastic Colorado campers. In Williamson County they found "great natural beauty so close at hand, and about to be lost." Linda saw the river as "a playground for college students and for the community."[68] But she accepted the dam's now-paramount local rationale: to bank water for future municipal use. "I can't see myself throwing myself across a tree to stop it and fifteen years from now having no water," she said in 1974.[69] Instead of protesting, she produced a movie.[70]

The San Gabriel River Crossings, filmed by a University of Texas graduate student, portrayed the North Fork and its four low-water crossings as they were before North Fork Dam stopped the river's natural flow. Grainy and a little hokey, the movie still washes the viewer into the river, taking poetic inventory of its treasures. The camera revels in the water, which reveals itself in hues of aquamarine, tourmaline, and sapphire—so clear one can see down to the river's bottom of smooth stone. Rapids gurgle and whoosh through hairpin turns, slowing at deep swimming holes where children splash and dive from great boulders. Canoeists paddle purposefully, caps pulled over ears. Fishermen cast flies for black bass. Bird trillings sweeten the air. Bluebells, spiderwort, primrose, and poppies tremble as the camera zooms in. Hikers gaze at steep white cliffs towering over Knight-Jenkins Crossing and amble through Crockett Gardens' old straw-

---

[67] Interview, Jeannine Fairburn, Nov. 14, 2001, Georgetown.

[68] Interview, Tom Graves, Nov. 15, 2001, McKinney, Tex.

[69] Interview, Linda Crawford Graves, May 1974, Georgetown.

[70] Diane Koenig, *San Gabriel River Crossings,* movie, circa 1976, Edward A. Clark Texana Collection (Southwestern University, Georgetown).

Linda Graves partaking in one of her favorite pastimes—birdwatching, circa 1995. *Courtesy Tom Graves, McKinney, Texas.*

berry fields. Boy Scouts pitch tents on a meadow and head off to explore, "going off like popguns in all directions." The effect is magical. "Every visit to the San Gabriel Valley creates a mood of transience," intones the narrator, "a feeling that the changes of moon and season, of weather and the work of man, could change this place in our absence, before we visit it again."

Linda Crawford Graves died of bone cancer in November 2001, but to this day her film compels viewers to reflect on one of her passions—the San Gabriel River's once vivid existence.

RIVER TWICE DAMMED

Don't talk, dig.

JOHN STEVENS, Chief Engineer of the Panama Canal

Without the relentless machinations of Wilson H. Fox and Owen W. Sherrill, neither of the San Gabriel River dams would have been built. Others were essential facilitators—Jake Pickle and Walter Wells leap to mind—but without Fox and Sherrill, both dam projects surely would have succumbed to the fierce opposition they inspired. Neither Fox nor Sherrill lived to savor his triumph. By the time the dams rose over the river each man had dreamed of dominating, both were dead.[1]

Between the time that the Corps of Engineers began acquiring land for the dams in the late sixties and the projects' completion a decade later, Williamson County had experienced other losses. Congressman Pickle, the dams' political mastermind and a legislative powerhouse, allowed the county to be dropped from his beloved Tenth District during the 1970 redistricting process.[2] Congressman Poage

---

[1] Wilson Fox died in February 1974, of a heart attack, on a bitter cold day while delivering a tax form to a client, according to his son, Dr. Jim Fox of Austin. See "Wilson H. Fox Services Today," *Austin American-Statesman,* Feb. 11, 1974. Sherrill died two years later, at 86. "Sherrill Services Today," ibid., May 6, 1976.

[2] Pickle needed to shed some of the rural parts of his district due to Austin's growth, and Williamson County got cut. Williamson County observers close to Pickle always

had taken Williamson County into the Eleventh District, rather in the spirit of a distant uncle's taking an orphan into the family out of a sense of duty, but with little enthusiasm.[3] Williamson County's loss of Pickle foreshadowed its ultimate embrace of the Republican Party after the 1980 election of President Ronald Reagan.

Granger successfully politicked to get Laneport Reservoir renamed Granger Lake, but it didn't help the agrarian market town.[4] By the time Granger Dam was dedicated in 1978, Granger was barely clinging to life. "Does the government have the right to come in and kill our town?" banker Truett Beard, who had taken over Ralph Moore's lakeside investment, wondered aloud at a town meeting.[5] Many of the once prosperous merchants of Davilla Street had shuttered their stores after losing their economic base—approximately 150 farm families who had cultivated the fabled Black Waxy, now gone to make way for Granger Lake. The elegant Storrs Opera House was razed to make way for a bank parking lot. Dozens of substantial family businesses—Mikulencak's Variety and Dry Goods, Granger Gin and Farm Supply, Grainger Grain, Inc., the old drugstore, even a popular beer joint—closed. The public school system lost a quarter of its

---

said that the congressman could have retained the county had he fought for it, but apparently he did not, due, it was thought, largely to the fact that the county never had contributed much money to his political campaigns. Or, perhaps Pickle discerned the coming "tilt" of suburban areas across the United States toward the Republican Party, a trend barely noticeable in Austin in 1970 but well developed by 1980.

[3] Poage's disdain for Williamson County's fractured political and business leadership was clear throughout the fifties and sixties. While he appears to have made a genuine effort to help individual Williamson County citizens (especially farmers) who needed assistance, his affection for the county and its political leaders was scant, as seen in his public discourse and private correspondence.

[4] Laneport Dam and Reservoir was officially renamed Granger Dam and Lake by an Act of Congress on January 3, 1975.

[5] "Granger fights for its life," *Granger News*, Nov. 8, 1973, Box 667, File 8, W. R. Poage Papers (Baylor Collections of Political Materials, Waco). Beard, who relished the memory of his old relationship with President Lyndon Johnson, and former Southwest Region Internal Revenue Service Director Frank Scofield acquired Ralph W. Moore's Texva Realty after Moore's death and reincorporated it as Gabriel Farms. Also see letter and enclosure, State Representative Dan Kubiak to D. L. Orendorff, Chief, Engineering Division, Corps of Engineers, Fort Worth, Nov. 26, 1973, with report by Bridgette Cavanaugh entitled "Granger Dam," ibid.

students, causing annual state funding to drop by approximately forty thousand dollars.[6] Ultimately the public school's tax base shrunk by 1.266 million dollars as the government converted 13 percent of the district's taxable property to tax-free status. As a direct effect, six teachers lost their jobs.[7] Granger's despair was palpable. "There is a pervasive sadness in this town," City Clerk Betty Hajda said in 1974, plucking nervously at imaginary lint on her blouse sleeve. "People are afraid."[8]

Taylor struggled against the tide of agricultural loss, as well, as it tried to cope with its eroded status as a key transportation center, after Interstate 35 bypassed it fifteen miles to the west. In 1974 it finally got the "loop" that "Son" Bland had wanted so badly, but not the "airline highway" to Austin it had sought since the 1950s. The 4.779-mile loop cost 4.3 million dollars.[9] The Soil Conservation Service's plan for a small dams network on the San Gabriel died after it became clear that the river would have two big dams, which would partially duplicate the small dams' benefits and permanently flood much of the land that the small dams would have protected.[10] Much of the San Gabriel

[6] A 1973 report written by Bridgette Cavanaugh and circulated by State Representative Dan Kubiak's office cited a loss of state school funding due to condemnation of private property through 1973 of $84,658. See letter, Kubiak to D. L. Orendorff, Nov. 26, 1973, ibid.

[7] "Acreage Acquired by the U.S. Government for Lake Granger Project," Jan. 1, 1976, Granger Independent School District, Granger.

[8] Interview, Betty Hajda, July 1974, Granger. The farm town's population fell from 1,256 in 1970 to 1,236 in 1980. In 1974 one in ten of the storefronts along Davilla Street were boarded up. In 2001 one in ten were in use. Also see letter, Herbert L. Sides to J. J. "Jake" Pickle, Apr. 26, 1967, Pickle Papers (Center for American History, University of Texas, Austin); letter, Pickle to Sides, May 2, 1967, 95-112-181, ibid. In the flat lands around Lake Granger, the expected increase in market value of real estate anticipated by the Army engineers, Poage, and Pickle did not occur.

[9] Interview, Chris Bishop, Texas Department of Public Transportation, Feb. 11, 2002, Austin.

[10] Letter, A. J. Wade to Poage, Feb. 27, 1968, Box 692, File 7, W. R. Poage Papers (Baylor Collections of Political Materials, Waco); letter, Poage to Wade, Feb. 29, 1968, ibid. In 1968 President Richard Nixon's administration chopped Soil and Water Conservation programs, lopping $120 million out of a $220 million budget, badly hurting efforts such as those by the Little River–San Gabriel Soil Conservation District to curb erosion and prevent flooding in their watersheds.

River, and the natural habitat it sustained, soon would be lost; what remained would never again flow freely and reinvigorate the Black Waxy with deposits of rich alluvial fertilizer during seasonal floods.

But the county gained financially. Round Rock shot into orbit, a rocket town aggressively pursuing a postindustrial, computer-based economy. Though its acquisition of Westinghouse Corporation did not turn entirely on Lake Georgetown's water, that water supply was very much the deciding factor in the city's subsequent netting of an impressive string of Blue Chip companies.[11] Overnight, it seemed, Round Rock led the county, growing 500 percent during the seventies, from 2,811 to about 15,000 citizens.[12] Georgetown grew too, though not nearly as spectacularly. The county seat seemed content, at least for the moment, to let Southwestern University, the county government, and Texas Crushed Stone's limestone-stripping operation provide most local employment. An ambitious young fellow named Bobby Stanton developed Serenada Country Estates west of Interstate 35 by buying up ranches and marketing his subdivision as a country haven near a lake. It was an enormous success. Soon Stanton was developing Georgetown's ritziest neighborhoods, all conveniently placed between Lake Georgetown and Interstate 35.

The county's population doubled from 37,305 in 1970 to 76,521 in 1980, making it the fastest-growing county in the Brazos River basin and the second-fastest-growing county in Texas. The growth was concentrated west of I-35, close to Lake Georgetown's anticipated water supply, in sprawling subdivisions that became Austin's suburban beachhead. In the first four years of the seventies, one-seventh of all county land west of the interstate, approximately forty thousand acres, passed from ranching into investor hands; land values west of the Balcones Escarpment rose from as little as thirty-five dollars per acre to one thousand dollars per acre.[13]

---

[11] Interview, Jim Hislop, July 1998, Georgetown. Hislop did say that Westinghouse counted on North Fork water to develop its planned "New Town" of 30,000.

[12] 1980 Census of Population, General Population Characteristics, Texas, 45–46; interview, Jim Hislop, July 1998, Georgetown. Round Rock officially counted 12,740 citizens in 1980, but some 15,000 hooked into city utilities.

[13] Field trip, author's notes, May 1974.

Road, River, and Ol' Boy Politics

Granger Dam—formerly Laneport Dam—was dedicated on December 1, 1978, a perfect crisp winter's day. The structure had cost American taxpayers 62 million dollars—nearly three times the 1958 projected estimate of 22.2 million dollars.[14] Congressman Poage, retiring after forty-two years in Washington, was the honored speaker. Until recently, Poage said, apparently forgetting his decades-long and ardent, though mostly futile, pursuit of a dam at Laneport, he could not see why this particular dam was sited where it was, but now he could see the logic of it. "These engineers are able to find spots where dams are needed," he proclaimed. Congressman Pickle, the man responsible for exhuming the dam from the Corps of Engineers' deep freezer, beamed at Poage, along with a phalanx of dignitaries from the Brazos River Authority and the Corps of Engineers. Granger's Ray Holubec, a former BRA director who owned two farms directly below the dam, won polite applause. J. Howard Fox, Wilson's and Henry's brother who had presided over the BRA board throughout much of the fight over Granger Dam, did not attend the ceremony.

No one mentioned that the BRA had not yet been able to sell the better part of Granger Lake's water, though it had been working hard to do so. Maj. Gen. Charles I. McGinnis, the Army's director of civil works, said that "someday" the lake would provide sixteen million gallons of water per day, "when it's needed."[15] Wilson Fox's dream of a revitalized Taylor piping reservoir water from Lake Granger to the city's new revenue-producing industries had not yet materialized.[16]

---

[14] Letter, Lt. Col. J. H. Hottenroth, Corps of Engineers, May 23, 1958, citing House Report 2247, July 1, 1954, p. 122 (National Archives, Southwest Region, Fort Worth); memo, Kevin McCarthy to Congressman Chet Edwards, Mar. 20, 1993, Corps of Engineers, Fort Worth. Also see "Poage, Officials Tour Granger Dam Site," *Taylor Daily Press,* Apr. 13, 1979.

[15] George Ferguson, "Granger Dam Dedicated Under Clear Bright Skies," *Taylor Daily Press,* Dec. 1, 1978.

[16] After Wilson Fox's death, Taylor leaders tried to convince General Manager Wells of the BRA that their city needed *all* of Lake Granger's water, thinking, as Georgetown leaders had thought, that Taylor could control and resell the water from the nearby reservoir. But, as in Georgetown's case, Wells "educated" Taylor against such a commitment. Wells feared the cities might not be able to make their payments on the water, and

North Fork Reservoir was not completed "simultaneously" with Granger Dam, as Owen Sherrill had insisted on and Congressman Pickle had promised, but the Corps did its best. Once money for the 38.8 million–dollar dam was flowing from Congress, the Army engineers efficiently erected the dam it had long argued would be a poor tool for controlling floods and storing water. On October 5, 1979, ten months after Granger Dam's gates opened, Lake Georgetown was dedicated. It seemed, somehow, a more auspicious occasion. Lady Bird Johnson got star billing, along with Congressman Pickle, who had not represented Williamson County for a decade. This time, Howard Fox attended the ceremony, along with Holubec, representing retired BRA directors. Also on the dais were Federal Judge Homer Thornberry, former congressman Poage, Congressman Marvin Leath, and Secretary of Labor Ray Marshall, who owned a nearby ranch.[17] Georgetown and Round Rock dignitaries were out in force, along with representatives of Williamson County's west-side entities that had not existed a few years before—the massive Anderson Mill and Brushy Creek municipal utility districts and the private Chisholm Trail Water Supply Corporation, all of which were jockeying for water supplies.[18]

Lake Georgetown's water was all under contract. In fact, Georgetown and Round Rock had each attempted to secure every last drop. In the end, the two cities split the reservoir's available yield at six million gallons a day, or 6,700 acre-feet a year each.[19] Pickle spoke of the political battle over the San Gabriel River dams. "In the compromise

---

with good reason. Alcoa's aluminum plant in Rockdale contracted for 5,000 acre-feet on Aug. 2, 1976. On April 16, 1979, Taylor led a consortium of small towns and water-supply corporations, contracting for 8,525 acre-feet of Lake Granger water—about half the reservoir's salable water supply of 16.2 million gallons per day. And in 1995 Del Webb Sun City contracted for 15 acre-feet. See *Final Environmental Impact Statement*, 1972, Section I, p. 5, and Brazos River Authority Federal Reservoirs Water Revenue, Fiscal Year 1989, Box 362 (Brazos River Authority, Waco).

[17] Joel Hollis, "North San Gabriel Dam is dedicated!" *Sunday Sun* (Georgetown), Oct. 7, 1979.

[18] San Gabriel Dam Dedication—North Fork Reservoir (Brazos River Authority, Waco).

[19] "Dow Chemical Company's First Request for Production of Documents and Information," ibid. In this 1989 statement of the BRA's water income, Lake George-

Don Scarbrough, left, and Congressman J. J. ("Jake") Pickle at the 1979 dedication of North Fork Dam. Courtesy Williamson County Sun, *Georgetown.*

we worked out," he told the crowd, "we assured everybody concerned that neither Georgetown nor Taylor could get one bucket of dirt or one drop of water ahead of the other. That's what I call working together—togetherness, Williamson County style!"[20]

It was an historic moment for Williamson County. The two dams transformed it as effectively as the Enclosure Laws altered England's pastoral environment and economic climate.[21] In October 1979, fifty-eight years after the 1921 flood, humankind controlled the San Gabriel River. Williamson County was a new place, reconfigured by half a century of maneuvering over the river's fate. Visionaries, specu-

---

town sales to Round Rock and Georgetown generated $215,357 for the BRA, while Lake Granger generated $.00—that is, not a cent—in income. The BRA had contracted with Georgetown on April 20, 1978, and Round Rock on May 2, 1978.

[20] Joel Hollis, "North San Gabriel Dam is dedicated!" *Sunday Sun* (Georgetown), Oct. 7, 1979.

[21] Georgetown School Superintendent Jack Frost recognized the historic importance of the dams and sent all history students to the dam dedication ceremonies. Don Scarbrough, "Passing Glance," *Williamson County Sun* (Georgetown), Oct. 7, 1979. Also see Garrett Hardin, "The Tragedy of the Commons," *Science,* 162 (Dec. 13, 1968), 1243–1248.

lators, boosters, engineers, land developers, water planners, and politicians all had a hand in the job.

In the process, they had nudged Williamson County from one world into another, from the agrarian society it had been to an incipient suburban mecca. The promise of bounteous water supplies courtesy of Uncle Sam's dams, and a federal interstate highway that had simply showed up, did the trick. In water-short Central Texas, Williamson County had water to spare.

Below: 1981 aerial view of the eastern portion of Lake Georgetown showing North Fork Dam, lower right. Most of the land above, or north of the lake to Andice Road (crossing the top third of the picture) was Roy Gunn's ranch. *Courtesy Pat Gunn Spencer, Georgetown.*

2005 aerial of same place. The Corps of Engineers did not allow private development around Lake Georgetown, maintaining a buffer zone that could be flooded in case of heavy rainfall or water storage emergencies. Notable developments on the old Gunn Ranch include Fountainwood Estates, center, and the H.E.B. complex, right. North of Andice Road (Williams Drive) is a small corner of Sun City Texas. *Courtesy Ercel Brashear, Georgetown.*

River Twice Dammed   241

Part Two

THE
ROAD

No man with a good car needs to be justified.

FLANNERY O'CONNOR, *Wise Blood*

HIGHWAY UTOPIA

> . . . it is something strictly American to conceive a space that is filled with moving, a space of time that is filled always with moving.
>
> GERTRUDE STEIN, *The Geographical History of America*

Seven years after Henry Ford's first Model T rolled off the assembly line, America's three and a half million car owners were yearning to explore the United States in their new vehicles. One thing held them back: outside the nation's cities, paved roads barely existed. Farmers suffered terribly from this deficit of decent roads, foundering in mud every time it rained. In 1916 Congress passed the path-breaking Federal-Aid Highway Act, which encouraged state governments to plan and construct a network of modern primary highways and offered to pay 50 percent of the cost.[1] A year later, the Texas Legislature created the Texas Highway Department, which outlined its dream scheme for roads to knit together the state's far-flung agrarian society.[2] Highway engineers mapped twenty-five primary highway routes, including one, originally named Highway 2, which came to be known as Texas's "Main Street," linking Dallas,

---

[1] Helen Leavitt, *Superhighway—Superhoax* (Garden City, N.Y.: Doubleday, 1970), 22–23; *Highways and the Nation's Economy,* Joint Committee on the Economic Report, U.S. Congress (Washington, D.C.: U.S. Government Printing Office, 1950), 13.

[2] "Highway Development," in Tyler, et al. (eds.), *The New Handbook of Texas,* III, 607–608.

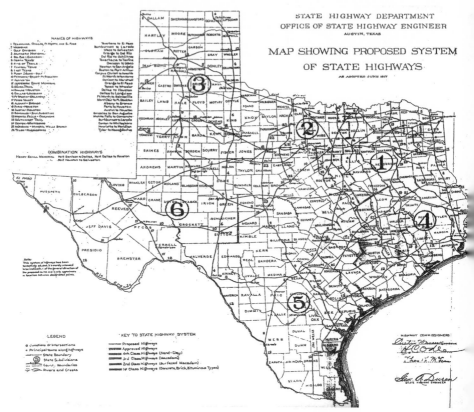

The Texas Highway Department's 1917 ideal highway system, which closely resembled the interstate highway system built in the 1960s. What would become Interstate Highway 35 was designated as Highway 2 in 1917, but in that design it passed through Taylor, rather than through Georgetown and Round Rock as it ultimately did when built. *Courtesy John Hurt, Texas Department of Transportation, Austin.*

Waco, Austin, and San Antonio.[3] It started at the Red River, headed south along the Balcones Escarpment, and ended in Laredo. Later, when it was actually built, it was called U.S. Highway 81.[4]

Eighty-five years later, most Texas drivers would instantly recognize the schematic Texas Highway 2 on a map. It was identical to what became Highway 81 and eventually Interstate Highway 35—

[3] Thomas E. Turner, "Main Street of Texas," *Texas Parade*, 13 (Apr. 1953), 17–19 (Baylor University Texas Collection, Waco).

[4] Map, "Proposed System of Texas Highways," June 1917, State Highway Department, Texas Department of Transportation, Austin.

with one exception. In Williamson County, the proposed Texas Highway 2 would have jogged east and coursed through Taylor, snubbing Georgetown and Round Rock. In 1917, that made perfect sense. Taylor was then the premiere market center between Austin and Waco; the Black Waxy from which it drew its wealth sustained prosperous King Cotton towns—Waco, Temple, Holland, Bartlett, Granger, and Taylor—lined up neatly along the Missouri, Kansas and Texas Railroad.[5] Taylor's reputation as the "world's greatest inland cotton market," its intersecting interstate rail systems, and a burgeoning industrial sector made it an essential cog in the state's market system. Seventy-five hundred people lived there.[6] Granger and Bartlett, ten and fifteen miles to the north, were booming, with nearly four thousand people combined. At least two thousand more farmed the Black Waxy within Taylor's market grasp.

By comparison, in 1917 Georgetown and Round Rock were villages. The county seat's twenty-eight hundred residents depended largely on the courthouse and a struggling Southwestern University for employment; Round Rock was a hamlet of nine hundred. Both towns serviced nearby farms and ranches, but these were lightly scattered compared to the county's densely cultivated east end.[7]

Texans wanted good roads. Bad roads were the farmers' bane, and farmers were the backbone of Texas's economy. The highway department's battle cry ricocheted across the state: "Get the farmers out of the mud!" And everyone else too. In 1910 a quartet headed by M. C. Cooke took off from Granger early one morning and drove a two-cylinder Buick five hundred miles to Plainview. The men camped and cooked their meals by the roadside and hugged the railroad tracks in case their car broke down. It didn't, but the journey took four days.[8]

[5] The Katy Railroad was the old Missouri-Kansas-Texas line, owned by Jay Gould and linking Central Texas blackland cattle and cotton interests with Fort Worth and Dallas markets. See Clara Scarbrough, 324.

[6] Herbert G. Willson, "Taylor, Black Land and Cotton," *Texas Magazine,* Feb. 1911, Houston; brochure: "Taylor and its' [sic] . . . Opportunities," Taylor Chamber of Commerce, 1940, "Williamson County Scrapbook" (Center for American History, University of Texas at Austin); Clara Scarbrough, *Land of Good Water,* 346.

[7] Clara Scarbrough, *Land of Good Water,* 346.

[8] Troy Worth, "Evaluating the Interstate," *Texas Highways,* 9 (Feb. 1962), 15.

A couple of years later, nineteen-year-old Margaret "Peg" Tegge eloped from her home near Granger with Circleville gin-owner Auburn Stearns to marry in Austin. It was a rainy day, and the nervous groom had to stop four times to repair punctures and pry the tires out of the mud. The thirty-mile trip lasted all day.[9]

In July 1919, after the Great War had ended, Gen. John J. Pershing dispatched a young American lieutenant named Dwight D. Eisenhower to lead a convoy of army vehicles across the country to demonstrate the military's capacity for quick movement. If Pershing hoped to dramatize the woeful state of United States highways, the demonstration succeeded. Eisenhower's caravan spent sixty-two days crossing the continent from east to west, sometimes traversing terrain with no roads at all. He never forgot the experience.[10] During the twenties suburban towns multiplied outside America's cities, and automobile sales exploded. Twenty-three million autos were on the road by 1923.[11] In Germany work started on a radically designed autobahn that allowed drivers to zoom along at speeds up to one hundred miles per hour on divided lanes. On this side of the Atlantic, America's first limited-access highway, the Bronx River Parkway, opened in the mid-twenties, inspiring New York's master builder, Robert Moses, to web his city with parkways and expressways.[12] But rural roads remained atrocious. In Williamson County in 1927, a bus carrying the Baylor University basketball team was "struck and torn to bits" at a railroad crossing in Round Rock by the International & Great Northern's Sunshine Special, on its way from Mexico City to St. Louis. "Ten of the twenty-one occupants [were] hurled into eternity in the twinkling of an eye," a reporter wrote. "Only the Carnage of a Battlefield Could Equal Scenes as Strong Men Weaken at the Sight Before Them," a newspaper headline read.[13] Texans were outraged, and Round Rock got a new arched bridge over its railroad

[9] Interview, Margaret Tegge Stearns, May 4, 1974, Taylor.

[10] David Osborne, "The Asphalt Bungle," *Inquiry* (Oct. 1981), 14.

[11] John Robinson, *Highways and Our Environment* (New York: McGraw-Hill, 1971), 48–49.

[12] Robert A. Caro, *The Power Broker: Robert Moses and the Fall of New York* (New York: Vintage Books, 1975), 144.

[13] *Williamson County Sun* (Georgetown), Jan. 28, 1927.

tracks. The Texas Highway Department focused on paving a network of rural roads, crucial to the Black Waxy farm economy of Williamson County.[14] The farm-to-market road was king. Texas highway engineers were considered heroes.

But the Depression killed auto sales and rural road building.[15] In 1934 architect Frank Lloyd Wright designed Broadacre City, a utopian alternative to the nation's existing rural, urban, and suburban patterns. Beautiful limited-access, six-lane freeways connected Wright's dense "high-rise" villages inhabiting an unspoiled agrarian landscape. In Wright's scheme, Broadacre City residents could choose between skyscraper homes in the town center or individual "organic" dwellings next to farm land. They could walk to work on the farm and walk to Beethoven concerts at night. To make all this workable, Wright thought, the U.S. government should give every American a one-acre plot of earth—and an automobile—as a birthright.[16] Wright never could sell his utopian dream, but his splendid vision of highways, whereupon drivers could never directly encounter oncoming traffic, probably influenced other transportation designers, like Norman Bel Geddes, whose model of a monumental limited-access, fourteen-lane superhighway crossing America was the hit of the 1939 World's Fair.[17] Certainly the German autobahn was part of both designers' thinking, but each gave it his own twist. President Franklin Roosevelt invested in highways mainly to boost employment, with no integrated transportation plan in mind. But after World War II,

[14] Texas highway builders, most of them trained at Texas A&M, astutely adapted the macadam road-building technique by using local materials, such as caliche or shell, or asphalt layered with tar, to form a weatherproof surface that would stand up to truck or auto traffic. The process was developed by the Scottish engineer John L. MacAdam. See T. U. Taylor, "County Roads," *Bulletin of the University of Texas,* Mar. 1890 (Center for American History, University of Texas, Austin).

[15] *Highways and the Nation's Economy,* 13.

[16] Frank Lloyd Wright, *The Living City* (New York: Meridian, 1958), 126–127, 146, 198–199; Robert McCarter, *Frank Lloyd Wright* (London: Phaidon Press, 1997), 243, 245–247. Originally Wright designed Broadacre City as part of a New York exhibition, but he continued trying to sell the concept for many years.

[17] Jeffrey L. Meikle, *Twentieth Century Limited* (Philadelphia: Temple University Press, 1979), 189, 194–195, 201–206. Geddes designed General Motors' Futurama exhibit, mounted in a building that looked like a giant carburetor.

during which the German autobahn had proven a military blessing, U.S. leaders began considering construction of a vast interregional highway network. They had the automobile and trucking industries and the expanding suburbs and deteriorating cities in mind, along with military needs. Congress passed the Defense Highway Act in November of 1941, asking each state to plan and coordinate such a system with its neighbors.[18] In 1944 Congress backed an interregional expressway system, but failed to agree on a workable plan, so the concept lay mouldering in highway engineers' offices.[19] In Texas and up through the Middle American states, one scheme blossomed in the popular imagination: a multilane, high-speed, Canada-to-Mexico expressway.[20]

During the Depression Williamson County's agrarian population shrank, as many farmers departed for city jobs. It shrank a little more during World War II, when the armed services called.[21] But the Black Waxy held most of its farming families. Most of the Czech-Americans near Granger and Taylor continued living on their land and near each other. The war helped Taylor, largely through the strength of Taylor Bedding Manufacturing Company, which supplied the U.S. Army with the bulk of its mattresses, made of Black Waxy cotton. Then, too, the International and Great Northern and the Kansas-Texas-Missouri rail lines met in Taylor and moved tons of agricultural products and heavy equipment.[22] Taylor was a center for meat shipping, as well, boasting three meat-packing plants and several cele-

---

[18] Mark H. Rose has exhaustively detailed the political development of the U.S. interstate highway system in his history, *Interstate: Express Highway Politics, 1941–1956* (Lawrence: Regents Press of Kansas, 1979). I relied on chapter two for much of this section.

[19] Rose, *Interstate*, 22–28.

[20] Letters, Gibb Gilchrist to Tim T. Warren, and Canada to Mexico Highway Association, July 14, 1932, G-G3, U.S. Highway System, 1931–32, Box 1185 (Texas Department of Transportation, Austin). Gilchrist was the Texas Highway Department's first highway engineer. The Canada-to-Mexico-Expressway theme recurs in highway department correspondence during the mid-1940s as possible interstate routes were being discussed. See, for instance, a letter from the Pryor, Oklahoma, Chamber of Commerce to the Texas Highway Department, Aug. 18, 1945, ibid.

[21] Clara Scarbrough, *Land of Good Water*, 345–346.

[22] The K–T–M rail line was affectionately called the Katy.

brated barbecue joints.[23] But some Taylor businessmen worried that after the war, their city's assets might erode.

These businessmen knew the highway department was drawing maps for a new four-lane divided highway, part of the proposed inter-regional system, that would pass through Williamson County, either along State Highway 28 (later to be known as State Highway 95) through Bartlett, Granger, and Taylor, or U.S. Highway 81, one of the state's primary trunk lines, through Georgetown and Round Rock, or possibly somewhere in between.[24] Williamson County movers and shakers fretted about what route the interregional would take, for it was clear that just as the railroads had created and destroyed towns, the superhighway could do the same. The Black Waxy was all Williamson County had of economic consequence. If the new interregional gave it a wide berth, no one could imagine the consequences.

Taylor leaders were keenly aware of these possibilities. In 1940 the Texas Highway Commission appointed a new chief engineer with ties to Taylor. His name was Dewitt C. Greer.[25] Greer had a special fondness for Taylor. It was a rail transport hub, and it seemed likely to favor highway development as well. With eight thousand people and a diverse industrial base, Taylor was the most important city within Austin's immediate orbit. Its business community was politically sophisticated, frequently inviting state bureaucrats to take an evening off to dine on T-bones and imbibe beer at the raffishly swanky Taylor Country Club. But Greer's Taylor links were personal too. He enjoyed dealing with Howard "Son" Bland Jr., the affable offspring of a pioneer founding family, whose chief mission in life became good roads—especially one linking Taylor directly to Austin.[26] The highway chief was also close to John M. Griffith Jr., who ran City

[23] "Taylor and its' [sic] . . . Opportunities," Taylor Chamber of Commerce, 1940, "Williamson County Scrapbook" (Center for American History, University of Texas at Austin), 7–7, 19.

[24] See "Official Map of the Highway System of Texas," Jan. 1, 1934, Box 1185 (Texas Department of Transportation, Austin); "Official Road Map, South Central Texas," circa 1940 (American Automobile Association, Washington, D.C.).

[25] Richard Morehead, *Dewitt C. Greer: King of the Highway Builders* (Austin: Eakin Press, 1984), 41.

[26] Unpublished obituary, "Howard Bland Jr.," (Taylor Public Library Archives).

National Bank and was a respected civic leader. Dewitt Greer's older brother, Marcus, had married Griffith's sister; the families were intimate.[27] Both of these connections would shape Taylor for decades—in the minds of Taylor burghers and in the ribbons of asphalt that were laid around the town.

[27] Telephone interview, Ed Griffith, Aug. 17, 2001, Taylor. Ed Griffith is the son of John M. Griffith. A fourth Griffith generation—Eddie Griffith Jr.—is now in charge of the still independent City National Bank. Also see *Central Texas Business and Professional Directory* (Austin: Centex Publications, 1950), 233 (Center for American History, University of Texas at Austin).

FOURTEEN

TAYLOR'S INTERSTATE?

We are gradually learning that the way a city grows, the
direction in which it spreads, and where it sickens and dies is
a factor not so much of zoning or real estate activity or land
values but of highways and streets and roads.

JOHN B. JACKSON, *The Southern Landscape Tradition in Texas*

The World War II years gutted Williamson County's rural roads,
especially east of the Balcones Escarpment, where county roads
thickly cross-hatched the Black Waxy as if by a web woven by a
slightly drunken spider. The highways were a mess. A trip from
Granger to Temple, a hop of 28 miles, was destined to break the axle
on a truck or ambulance.[1] State Highway 95 was a ruin, turning to
sticky mud when it rained, or splitting into deep hard fissures in the
heat of summer. Taylor wanted Texas 95 fixed. Taylor also plugged
relentlessly for an "airline" route to Austin—a straight shoot, rather
than the existing circuitous path that dragged cars and trucks across a
one-lane bridge and down tiny Round Rock's Main Street.

No one knew where the interregional, or the Canada to Mexico
superhighway, as it was often called, would slice through the county,
assuming it got built. For that matter, no one could imagine what
such a highway might look like—especially its sheer magnitude—few
having traveled to Europe and laid eyes on an autobahn. In 1944 the
county's one officially designated primary highway was United States

---

[1] Hearing, Texas Highway Commission, Judge Sam V. Stone, Williamson County,
Apr. 13, 1943, Box 1329 (Texas Department of Transportation, Austin).

81, which sent traffic through the hearts of Georgetown and Round Rock. But a strong case for an alternative interregional route could be made—from Temple to Taylor to Austin—straight through the populous Black Waxy, where most of Williamson County's people lived and money got made.

Only a handful of people in Williamson County even knew that a superhighway was on the drawing board. Talk of interregionals was limited mostly to Washington government circles, Texas highway engineers, and people whose business involved trucking, automobiles, or new housing. In the forties, and even through the fifties, the average Williamson County citizen valued local farm roads above all else. There were hundreds, built and maintained by a cash-strapped county government. After all, few drove to Austin to work. A trip to "the city" was reserved for weekend adventures, traveling salesmen, and honeymooners. Traffic was scant. The most congested road in Texas lay just northeast of downtown Fort Worth, where State Highway 121 intersected State Highway 183; average daily volume there was 32,150 vehicles. Austin's highest traffic count was 6,287 vehicles a day at the Colorado River bridge.[2] At least 70 percent of Williamson County's thousand-mile road system lay east of the Balcones Escarpment, converting the small farms owned by Czech, German, and Swedish Americans into rural ethnic neighborhoods as clearly defined as any grid city's.[3] A 1940 road map of eastern Williamson County shows a tight pattern of more than two hundred squares, triangles, and rectangles of agricultural land bounded by county roads, reaching ultimate density on the Black Waxy. West of U.S. 81 fewer than a dozen county roads existed. These followed the flow of the San Gabriel's North Fork and Brushy Creek, and colonized a few farm plots between Florence and Andice.[4] It was all the county could do to keep the roads passable. Texas counties were pressed even harder after passage of the

[2] *Texas Highways Fact Book* (Austin: Texas Highway Department, 1959 (Center for American History, University of Texas at Austin).

[3] Telephone interview, Jerry Mehevec, a longtime former Williamson County commissioner who represented road-dense eastern Williamson County, Jan. 27, 2002, Taylor. Mehevec died of cancer in early 2005.

[4] Map, Williamson and Lee Counties, No. 9. July 31, 1944, Docket No. 4, State Highway Commission, Box 1327 (Texas Department of Transportation, Austin).

Road, River, and Ol' Boy Politics

Dewitt C. Greer. *Courtesy
Texas Department of Transportation, Austin.*

John M. Griffith.
*Courtesy Eddie Griffith,
Taylor.*

1941 Federal-Aid Highway Defense Act, when the Texas Legislature decided that the *counties*, not the state, should foot the bill for the proposed interregional system's considerable right of way. It was the only state in the Union to insist that counties pay a share of the coming interstates.[5]

In Taylor, while Wilson Fox led the pro–Laneport dam forces, Howard "Son" Bland led a gentle assault on the State Highway Department, not only to construct what amounted to a new State Highway 95 between Taylor and Temple, including replacing the "nightmare" bridge at Circleville, and to build a new airline highway to Austin, but also for a "bypass" loop highway around Taylor and relocation of State Highway 102 between Taylor and Lexington, which amounted to a new road.[6] Bland peppered Chief Engineer Dewitt Greer and other road engineers with cheerful but persistent letters until his death in 1952.[7] Afterwards the Taylor–Austin airline highway was a lost cause.[8] But the highway department did eventually build a four-lane divided-highway loop around Taylor at a cost of 4.3 million dollars, greatly enhancing Taylor's reputation within the realm of the highway lobby.[9] Bland's long-sought Highway 95 improvements, including a new bridge at Circleville, were completed in the late fifties.[10]

[5] "A Program for Texas Highways," 1959, Texas Research League (Center for American History, University of Texas, Austin).

[6] Texas Highway Commission meeting, Advance Summary, July 31, 1944, State Highway Commission, Box 1327 (Texas Department of Transportation, Austin).

[7] Letters between Greer and Howard "Son" Bland, 1950–1951, Box 1511, General Files, State Highway Department (Texas Department of Transportation, Austin). Bland died of a heart attack at the age of sixty on January 3, 1952, leaving Taylor bereft of its "all-time outstanding citizen," according to the *Taylor Daily Press*, Jan. 3, 1952.

[8] Taylor floated the idea of an "airline" highway to Austin again in April 1964, but was shot down at the highway department. By that time, Taylor leaders feared losing not only Interstate 35 to Georgetown, but the Laneport reservoir as well. See Wray Weddell Jr., "Travis [County] Cool to Airline Highway," *Austin American-Statesman,* Apr. 14, 1964, p. 8.

[9] The exact figure, according to the Texas Department of Transportation, was $4,323,370. See Griffin Smith, "The Highway Establishment and How It Grew," *Texas Monthly* (Apr. 1974), 84, 91–92.

[10] "Dallas Firm Bids Low on Circleville Bridge," *Taylor Daily Press*, Jan. 11 1956; "Re-Routing of Highways through Taylor Proposed," ibid., Feb. 24, 1956; "On-The-Ground

Across Williamson County, one result of all this talk about "super-highways"—with references to "loops" and "bypasses" and "express-ways"—was confusion. There were so many schemes and rumors in the mill—a new interregional highway that might run through Taylor or Georgetown or somewhere in between; a combined highway and dam bridging Georgetown if that became the interstate's route; a new toll road, whose path remained a mystery, connecting Dallas and San Antonio; and Taylor's ambitious bundle of projects, including a new highway to Austin—that the confusion became part of the story.[11]

The confusion was compounded by the county's ongoing battle with the State of Texas over right-of-ways for the new interregional highway, which under state law the county had to purchase. From 1944 until 1956, County Judge Sam V. Stone refused to buy a foot of right-of-way until he was certain of the interregional's route through Williamson County—not at all an unreasonable position. By 1954 every other county on the interregional's route between Dallas and San Antonio had completed its right-of-way purchases, but Judge Stone held out, earning the wrath of the Texas Good Roads Association and attracting negative press attention across the state.[12] Still Stone continued to resist buying right of way for Interstate 35 until 1956, after the federal government agreed to pay 90 percent of the cost of the interstate system. The Texas Highway Commission then ruled that Texas counties would not have to pay for right-of-way after all; the federal government would take care of it.[13] Throughout the standoff Stone argued that Williamson County did not know the precise route of the interstate, that even if the route were known the

---

H'way Surveys Begin Here," ibid., June 6, 1956; Griffin Smith, "The Highway Establishment and How It Grew," *Texas Monthly* (Apr. 1974), 84, 91–92; interview, Tom Bullion, May 12, 2000, Taylor.

[11] The proposed toll road supposedly would have split the distance between Georgetown and Taylor. It popped up frequently in newspaper accounts in the mid-1950s, but then vanished from public discourse.

[12] Letter, Ike Ashburn to Sam V. Stone, Sept. 29, 1954, Box 1511, General Files, Williamson County, gtx, Texas Highway Department (Texas Department of Transportation, Austin); letter, Stone to Ashburn, Sept. 30, 1954, ibid.; letter, Dewitt C. Greer to Stone, Oct. 1954, ibid.

[13] Texas Research League, "A Program for Texas Highways," Austin, 1957 (Center for American History, University of Texas, Austin).

Taylor's Interstate?

David Fontaine Forwood, founder of Taylor Bedding Manufacturing Company. *Courtesy Taylor City Library, Taylor.*

county did not have half a million dollars to buy three hundred feet of right-of-way along the interstate's path, and that "the people of Williamson County would not even think of passing" a bond to pay for it. "This is not a local road," he insisted. "It is designed as a national highway and for national defense. . . . This is not a local proposition at all. Instead of helping the towns, the new highway will carry traffic *around* them."[14]

Judge Stone's refusal to cooperate saved his county a great deal of money, but it also drew attention to the fact that the interstate's route was not fixed through Williamson County—though by the late forties highway engineers agreed it would more or less follow U.S. Highway 81. But local confusion lingered about the final route, especially in Taylor, whose leaders had long believed that Taylor would lie on the path of the ultimate superhighway between Dallas and Austin. In the midst of postwar optimism about creating a modernized Williamson County—one dripping with crystalline lakes and wealthy farmers and industrial riches and jet-shaped cars swooping to Austin—the multitude of new highway possibilities during the forties and fifties hopelessly muddled the truth about whether Taylor did or

[14] "Thornberry Supports County H'way Problem," *Taylor Times*, July 21, 1955. Judge Stone was widely respected in Williamson County and was easily reelected for many years. Part of his winning political philosophy, according to Georgetown banker Jay Sloan, was to attend every funeral in the county. Interview, Jay Sloan, Sept. 11, 2000, Georgetown.

Road, River, and Ol' Boy Politics

Taylor Bedding foreman Herbert Meiske inspects bleached cotton as it emerges from dryers. The company invented the process of fireproofing cotton mattresses and insulation, making it an industry leader for decades. *Courtesy Taylor City Library, Taylor.*

David Fontaine Forwood Jr., left, and an unidentified man watch as four thousand mattresses per day are "Taylor made" for American troops during World War II. The young Forwood was killed flying a private airplane near Taylor. *Courtesy Taylor City Library, Taylor.*

did not shove the national defense highway fifteen miles off its original track, over to the lightly settled western side of the county.[15]

Urban myths powerfully affect how people feel about places. Such a myth dogs Taylor, and occurs in several forms. When talk of the interregional highway first surfaced, so the story goes, several of Taylor's leading lights decided that a limited-access expressway would hurt the town. It would punish farmers, cutting up their fields and breaking road connections. It would threaten Taylor Bedding Manufacturing Company, maker of the famous Morning Glory mattress, which required a healthy supply of low-wage workers. It had them bottled up in a nearby neighborhood composed of African Americans and Mexican Americans. For these people, commuting to Austin for higher-paying jobs was not an easy option. But with a free chute to Austin, who knew what might happen? The city's cheap labor pool might vanish.[16] Several Rice's Crossing farmers felt that a four-lane divided highway with few access points would ruin their ability to get to their crops. A state highway was one thing; an interstate quite another. It was huge— a three-hundred-foot-wide barrier.[17] A farmer might face a twenty-mile drive to get from one end of his field to the other, unable to "cut across" an interstate highway. In the two versions of Taylor's urban myth, either Taylor Bedding founder David Fontaine Forwood or

[15] It helps to remember who Williamson County's congressmen were during this period: Lyndon Baines Johnson, a proven winner of federal largesse, and Homer Thornberry, who was no slouch in Congress himself. That sort of representation, coupled with Texas's powerful congressional delegation, gave District Ten's constituents reason for optimism.

[16] Interview, Municipal Judge James Miles, Sept. 21, 1999, Taylor. Miles's story, which he got from several Taylor old-timers, suggests a similar circumstance to that related by David Montejano in his excellent *Anglos and Mexicans in the Making of Texas, 1836–1986* (Austin: University of Texas Press, 1987), about how South Texas Anglo farmers tried to limit the mobility of their Mexican workers. For fear of their laborers deserting the fields at the wrong time, the farmers consistently opposed the "good roads" movement and discouraged car ownership by Mexicans, preferring to haul them around in trucks. See Montejano, *Anglos and Mexicans in the Making of Texas,* 200–201.

[17] Telephone interview, John Hurt, Texas Department of Transportation, Feb. 24, 2004, Austin. The footprint of an interstate can range from two hundred to six hundred feet wide, depending on the number of lanes built, access roads, and location. In the open countryside where land was inexpensive in the 1960s, the highway department tended to build on the wide side.

Rice's Crossing grower Mahon Garry called on Taylor banker John Griffith to ask his help in getting his highway czar friend Greer to kill the interstate route near Taylor. That, supposedly, is what happened.

Yet another version of the story is so delightfully outrageous that it has become one of Williamson County's cherished courthouse myths: Taylor's powers-that-were so detested the notion of an interregional highway that they threatened to "steal" Georgetown's courthouse. Georgetown's powers-that-were had to agree not to put up a fight in Austin if the interregional ran through their town, along U.S. 81—otherwise, Taylor would take the courthouse away from Georgetown by forcing a county-wide election, which Taylor could win.[18] That move supposedly shifted Interstate 35 fifteen miles west of Taylor, to U.S. Highway 81 through Round Rock and, to the north, Georgetown.

Does Taylor's urban myth reflect a semblance of reality? Dewitt Greer's records bear no hint of any chat with banker Griffith that might have changed the path of Interstate 35.[19] Of course, Greer and Griffith probably would not have written each other on such a matter. If Griffith did approach Greer, he would have picked up a phone and called him, or simply visited with him over dinner.[20] Virginia Forwood Lawrence, heir to Taylor Bedding, thinks the story that her father blocked Interstate 35 from coming through Taylor is bunk. "He would have told me," she said of her father. "He trained me to take over the business and he told me everything. I've never heard the story."[21] On the other hand, Mahon Garry's son, Mahon "Buzz"

[18] Interview, Gene Fondren, Mar. 9, 2000, Austin. Fondren, formerly Wilson Fox's law partner, said that the respected Rice's Crossing farmer Mahon Garry was the lynchpin for opposition among the farmers. Fondren, now retired, represented Williamson County in the state legislature and for decades was chief lobbyist for the Texas Automobile Dealers' Association. He remembers Williamson County District Clerk Stiles Byrom as being the source of the story about Taylor threatening to "steal" the courthouse.

[19] Dewitt C. Greer Papers (Texas Department of Transportation, Austin). A thorough search produced no scrap of evidence along these lines.

[20] Interview, Ed Griffith, Aug. 17, 2001, Taylor. "I've heard rumblings about this rumor, but not from Dad," he said in a telephone conversation. "It's certainly possible. I just don't know." Dewitt Greer's official Williamson County correspondence from that era does not reveal any letters between Griffith and himself.

[21] Interview, Virginia Forwood Lawrence, Mar. 28, 2000, Taylor.

Garry, heard the story "all my life," but "I never heard Daddy talk about it." The version his father told him had Forwood blocking the highway.[22] A longtime Taylor city attorney, Tom Bullion, doesn't believe it. "If there had been a big discussion in the city I would have known about it, and I don't remember any," he said. "But if they did turn it down, they made a terrible mistake."[23] A lifelong civil engineer who came to Austin in 1946 with the U.S. Bureau of Roads to work on the interregional project thinks the interstate route always followed U.S. Highway 81. "I don't think another route was ever considered," said Ralph Rich, "unless it was in someone's mind."[24] Then there is the nagging question, Why on earth would Taylor leaders kill the interstate while officially pressing for a direct airline highway to Austin? Or were there two opposing camps in Taylor?

Nonetheless, there is some hard evidence that Taylor did rate as a serious prospect for the interregional route, at least during the earliest thinking about the road. During a 1944 Texas Highway Commission hearing, Dewitt Greer and Judge Stone kicked the subject around as part of a larger discussion about where right-of-ways were likely to occur and whether the highway department should build a new Taylor-to-Austin airline highway if the interregional was going to connect the two cities anyway. The proposed interregional between Austin and Waco "would cross U.S. Highway No. 79 at a point near Taylor," linking Taylor directly to Austin, a highway department memo informed Texas Highway Department commissioners before a public meeting—perhaps the most conclusive document on the matter.[25] During the public session, the following conversation clearly indicated that no final decision had been made on the interstate high-

---

[22] Telephone interview, Mahon "Buzz" Garry, Apr. 5, 2000, Austin.

[23] Interview, Tom Bullion, May 12, 2000, Taylor. But Bullion did not arrive in Taylor until April 1956, and a rough version of the U.S. Highway 81 interstate route was determined in the late 1940s. In May 1949 Henry Fox wrote a piece ("Super-Four-Lane Bridges at Georgetown Without Holes in Them to Solve Dam Situation Is Reader's Suggestion") in the *Taylor Times,* reprinted in the May 22, 1949, *Williamson County Sun* (Georgetown), citing the interstate route as following U.S. 81.

[24] Interview, Ralph Rich, Mar. 31, 2000, Austin.

[25] Memo, Staff to Texas Highway Commission, July 31, 1944, Number 4, District 14, Box 1327 (Texas Department of Transportation, Austin).

way's path through Williamson County. But the map used by Chief Engineer Greer to illustrate the "Proposed Interregional" highway marked the route as passing south from Temple to Bartlett, Granger, and Taylor, crossing State Highway 79 about three miles west of Taylor at Frame Switch, and then heading straight into Austin. Here was the exchange:[26]

> MR. GREER: From Temple to Austin we have been studying the matter of a regional highway; we find we would serve about the same number of people and the same travel distance to Austin [by bringing it through Taylor].
>
> JUDGE STONE: That 79 traffic comes through there.
>
> MR. GREER: Our idea was when and if an express highway is built through it should take the place of one of these roads, otherwise we would have three highways.[27] The direct route is up through here [indicating on map].[28]
>
> JUDGE STONE: Where does that proposed interregional highway cross the I&GN [railroad]?
>
> THE CHAIRMAN: It would be way west of Hutto, it wouldn't miss Georgetown by more than five or six or seven miles.[29]
>
> JUDGE STONE: Would you expect to serve Granger? You know you have a road proposed on up to Temple. Does this interregional strike Temple under the plan now? Then it would take the place of your road from Granger in there, wouldn't it?
>
> MR. GREER: It ought to be either one way or the other.
>
> THE CHAIRMAN: In my opinion it [the interregional] will stay over on this other location.

---

[26] Transcript, Texas Highway Commission hearing, July 31, 1944, pp. 20–22, ibid.

[27] The three highways Greer was referring to were U.S. 81, State Highway 95, and the interregional highway.

[28] Greer's map shows a new Granger-Taylor-Austin highway in red marked "Proposed Interregional."

[29] Brady Gentry was chairman of the Texas Highway Commission, though he is not named in the transcript. His description of where the interregional would cross State Highway 79 does not square with Greer's map, which shows it crossing halfway between Hutto and Taylor. Had it crossed where Gentry indicated, it would have struck south Williamson County almost exactly where the I-35 "reliever" highway, State Highway 130, or 130 Toll, will pass, slightly west of Hutto.

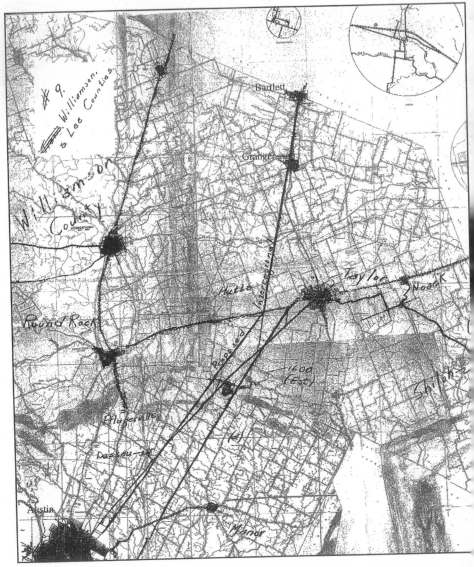

Dewitt C. Greer's 1944 sketch, on a state road map, of the "Proposed Interregional" highway through Williamson County. His route followed Highway 95
south from Temple, passed through Bartlett and Granger, and crossed Highway 79
three miles west of Taylor at Frame Switch. He also penciled in two possible "airline highway" routes from Taylor to Austin. *Courtesy Texas Department of Transportation, Austin.*

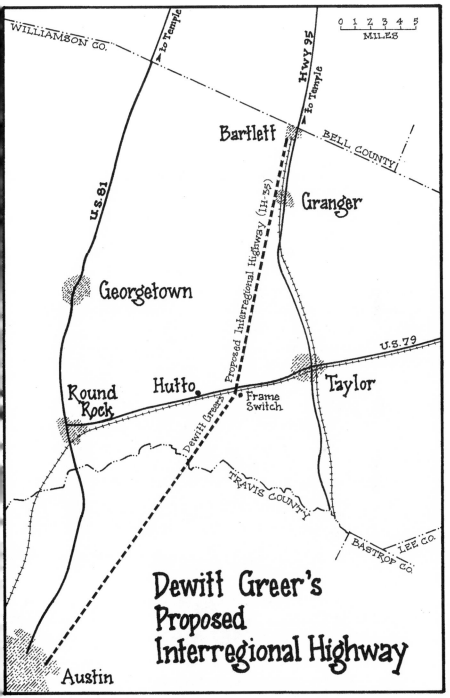

Simplified version of Dewitt C. Greer's 1944 map of "Proposed Interregional"
through Williamson County. *Drawn by Kristen Tucker Pierce.*

JUDGE STONE: On 81?

THE CHAIRMAN: In my opinion it will. You have Belton and Georgetown on this route and on the other route you only have Taylor.

MR. WILLIAMS: And Belton.

The chief engineer's map, which appears to have been drawn by Greer himself in red ink, showed no indication of a proposed interregional along U.S. Highway 81 through Williamson County.[30] What Greer's map shows is a Temple-to-Taylor-to-Austin interregional highway that could have become Interstate 35. Today a few Texas Highway Department old-timers remember Taylor's lost interstate. "Oh, yes," said the chief spokesman for the Texas Department of Transportation in Austin when the subject was raised. "Originally, Interstate 35 went through Taylor."[31]

Does it matter? In the end, the interregional highway eluded Taylor. Thus, it might be argued, it is a nonstory. But the new superhighway changed everything it touched, and much it didn't. Had it brushed by Taylor, as Greer's 1944 map indicated it would, Taylor, rather than Round Rock, would likely have become Williamson County's leading city. Today Georgetown might be worrying about decline rather than grappling with growing pains.[32]

Meanwhile a shift of seismic importance to Texas's agrarian core took place. In the late 1940s the highway department's interest in farm-to-market roads started to wane. In 1946 the Texas Legislature created a dedicated Highway Fund, which meant that the road engineers never had to go begging again.[33] For all practical purposes they

[30] Map, Williamson and Lee Counties, showing a standard 1944 road map with Dewitt Greer's proposed interregional highway and "direct" highways inked in between Austin and Taylor.

[31] Interviews, John Hurt, Mar. 22, 2000 and Jan. 24, 2002 (Texas Department of Transportation, Austin).

[32] It is possible that Hutto would have flourished more spectacularly than Taylor had I-35 passed between them, but that seems doubtful, given the political and economic sophistication of Taylor leaders during the fifties and sixties.

[33] The Highway Fund was the brainstorm of the Texas Good Roads Association. It put a percentage of the gasoline tax into a fund that could never be used for anything but building highways, "liberating," as Griffin Smith wrote in *Texas Monthly*, the High-

could choose their own projects. By 1947 energy that had once focused on rural transportation needs was refocused on more dramatic projects, such as urban expressways and the interregional system that some day, the highwaymen hoped, Washington would give them the money to build. The days when farmers were kings of the road were finished, though few realized it. Attempts in the 1947 and 1949 legislatures to protect the state's farm-to-markets failed. Afterwards the politicians pretty much left the engineers alone.[34] Chief Engineer Greer's reputation for honesty and technical excellence maintained the "liberation" of the Texas Highway Department from the usual political bothers. The superhighway had won.

If Taylor's power brokers did manage to shift the interregional highway fifteen miles west, they pulled off a coup that scores of larger and more powerful American cities failed at when they tried to alter *their* interregionals' routes.[35] It seemed to parallel how Taylor leaders successfully lobbied for a wildly unpopular flood-control reservoir, which a small number of heavy hitters thought indispensable to Taylor's future. With regard to both federal works projects—the dam and the interstate—Taylor believed it could alter them to suit its needs. Georgetown too, led by the unstoppable Owen Sherrill, overcame the Army Corps of Engineers' (and Taylor's) opposition to get its own dam built.

In short, and curiously, Williamson County's political and business elites *expected* to control and shape their futures. They almost uniformly believed in the American dream of small-town "can-doism," buttressed by political "pull" and big government assistance. There were few doubters; the doubt crept in later. At mid-century the

---

way Department to pick and choose its projects. Smith, "The Highway Establishment and How It Grew," *Texas Monthly* (Apr. 1974), 82, 84, 86.

[34] Ibid., 84, 86.

[35] At the proposal stage, the interregional's route caused consternation in nearly every city it approached. If it ran through the center of a large city, as it usually did, the neighborhoods nearly always objected; if it bypassed a small town, as it usually did, business interests fretted that the downtown would die. In Texas, Galveston, and San Antonio residents heatedly objected to the loss of old neighborhoods and cherished parks, and landowners tied the project in knots over the Interstate 35 route through Waco, which sliced right by Baylor University and extinguished the vigor of a thriving U.S. 81 strip.

power was local, in the hands of the muscled elite, and to hell with state and regional and federal plans, unless they could be turned to local advantage.

The point of Taylor's urban myth is that, whether true or not, fifty years later Taylorites accept it—or believe it easily *could* have happened. Most believe in the power of their former leaders and lament the loss of that power. They may shake their heads at the thought that their city's legendary public servants doomed a rich vein of economic development, but still they tell the tale with glee. Shortly before she died, Ruth Mantor, Taylor's most revered elder citizen, didn't know the truth of the tale, but she approved of the outcome. "Thank goodness it didn't happen," she said of Interstate 35 coming to Taylor.[36] Others sense Austin nipping at Taylor's outer flanks, hopscotching along Highway 79 from Hutto. When the megalopolis arrives, they think, Taylor will have been well served by the delay afforded by Interstate 35's wide berth; Taylor saw, and learned from, what happened to Round Rock and Georgetown.

In this vision of the future, Taylor's unrenovated historic homes and leafy neighborhoods become perfect candidates for gentrification, along the lines of Austin's Hyde Park or Georgetown's Old Town. But Taylor has always held industry in high regard and will continue to attract solid industrial concerns. The Black Waxy will convert nicely into inexpensive "starter-home" subdivisions; farmers can finally cash in. It will be a healthy mix, a small city topping out at fifty thousand people. I-35's traffic problems, noise, accidents, crime, ugliness, and pollution will be safely distant. Austin will swallow Round Rock and Georgetown, and Taylor will dominate Williamson County once again. In this vision of Williamson County's future, the power of Taylor's old ruling elite prevails.[37]

[36] Interview, Ruth Mantor, Oct. 29, 1999, Taylor. The Texas Legislature named Mantor a "Texas Treasure" about a year before her death.

[37] This paragraph is drawn from dozens of interviews with Taylor residents, county watchers, and my own observations. In the end, it is an optimistic scenario for east Williamson County in the year 2020.

FIFTEEN

ROUND ROCK'S "DESIRE LINE"

We do our engulfing in the name of progress; nothing must impede the "wheels of progress," and nothing does. . . . Like lava from Etna, this pressure overflows the countryside, filling in the meadows and marshes, felling the woodlands, forcing the brooks underground. Nothing is impregnable.

EDWARD WEEKS, *In Friendly Candor*

Sometimes a hero spontaneously rises from a community. Round Rock's hero was Louis N. Henna. In 1956 Henna owned two automobile dealerships and a great deal of real estate. He was lord and mayor of his town. Henna had moved up the hard way. During the Depression he started working for a filling station, and in 1932 he borrowed money from General Motors to start an auto dealership. Company officials were horrified when they realized they had agreed to bankroll a seventeen-year-old boy, but Henna made the deal stick.[1]

He was a brilliant promoter. In 1938 he erected what was said to be the "world's largest road sign." It read,

HENNA MOTOR CO.
Round Rock
Best place in Texas to buy a Chevrolet

---

[1] Interview, Billie Sue Henna Cariker, Sept. 14, 2000, Round Rock; interview, Jay Sloan, Sept. 11, 2000, Georgetown; interview, N. G. "Bunky" Whitlow, Sept. 14, 2000, Round Rock. Whitlow says Henna was 19 years old when he started the dealership.

Louis N. Henna. *Courtesy Louis Henna Jr., Austin.*

and faced U.S. 81 where drivers could goggle at it. To emphasize the sign's size, Henna parked a 1937 Chevy right in front of it, reducing the large, boxy automobile to the visual equivalent of a child's toy.[2] The press dubbed him a "modern Midas," but what Round Rock most admired about Henna was his generosity. When the little town needed a dentist, he built a modern dental clinic, advertised in the *Texas Dental Journal,* and recruited one. When the Lutheran orphanage shut down, Henna, a God-fearing Baptist, deeded over 112 acres of farm land he owned and gave two hundred thousand dollars to build the Texas Baptist Children's Home at the intersection of U.S. 81 and 79. When it opened in 1950, sixty children of all religious faiths moved in.[3]

[2] Noel Grisham (ed.), *Round Rock Texas U.S.A.!!!* (Round Rock, Tex.: Round Rock Kiwanis Club, Sweet Publishing Co., 1972), 38. Henna's widow, Billie Sue Henna Cariker, grinned when recalling that sign. "Mr. Henna said there was enough board feet of wood in that sign to build a five-room house!" she said. Interview, Billie Sue Cariker, Sept. 14, 2000, Round Rock.

[3] Jimmy Banks, "Louis Henna: Hobby: Orphans," *Texas Parade,* 11 (Feb. 1951), 30–31 (Texas State Archives, Liberty); "The Texas Baptist Children's Home, Round Rock," *Central Texas . . . Business and Professional Directory"* (Austin: Centex Publications, circa 1952 (Center for American History, University of Texas at Austin); interview, Billie Sue Henna Cariker, Sept. 14, 2000, Round Rock.

Road, River, and Ol' Boy Politics

Louis Henna erected "the world's largest road sign" advertising his Round Rock car dealership. *Courtesy Louis Henna Jr., Austin.*

Texas Baptist Children's Home was built largely because of the generosity of Louis Henna. At the home's dedication, Henna, sitting, second from right, observes the ceremony with his wife, Billie Sue Henna. *Courtesy Louis Henna Jr., Austin.*

"People just swarmed to Louis Henna. He was like a magnet," recalls an admirer. "Before anybody did anything, they'd say, 'Let's go down and talk to Mr. Henna.' If Mr. Henna was not for it, I wouldn't advise you to try to do it."[4]

In June 1956 the fifteen-year debate over the form and funding of the long-proposed interstate highway system was resolved. It helped that a popular President Dwight D. Eisenhower was pushing for it on military grounds, which seemed freshly relevant with Soviet aggression on the rise.[5] Congress gave the system a new moniker—the National System of Interstate and Defense Highways—and gave something to all the disputing lobbies (trucking, rural, urban, and military). But mostly, Congress provided money. The key factor that lubricated the final bill's passage was the federal government's pledge to pay 90 percent of the cost. The state or local governments would pick up the rest. It was thought that the 41,000-mile interstate system would cost twenty-five billion dollars and could be built in ten years.[6] A supporter of the interstate legislation justified it with what became mainstream thinking: "America lives on wheels, and we have to provide the highways to keep America living on wheels and keep the kind and form of life that we want."[7]

Cars and expressways were seen as tickets to freedom critical to the American democratic way; it was unthinkable to question them.[8] Even more than the rest of the country, Round Rock was obsessed with cars. In the fall of 1956 the town of fourteen hundred had three car dealerships—Henna "No Stuttering We Trade" Motor Company,

---

[4] Interview, N. G. "Bunky" Whitlow, Sept. 14, 2000, Round Rock.

[5] Most people credit Eisenhower for thinking up the interstate system, but it was an old idea, supported by Presidents Franklin Roosevelt and Harry Truman.

[6] Rose, *Interstate*, 92–94. Ultimately, the cost exceeded $100 billion. The system was mostly, though not entirely, completed by the mid-1980s, when many parts of it were becoming overwhelmed by traffic. See David Luberoff, "A Tale of Two Tables," *Governing*, 10 (May 1997), 80.

[7] Rose, *Interstate*, 69. The secretary of the treasury, George M. Humphrey, made this comment on May 2, 1955.

[8] "Texas Highways: 2000 A.D.," *Texas Highways*, 9 (May 1962), 13–15; speech, Dewitt C. Greer, "The Future of Highways and Highway Transportation," Pacific Regional Conference International Road Federation, Sydney, Australia, Mar. 3, 1961, reprinted in *Texas Highways*, 8 (Mar. 1961), 15.

"Rocks-in-His-Head" Todd Motor Company, and Henna Chevrolet.[9] A dozen filling stations, automotive repair shops, and eateries lined the Highway 81 strip, locally named Mays Street, which defined the tiny downtown business district and arched over the railroad tracks where half Baylor's basketball team had died in 1927.[10] U.S. Highway 79 was the county's major east-west link, tying Taylor to the northern edge of Round Rock and dead-ending into Highway 81. Louis Henna owned two corners of that intersection, having given the third corner away for the site of the Baptist Children's Home.[11] The new interstate's route would make or break people, whether they ran a Texaco gas station or owned hundreds of acres of ranch land. Round Rock leaders knew what they wanted: the interstate should run right through town on Mays so their traffic-dependent city would not fail.[12]

Henna didn't own the local newspaper. He didn't need to. The four-page *Leader* was dominated by car, truck, oil, and gas ads.[13] In a town and in a time like this, newspaper ads carried great weight. The autumn months of 1956 were harsh times for Williamson County. The drought had forced Congress to extend the Soil Bank Act, which paid farmers not to plant crops and gave emergency relief to ranchers. Cattlemen celebrated as Congressman Thornberry explained the act's extension as "an effort to help farmers and ranchers hold onto their foundation breeding herds."[14] Round Rock's leading bank, Farmers State, had 1.65 million dollars in deposits; the beloved Round Rock

---

[9] Henna Motor Company was owned by Billy Henna, Louis's brother, but most people in Round Rock felt that Lous was the dealership's guiding hand. Louis Henna moved Henna Chevrolet to north Austin on Interstate 35 in 1966; Todd Motor Company, owned by Jesse Todd and financed by Georgetown banker Grogan Lord, was sold and became Leigh Motors. That dealership now sits empty.

[10] Mays Street was named for L. M. Mays, a cattleman who helped establish Greenwood Masonic Institute, or Round Rock Academy. Clara Scarbrough, *Land of Good Water*, 251.

[11] Interview, Bill Todd, Sept. 14, 2000, Round Rock; interview, N. G. "Bunky" Whitlow, Sept. 20, 2000, Round Rock; interview, Billie Sue Henna Cariker, Sept. 14, 2000, Round Rock.

[12] Interview, N. G. "Bunky" Whitlow, Sept. 14, 2000, Round Rock.

[13] See *Round Rock Leader*, Aug. 30–Nov. 22, 1956.

[14] Ibid., Sept. 29, 1956.

Cheese Company (a favorite stop for U.S. 81 wayfarers) announced it was closing for good. Robertson Steak House served Thanksgiving dinner for $1.25; the Clay Pot Cafe dished out a half-pound barbecue steak or southern fried chicken, pear and cheese salad, stuffed potatoes, English peas in cream sauce or Corn O'Brien, hot rolls and butter, coffee or iced tea, and peach short cake for eighty-five cents.[15] The man who would later become Round Rock's leading sleight-of-hand artist at attracting industry, Norman G. "Bunky" Whitlow, was the high school principal. One day that autumn Whitlow introduced his students to a new "magic machine"—a tape recorder. "A whirling spool of magnetic tape is giving entertainment at lunch . . . and adding spirit to activities at Round Rock High School," the *Leader* reported on page one.[16]

Readers of the *Leader* might have missed another item that fall, since it was buried on an inside page, surrounded by ads and school news from Jollyville in south Williamson County. On November 8, 1956, the newspaper published a letter that Helen Irvin, president of the Round Rock Historical Association, had sent to Dewitt Greer at the Texas Highway Department. In it she had expressed the group's dismay at the proposed routing of the new interregional highway "through the portion of town known as 'Old Round Rock' or 'Old Town.' " On behalf of her organization, she asked that the highway department resurvey the interstate's possible routes, "looking to the rerouting of U.S. 81 outside the western corporate limits of the City of Round Rock." She and her group wanted the road engineers to move the interstate half a mile west of its planned route, swinging it around residential Round Rock. She suggested that the *Leader's* readers sign a petition to be sent to the highway commission.[17] It was all very low key and, except for two angry letters to the editor opposing the historical group's efforts, the matter never merited another mention in the town's only newspaper.

The Round Rock Historical Association's frontline troops were townswomen, but Col. W. Ross Irvin, a retired World War II cavalry-

[15] Ibid., Oct. 18, 1956.
[16] Ibid., Nov. 1, 1956.
[17] Letter, Mrs. W. R. Irvin to D. C. Greer, Oct. 21, 1956, Box 1329, Texas Highway Commission (Texas Department of Transportation, Austin).

man and civil engineer, inspired and commanded its maneuvers.[18] In 1950 he and his wife, Helen, had purchased the imposing hundred-year-old Washington Anderson House, its limestone blocks chiseled by a mason imported from Sweden. The "Wash" house stood two stories tall and overlooked Brushy Creek.[19] The Round Rock Historical Association's board listed the wives of many of the city's leading "establishment" businessmen along with the wives of several retired army officers, veterans of battles in the Philippines and the Far East, who had followed the Irvins to Round Rock in the early fifties.[20] The group, known as the "army colony," came almost en masse, investing money and energy in the buying and restoring of 1850-era limestone "wrecks" as romantic retirement homes. In addition to the Irvins, Gen. T. F. and Mildred Wessels purchased and renovated eight acres and a Spanish Colonial–style house on Brushy Creek near the Highway 81 and Highway 79 intersection; Col. William N. and Nan Todd restored the 1934 "Mexican Schoolhouse" perched on College Hill above Brushy Creek, and Col. Alex B. and Ruth MacNabb resurrected an 1855 four-room cottage, with fifteen-foot ceilings and four fireplaces, on the south bank of Brushy Creek just opposite the rock for which the town was named. Old Town had captivated them and they invited their friends to visit. Their friends fell under the spell and spread the word about this gracious little place.

A scheme evolved.[21] Williamsburg, Virginia—Texas style—was the model. The raw material existed. West of today's "Old Round Rock" a village of stone buildings built mostly in the 1850s sprawled along Brushy Creek and on hilly dirt roads. The "Old Town" (originally called Brushy Creek) had flooded in 1900, and the village had ceased

[18] Interview, Bill Todd, Sept. 14, 2000, Round Rock; interview, Harriet Irvin Rutland, Jan. 14, 2005, Austin.
[19] Interview, Bill Todd, Sept. 14, 2000, Round Rock. "Wash" Anderson fought for Texas independence from Mexico in the battle of San Jacinto.
[20] The group's president was Helen Irvin, wife of Col. W. Ross Irvin. Vice president was Mrs. T. E. Nelson, who was married to the powerful banker and landowner. Other officers and directors included Mrs. W. J. Walsh, Mrs. L. S. Landrum, Mrs. H. N. Egger, and Mrs. Robert Carlson—all wives of influential men in town. It is said that Helen Irvin approached Louis Henna's wife, Billie Sue Henna, and invited her to be the group's president. Mrs. Henna is said to have declined. The story cannot be confirmed.
[21] Interview, Bill Todd, Sept. 14, 2000, Round Rock.

to be a place where the wealthy lived or conducted business. It was not until 1956, when the "army colony" came along that Brushy Creek experienced a renaissance. The place boasted a romantic history. The Great Seal of Texas and the Texas Archives had been secreted away there for a time, when Houston threatened to snatch the capital away from Austin. Two celebrity desperadoes had sauntered down "Old Town's" streets: bank robber Sam Bass ("Robin Hood on a Fast Horse") died at the Round Rock Hotel after a gun battle; and John Wesley Hardin, son of a circuit-riding Methodist minister, attended Greenwood Masonic Institute in Old Round Rock before choosing a career path that included murdering twenty-seven men.[22] As for a supply of tourists, Austin was nearby. And the new interstate—*if* it safely circumvented Old Town—would provide a bountiful supply.

"In the years to come," Helen Irvin wrote the Highway Department's chief engineer,

> Round Rock can be made a mecca for tourists due to the happy circumstance that most of the buildings in Old Town can be restored as homesites of an antebellum era. Texas has few areas so rich in tangible historic assets as Round Rock. Businessmen in Williamsburg, Virginia and Natchez, Mississippi do not have to be told today of the monetary reward that such areas bring. Our western neighbor, Taos, New Mexico, has been particularly successful in extracting the tourist dollar. ... Round Rock's charm, like that of Williamsburg, Virginia, lies in the character of the village as a whole. With a large expressway booming through its midst, the character of the village will change, in fact, die.[23]

She appealed, artfully, to Greer's masculine engineering mind. She wrote of tourist dollars and incentives and safety features and structural engineering, and then she massaged the chief engineer's ego. "Please remember that we are women and not engineers!" she flut-

---

[22] Letter, Mrs. W. R. Irvin to D. C. Greer, Oct. 21, 1956, Box 1329, Texas Highway Commission (Texas Department of Transportation, Austin). Also see Clara Scarbrough, *Land of Good Water,* 289, 298; Grisham (ed.), *Round Rock Texas U.S.A.!!!,* 4.

[23] Letter, Mrs. W. R. Irvin to D. C. Greer, Oct. 21, 1956, Box 1329, Texas Highway Commission (Texas Department of Transportation, Austin).

Col. Ross and Helen Irvin in the late 1950s, relaxing in their historic home on Brushy Creek in Round Rock. *Courtesy Harriet Rutland, Austin.*

tered.[24] What she did not mention but clearly knew was that early in 1956, *before* the Interstate Act was passed, Texas Highway Department officials had met with Mayor Henna and a small group of Round Rock businessmen to determine the new interstate's route. The Round Rock men wanted the interstate to slice straight through downtown, but the engineers advised against it, since its approximately three-hundred-foot swath would gut the business district. A consensus, including both the engineers and Round Rock leaders, had formed that the best solution would be to keep the interstate as close to downtown as possible without destroying the business district. That would drive it through Round Rock's Old Town, several

[24] Ibid.

Col. W. Neely
and Nan Todd.
They restored
Round Rock's
"Mexican
Schoolhouse" in
the mid-1950s.
*Courtesy Col.
William ("Bill")
Todd, Round
Rock.*

blocks west of Highway 81. This route collided with the army colony's restored homes and the Round Rock Historical Association's plans for a Texas Williamsburg. But it kept the interstate from completely bypassing the town, as it would at Georgetown. The highway department's "preferred route" stuck to Highway 81 until just north of its intersection with Highway 79, where it veered a few blocks west, avoiding downtown, but bisecting Old Town.

The highway department's choice was hardly unique; this sort of giant concrete pathway regularly carved through American cities throughout the interstate system's quarter-century buildout. At mid-century, with little experience to guide them, highway engineers and city planners believed it best to thread interstates right through the hearts of big cities and "bypass" smaller towns. The Texas Highway Department had established what it called a "traffic desire line" to help make that decision. The rule of thumb was that if a city had 150,000 residents, the interstate would split the downtown district.

The native fieldstone "Mexican Schoolhouse" was built in 1934 by Texas Relief Commission workers. It had one enormous room. From the 1920s through the 1940s, most Texas Hispanics attended segregated schools. *Courtesy Col. William ("Bill") Todd, Round Rock.*

After restoration in 1955, the Todds christened their College Hill home "Double N Acres." It had four fireplaces, a new garage addition with living quarters, and fieldstone patios which became focal points for entertaining. *Jean Todd Photo, Courtesy Col. William ("Bill") Todd, Round Rock.*

Smaller cities were to be "skirted."[25] If a smaller city didn't object, however, or was powerless, such as Williamson County's Jarrell, the engineers might split even tiny villages. The process was more fluid than advertised. In Texas the highway department generally did what it wanted to do.

[25] The theory posited that the interstates would pull suburban drivers into the cities'

This 1853 home, photographed in 2005, nestles in an oak grove south of the Brushy Creek "round rock" for which the city is named. The home was first restored in the 1950s by Col. and Mrs. Alexander B. McNabb. In 1972 Edward Don Quick bought it while his wife Jeanie was giving birth to their first child. She re-restored it; today it is the only "army colony" home that remains a private residence. *Mark Ashley Photo, Courtesy* Williamson County Sun, *Georgetown.*

Helen Irvin requested a special hearing before the Texas Highway Commission. She and Colonel Irvin lined up some impressive support. They got a thumping endorsement for their cause in the *San Antonio Express,* triggering letters to Greer from "indignant" and "shocked" readers from San Antonio and elsewhere.[26] The Irvins also

---

central business districts if the highways ran close to the cities' cores, and that interstates would rehabilitate the poor and/or minority neighborhoods through which they often ran. Both theories proved witless. Numerous books have analyzed the results of these theories, focusing mostly on urban areas, among them Albert Benjamin Kelley, *The Pavers and the Paved* (New York: D. W. Brown, 1971); Lewis Mumford, *The Highway and the City (Essays)* (New York: Harcourt, Brace & World, 1963); Leavitt, *Superhighway—Superhoax* (New York: Ballantine Books, 1970); Tom Lewis, *Divided Highways: Building the Interstate Highways, Transforming American Life* (New York: Viking, 1997); and John Robinson, *Highways and Our Environment* (New York: McGraw-Hill, 1971). Also see "The Public Eye," *Texas Highways,* 7 (Aug. 20, 1960), 20.

[26] Lucile Stewart Krisch, "Work Threatened," *San Antonio Express,* undated but

got strong backing from the Austin Heritage Society and briefed a committee of the Texas Historical Survey, which apparently never took a stand.[27] Helen lobbied Williamson County's powerful Judge Sam Stone, who wrote a letter on her behalf to the Highway Commission's chairman hinting there might be legal problems with the interstate's right-of-way process in Round Rock.

"Before the Federal-Aid for the procurement of right-of-way was approved," Stone wrote Greer, "your Interstate Highway Engineer, Mr. Travis Long had laid out and prepared deeds for this route through Round Rock, changing the present route of Highway No. 81 from the business district to a residential section of the village, which passes through a recently improved area, intersecting two small parks and passing very near several old land marks, some of which have been restored and passing through some which likely will be restored if said right of way does not interfere."[28] Although he never acted on it, the judge suggested that the highway department was out of order, possibly acting illegally, in establishing the Interstate 35 route through Round Rock.

Judge Stone added that he felt certain the Highway Commission would want to hear the Round Rock "committee of ladies . . . who are so interested in this location." A hearing was set for November 21. Round Rock's City Council sent a strong resolution favoring the proposed route, and the Chamber of Commerce sent a surprisingly weak resolution that might have meant anything.[29] Ed Bluestein, chief engineer for District 14, who was in charge of building the

---

obviously written in November 1956, with attached letters to Greer and his replies.

[27] Letter to the Editor, *Round Rock Leader,* Nov. 8, 1956; Krisch, "Work Threatened," *San Antonio Express*, undated.

[28] Letter, Judge Sam Stone to E. H. Thornton Jr., Oct. 31, 1956, Box 1329, Texas Highway Department (Texas Department of Transportation, Austin).

[29] Letter, Louis N. Henna to Dewitt C. Greer, Nov. 14, 1956, Box 1329, Highway Commission File, Williamson County folder 3 (Texas Department of Transportation, Austin); Minutes, City of Round Rock, Nov. 13, 1956, ibid.; Letter, Round Rock Chamber of Commerce to Greer, Nov. 14, 1956, ibid. The Chamber's resolution dodged the central issue. It read, in its entirety, "A motion was made and seconded that we go on record that the Chamber of Commerce cooperate with the City of Round Rock and with the Texas Highway Department in securing a route through Round Rock for the new highway." Ibid.

Round Rock's "Desire Line"

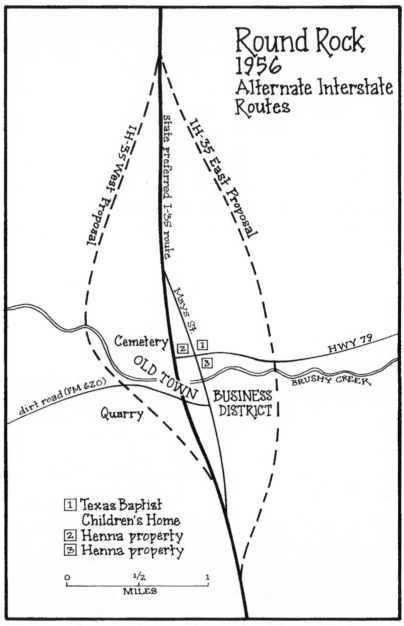

Round Rock
1956
Alternate Interstate
Routes

IH-35 West Proposal

state preferred I-35 route

IH-35 East Proposal

Mays St.

Cemetery [Z] [1]

[3]

OLD TOWN

HWY 79

BRUSHY CREEK

dirt road (FM 620)

Quarry

BUSINESS
DISTRICT

[1] Texas Baptist
    Children's Home
[2] Henna property
[3] Henna property

0   ½   1
           MILES

In 1956 Round Rock was split between those wanting to develop "Old Town" as
an historic residential and tourist attraction, and those wanting to keep the interstate highway close to downtown. A preservation group tried to shift Interstate
Highway 35 half a mile west of its present route. *Drawn by Kristen Tucker Pierce.*

interstate through Williamson County, shot off a letter to Greer defending "what we think to be a splendid location" through Round Rock. At first, Bluestein explained, the route was supposed to go *east* around Round Rock, but "rapid and costly development in that area precluded a route as originally planned." Now, bowing to a "minority group" to loop the highway west of city limits "would involve us in innumerable complications." He noted that engineer Long had met with the Round Rock City Council "and a group of business men . . . several months ago at a time when he presented them with a right-of-way map and deeds officially requesting them to acquire the right-of-way. That was the policy at the time. The City Officials and business men almost unanimously approved our location. . . . This group is certainly not in concurrence with the group opposing our proposed location, this latter group being largely retired military personnel presently living in the Round Rock area."[30]

In a postscript, Bluestein added the clincher. "Incidentally," he wrote Greer, "our proposed route . . . through the Round Rock area has already received approval of our Road Design Division and the United States Bureau of Public Roads." The Bureau of Public Roads was the ultimate authority. In other words, the deal was done. It could be undone, but that would cost time and money, assets that Chief Engineer Greer hated wasting.

———

For all practical purposes, the Texas Highway Commission's hearing on I-35's Round Rock footprint was a put-up job. As always, a highway department memorandum prepared commission members for what to expect. It squeezed the Round Rock Historical Association's rationale for protesting into four short lines. Then the typed memo continued for another full page, summarizing Bluestein's arguments, reprinting the Round Rock City Council's and Chamber of Commerce's resolutions, and detailing the advantages of the depart-

---

[30] Letter, Ed Bluestein to Dewitt C. Greer, Nov. 14, 1956, ibid. The claim that the route of I-35 had been changed to avoid "rapid and costly development" in east Round Rock was decidedly strange. If there *was* development at that time, it escaped the notice of even the banking community, including N. G. "Bunky" Whitlow of Farmers State Bank, who is legendary for his knowledge of the community.

ment's proposed route. It would, the memo stated, utilize investments previously made in U.S. Highway 81, adequately combine service to local businesses and through traffic, and connect easily to U.S. Highway 79. The memo concluded with a recommendation "that the [Round Rock] delegation be carefully heard and the matter be reviewed again in executive session, leading to a decision to reaffirm the route as established by the District Engineer and his staff."[31]

Considering the hurdles she faced, Helen Irvin, flanked by Nan Todd and Milly Wessels, scored some impressive points. Chairman E. H. Thornton Jr. of Galveston repeatedly reminded Irvin that this day's hearing would not count; whatever emerged from the proceedings, he said, another hearing would be held—one inviting all Round Rock citizens to participate. That would be wonderful, Irvin replied. Her group welcomed a public hearing during which all possible routes—east skirting Round Rock, west skirting Round Rock, and through the middle of Round Rock—could be discussed by everyone.

She presented a petition signed by sixty-six people, a surprising number of them prominent Round Rock businessmen (or their wives), who in Irvin's words "dislike the present proposed routing of U.S. 81 and the way it has been forced upon us without discussion. ...We do not feel fair consideration has been given to all sides. We do know that your engineers and our Mayor, Louis Henna, and the City Council agreed on the present route. However, interested citizens did not have the opportunity to voice their opinions."[32]

She reminded commissioners of the highway department's written policy of bypassing small towns. Georgetown was larger than Round Rock; it would be bypassed. "We most earnestly request a bypass," she said. The interstate's current routing, she said, would "render imprac-

[31] Memorandum to Highway Department Commission, Nov. 21, 1956, Williamson County, District 14, ibid.

[32] Speech, Helen Irvin, Nov. 21, 1956, ibid. The petition was something of an afterthought. Irvin told commissioners she had been "surprised" at the response. If her group had gone from house to house, she thought there would have been "hundreds" of signatures. As it was, the petition was signed by many of the most powerful people in Round Rock. The names included Mr. and Mrs. Will Wilson, Mrs. Tom E. Nelson, Petrenella McConico, Dick Mayfield, Howard and Lee Nora Bible, Ann Abell Behrens, Mr. and Mrs. Charles W. Prewitt, Mr. and Mrs. L. S. Landrum, and Mr. and Mrs. H. N. Egger—most of them old Round Rock families.

ticable the restoration of many of the historical buildings and homes of Round Rock. Most of these were erected in the 1850's, and Round Rock is one of the few places in Texas that has as many of these buildings. The Heritage Societies of Texas, the Daughters of the Republic of Texas, the State Historical Survey, and many other organizations are trying to preserve these historical sites . . . and to make them known to scholars and tourists."

Further, the proposed expressway, she testified, would require "either underpasses or overpasses in dealing with two railroads, and at least two roads within the City of Round Rock, and a four-lane bridge over Brushy Creek." It would destroy two churches, the American Legion Building and "ruin" the Legion Grounds and Old Settlers Park—all easily avoided. Irvin listed ten "advantages" of a western interstate bypass around Round Rock: lower cost of land, no high bluff to bridge on north entrance over Brushy Creek, no town lots to purchase, no expensive buildings to condemn, no damage to historical sites, level terrain, greater industrial expansion allowed, no town streets to consider as crossways, one railroad crossing eliminated, and the general expressway policy of bypassing small towns honored.

She was correct on every point. The chief problems with the western bypass option were that U.S. Highway 79 would need to be extended roughly half a mile and that Round Rock businessmen were generally terrified of any plan to move the interstate away from the town's center, for fear they would lose business. Certainly, the potential value of Henna's Highway 79–Highway 81 intersection property, undeveloped save for the Children's Home, would be greatly diminished.

Though the commissioners had apparently come prepared to rubber-stamp Greer's plan, Irvin's presentation sparked what seems to have been genuine sympathy on the part of at least some members. Herbert C. Petry Jr. of Carrizo Springs seemed almost smitten with Irvin, intervening several times to soften Chairman Thornton's harsh language and restating Irvin's points more forcefully to his fellow commissioners. The fact that the delegation was composed of "ladies" may have cut both ways. Thornton treated the protesting women as if they were thick-witted children; Petry galloped to the rescue like a gallant defender. Irvin herself implied that the status of the women in her group did not quite measure up to that of their husbands, explaining to Chairman Thornton, "it was hard for us to come today; when we know that the City Council and the Mayor have approved the proposed route it makes it embarrassing for our husbands to protest because they hate to get involved, but we did not feel we had a fair voice in the matter before it was decided."[33]

The commission retired to executive session. If another public hearing was held later, no record exists of it. Dewitt Greer's file on the Round Rock interstate route ends with Helen Irvin's futile plea.

————

Twelve years later Interstate 35 sliced through Old Town, cutting a deep gash through College Hill, which rose from Brushy Creek's north bank.[34] The army colony's vision of a Texas Williamsburg was long dead. (Other small towns did successfully pursue that vision: Salado and Fredericksburg, for example.) "The highway ended up running right smack through my Daddy's front yard," lamented Bill Todd, son of Col. William N. and Nan Todd. The "Mexican schoolhouse" his parents restored was not particularly old, but it was pleasing, a long limestone structure lined with multipaned floor-to-ceiling windows. The Todds constructed an L-shaped addition that opened onto an enormous shell-encrusted limestone patio.[35] "It was the most beautiful place," Todd said, his eyes misting. "Sitting there, you could see these beautiful sunsets. You could see from our house to Fort Stockton, except for the barbed wire. There was nothing in between."[36]

Most evenings, cocktails were served on the patio. "Travis Long, the district engineer who was in charge of building the interstate, came by our house regularly at five o'clock," Todd said. "He knew he could get a drink there. My father and he became pretty good friends. All our friends had wicker baskets they packed with food, and every evening, they would congregate on somebody else's patio. It was a very pleasant little town in those days." Bill Todd's young family lived

[33] Transcript, State Highway Commission, Nov. 21, 1956, ibid.

[34] I-35 was not completed through Round Rock until 1968, the last stretch completed in Williamson County.

[35] Grisham (ed.), *Round Rock Texas U.S.A.!!!*, 122. The original "Mexican School" was built in 1934 by Texas Relief Commission workers.

[36] Interview, Bill Todd, Sept. 14, 2000, Round Rock.

at the Schoolhouse for a time. "We just walked to town," he remembers. "It was about a mile. Our kids used to traipse down Highway 620 with two little dogs chasing them." It was a sleepy, two-lane country road then, producing more dust than traffic.

After it became clear that they would soon have an interstate highway practically in their back yard, the Wessels sold their home to the First United Methodist Church, which built several structures that wrapped around the house, concealing it from highway drivers. A cluster of magnificent oak trees, including Council Oak, said to be the largest in Williamson County, shelters the old patio. It is easy to see why the house attracted the Wessels, who had lived in the Far East: it resembles the colonial bungalows of South Australia's Barossa Valley.

The Irvins lived at the "Wash" (rechristened "El Milagro"), with its bass-stocked Brushy Creek lake and strolling peacocks, until they died. Their daughter, Harriet Rutland, inherited it. Later she bought and restored several Old Town structures.[37] The "Wash" is now the keystone building in an office complex called Heritage Center on U.S. Highway 79.

In the early 1970s the MacNabbs sold their 1855 Brushy Creek cottage to Edward Don Quick, a commercial real estate broker who grew up hunting arrowheads along Brushy Creek. Quick purchased it for his wife, Jeanie, while she was in a hospital giving birth to a son. The Quicks live there still, lovingly tending the only one of the army colony's restored homes that survives as a residence. It is a magnificent place, its back lawn rolling down to Brushy Creek.

---

[37] Paul L. Daley, "Round Rock's St. Charles Inn Brings Back Stagecoach Days," *Austin American-Statesman,* July 26, 1970. The Irvin's daughter, Harriet Rutland, went into an unusual form of real estate, buying dilapidated buildings, restoring and reselling them. Among her projects were Old Town's St. Charles Inn, Rose House, and the old post office. She was living at home when her parents started protesting the I-35 route, but she was also working for the public information director of the Texas Highway Department. "I was Betty Harriet Philips then, and I didn't mention who my parents were. My boss had me typing replies to my own father," she said. Interview, Harriet Irvin Rutland, Jan. 14, 2005, Austin. "It was all politics," she said of the engineering decision on where to place I-35 through Round Rock. "Judge Sam Stone and Louis Henna wanted it where it was." Ibid. Clearly that was true with regard to Louis Henna, but there is no hard evidence of Judge Stone's favoring the highway department's "preferred" route. If anything, his letters indicate a tilt toward the Irvins' position.

After Bill Todd's mother died, he sold her home and land at the intersection of I-35 and State Highway 620 to the Comfort Inn motel chain. "I am an evil man," he said with a grimace of pain. "I let them demolish that thing. You couldn't move it. It was a beautiful house. It was just in the wrong place."[38] In the motel's back parking lot one catches a whiff of what Colonel Todd and I-35 builder Travis Long must have admired as they faced west, sipping drinks late in the day on College Hill: dry wall stone terraces sloping down toward the creek, an ancient and dying oak tree, a pungent bed of wild iris, a fragment of flagstone patio. On the front side of Round Rock's Comfort Inn, one stands and watches the daily product of Interstate 35—154,000 vehicles speeding by.

[38] Interview, Bill Todd, Sept. 14, 2000, Round Rock.

THE GOLDEN EGG

The new superhighways created tremendous opportunities for
land development in the remote hinterlands of big cities. An
unthinkably long commute on old country roads now seemed
reasonable on the freeway.

JAMES HOWARD KUNSTLER, *Geography of Nowhere*

Ａll across America the 1956 creation of the National System of
Interstate and Defense Highways opened a Pandora's box of
possibilities for cities, property owners, and entrepreneurs. As
details of the new highway system percolated from the U.S. Bureau of
Public Roads to state highway departments to city officials' offices
and chambers of commerce, the gas station, motel, and cafe owners
whose enterprises depended on passing traffic started writing letters
to anyone they thought was in the know. Most expressed concern—
deathly concern—about the proposed interstates' precise routes.
Could the mom-and-pop operations continue harvesting greenbacks
from passing travelers with the new interstate system? If not, they
wanted it changed.[1]

In Texas, highway engineers had pondered interstate routes since
the mid-1940s. Originally the U.S. 81 "superhighway" route through

---

[1] Senator Lyndon B. Johnson received heartrending letters from his Texas constituents, mostly small-business owners, who feared the new interstate routes would ruin
businesses that had taken them lifetimes to build. One involved a West Texas motel
owner who had scrimped and saved to purchase and upgrade a motel that the proposed
interstate would miss by a mile. See 1958–1960 Senate Case and Project Files, Boxes 603,
677, 774 (LBJ Library, Austin).

Austin—now Interstate 35—would have cut through the heart of the University of Texas campus, but that idea was quickly abandoned.[2] In Williamson County, as we have seen, road engineers first thought the interstate might go just west of Taylor. But by late 1956 the "U.S. 81 Superhighway" route—Interregional Highway 35—was firm. In Williamson County, it would follow U.S. Highway 81 except at Georgetown, where it looped about a mile west of the downtown courthouse, and at Round Rock, where it paralleled Highway 81 six blocks west of downtown, cutting through Round Rock's historic Old Town.[3]

U.S. Highway 81 was Georgetown's and Round Rock's lifeline to outside money. It wasn't much, but it was nearly all either town had.[4] Unlike Taylor, Georgetown and Round Rock had no fabulous Black Waxy Prairie sustaining thousands of farm families, no big industry, no interstate rail junction, no nearby oil fields. Good farm land existed east of Georgetown and Round Rock, but it did not compare to the Black Waxy, which kept a dense agricultural population in business. West of the Balcones Escarpment towns, several score ranching families wrestled with the prospect of default during the terrible drought of the fifties. Thus, in Georgetown and Round Rock, the money brought to town by highway traffic was important. Filling stations studded Highway 81 through both cities; each had eateries and drive-ins geared toward passing traffic: Round Rock touted Robertson's Steak House and the Clay Pot, while Georgetown had the L&M Cafe and the King Cole. A Georgetown man recalls "crickets piling up so high in front of the L&M and King Cole cafes they had to get dump trucks to haul them away. The cafes were open all night and their lights drew those crickets, as well as travelers. Highway 81 was the backbone of the United States."[5]

[2] Kelly Daniel, "Vintage Plan Shows What I-35 Might Have Been," *Austin American-Statesman,* Nov. 28, 1999, p. B-1.

[3] Right-of-way costs may have been one factor in keeping "loops" to a minimum in Williamson County. Once Texas counties were absolved from acquiring interstate right-of-way, it became more economical for state engineers to mimic established routes of U.S. highways, where right-of-way already existed, than to march across virgin territory.

[4] According to the Texas Highway Department, Georgetown had 4,951 residents in 1950; Round Rock had 1,438.

[5] Interview, Paul Hindelang, July 2, 1998, Georgetown.

Many feared that the loss of Highway 81 traffic to the new interstate would kill their towns. In Round Rock, as we have seen, the highway department scotched a local proposal to ram I-35 right through the heart of the business district along Mays Street. In Georgetown a leading industrialist said, "when the local folks first heard that the interstate was going to loop around Georgetown, they became frantic. . . . There was a strong contingent that thought it should go on both sides of the Courthouse—Main Street for northbound traffic and Austin Avenue for cars heading south. Of course, nobody in Georgetown had *seen* an interstate. It was a foreign concept to everybody. Their real concern was that the world was going to pass them by."[6]

Williamson County ranchers and farmers grimly discussed the federal farm policy, which paid cattlemen to slaughter herds and farmers to refrain from planting crops.[7] They told macabre jokes about the coming interstate, which was viewed as a killer of small towns. When Georgetown's Sam Brady told friends he had bought property on Highway 81 for a new insurance office north of the Courthouse Square, they advised against the plan.[8] "You don't want to go moving down there," hooted "Jelly" Caskey. "When the interstate comes in, we're going to have to plow up Austin Avenue and put it in the Soil Bank!"[9]

On October 1, 1959, the Bureau of Public Roads declared the Interstate 35 route through Texas official.[10] One essential element of the plan, however, remained somewhat fluid: its interchanges and ramp exits. If I-35 was replacing the "backbone of Texas," the new superhighway's interchanges—where drivers would enter, exit, and cross the expressway—were like a backbone's vertebrae. If the vertebrae were packed too tightly together, the spine might fuse, clogging traffic and defeating the chief purpose of the interstate—to move cars

---

[6] Interview, Bill Snead, July 2, 1998, Georgetown.

[7] In Central Texas, farmers and ranchers not only had to contend with the worst drought in memory but also foot-and-mouth disease, which had infected Texas cattle. Many herds had to be destroyed.

[8] Through Georgetown, U.S. Highway 81 is named Austin Avenue.

[9] Interview, Sam Brady, July 7, 1998, Georgetown.

[10] Interview, Morton Broad, Mar. 23, 2000, Austin.

quickly and safely. But if the vertebrae lay too far apart, the backbone would become a barrier, forming an impermeable wall through rural areas, cutting farms, ranches, and towns into unusable bits and blocking economic development. The negative impact of this interstate "wall" was somewhat mitigated by Dewitt Greer's devotion to access, or frontage, roads parallelling Texas's interstates—a quirk shared by few states in the Union—but still, in a largely rural state, the interstate system was seen as an out-scaled intrusion on agricultural areas.[11]

If one ponders a 1959 Williamson County map, it becomes obvious that highway planners were planning interchanges and ramps to serve existing communities, but without much thought about the growth that would be triggered by the interstate and its exchanges. The number of interchanges was strictly limited to allow for speedy traffic and to discourage accidents, and because of their scarcity, an interchange meant potential money in the bank. In 1959 the approved official interchanges were marked on the designated I-35 route.[12] But that didn't stop people from trying to alter them. Sometimes, armed with sufficient political pull and a compelling argument, they succeeded.

One of the first petitioners to write Senator Lyndon Johnson about interstate exchanges was Dion VanBibber, the fabled Salado restaurateur who had turned an historic log cabin into a dining legend. His place, the Old Stage Coach Inn, had been written up in *Life* and *Gourmet* magazines.[13] In the fifties it was the only culinary bright

---

[11] Don Fairchild, "IH 35: Man's Effort to Join Nation by a Highway," *Austin American-Statesman,* Jan. 10, 1971; Kelly Daniel, "Frontage Roads to Become Texas Highway Relics," *Austin Statesman,* Aug. 2, 2001, p. A-1.

[12] The originally planned I-35 interchanges in Williamson County included (running north to south): one in Jarrell, one at Corn Hill, others leading to Theon, Walburg, Florence, and Andice (which became Williams Drive in Georgetown), one to Georgetown Airport (Lakeway Drive), one in Georgetown at Highway 29, one south of Georgetown at what was then Owen Sherrill's land, one at Texas Crushed Stone, one north of Round Rock at Sam Bass Road, one at Highway 79 in Round Rock, another at FM 620 in Round Rock, one south of Round Rock close to Gattis School Road, and one at FM 1325. An underpass was built at McNeil Road with no full access to I-35. See "Land Purchase Will Begin for Interregional Hwy 35," *Williamson County Sun* (Georgetown), Sept. 14, 1961. Conspicuously missing in Georgetown was an exit at Leander Road.

[13] Letter, Mrs. Claud C. Westerfeld to Senator Lyndon Johnson, May 3, 1958, Senate

Road, River, and Ol' Boy Politics

spot (above the diner level) between Dallas and Green Pastures in Austin. The Inn sat a few feet from U.S. Highway 81 in the middle of the village of Salado. Except for the highway, Salado was a relic from the nineteenth century, when it had been founded as a stagecoach stop where teams of horses could be changed and travelers could fortify themselves with food. Williamson County residents loved driving up to Stage Coach Inn, a few miles north of the county line, to dine on shrimp cocktail, tomato aspic, prime ribs of beef, corn fritters, and "Strawberry Kisses" under the great sprawling oaks of VanBibber's romantic patio.

Like most highway restaurant owners in the United States, Van-Bibber was worried about the interstate. "The superhighway is truly a wonderful institution," he wrote. "Also, on occasion, it can be very ruthless."[14] At Salado, it had been determined, there would be no exit from which motorists could easily reach the Old Stage Coach Inn from Interstate 35. VanBibber had corresponded with the engineer in charge of I-35 through Bell County before concluding that his case was "hopeless."[15] Then he appealed to Senator Johnson. Originally the Texas Highway Department had recommended an exit ramp at Salado that would have served Old Stagecoach Inn nicely, but the U.S. Bureau of Public Roads had deleted it from the national plans. It had been a policy decision. The Bureau took the position that no access ramps or connections should "serve any privately owned installation" because "this would encourage other businesses similarly situated to request ramps and connections."[16] Time was now of the essence; I-35 was about to open through Salado.

VanBibber laid out the situation for Senator Johnson, adding a personal note:

Case and Project 1958, Container 603 (LBJ Library, Austin); Dion VanBibber to Johnson, Mar. 26, 1958, ibid., Container 603 (LBJ Library, Austin). Westerfield alluded to the Bell County restaurant's national reputation. I remember being impressed by the framed articles hanging in the entryway when my family ate there in the 1960s.

[14] Letter, Dion VanBibber to highway engineer Thomas C. Collier, Mar. 12, 1958, ibid.

[15] Ibid.

[16] Letter, Thomas C. Collier to Dion VanBibber, Mar. 5, 1958, ibid.; F. C. Turner to Johnson, May 21, 1958, ibid.

The Golden Egg

You probably do not remember me but I am sure you remember my establishment. . . . For, I recall that you had lunch there with friends not long after your illness. I do hope that you remember it favorably for it is in serious trouble. . . . Be sure that this is not a whim nor the wail of one who sees a hundred thousand dollar investment going down the drain but also the urgent cry for help from an old historic place that hopes to carry on for many years to come.[17]

A few weeks later Interstate 35 opened in Bell County, and the restaurateur reported to Johnson:

Immediately . . . north-bound patrons of the Inn were reduced to a mere trickle of bewildered, unhappy and often indignant guests who were obliged to pass the entrance and detour back. This condition prevailed until some more forthright travellers decided to leave the Expressway and cross to the access road, via the ditch, at the point where I have requested that a ramp be constructed. Business returned to almost normal. However, on this rainy week end when such a crossing was impossible, business was again reduced to fifty per cent of normal. It would certainly seem that this is *prima facie* evidence that such a ramp is desirable and necessary.[18]

The day he received this epistle, Johnson whipped off a message to the U.S. Bureau of Public Roads. We do not know what he said, because the letter is not in his archives.[19] But the Bureau replied to it with terrific news. Suddenly the Texas Highway Department had targeted "development of the interstate highway" in the Salado area with freshly obtained Federal-Aid Highway Act funds.[20] Salado

---

[17] Letter, Dion VanBibber to Johnson, Mar. 26, 1958, ibid.

[18] Letter, Dion VanBibber to Johnson, Apr. 14, 1958, ibid.

[19] The letter must have been written because it was referred to twice in subsequent correspondence, including a letter from F. C. Turner, deputy commissioner of the Bureau of Public Roads, to Johnson dated April 24, 1958, in which Turner refers to Johnson's April 15 letter. Ibid.

[20] Letter, F. C. Turner to Johnson, ibid.

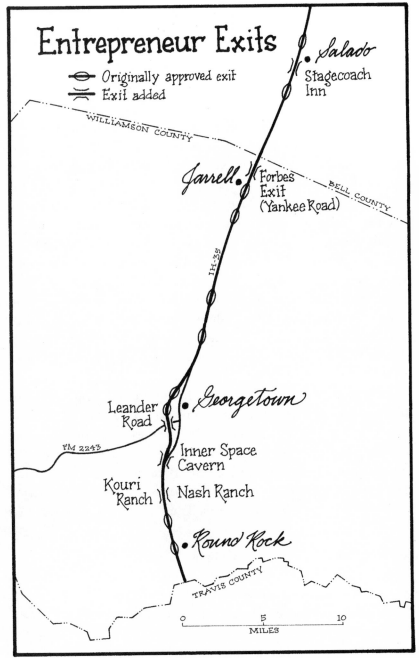

Entrepreneur Exits

- ⊖ Originally approved exit
- ≋ Exit added

Salado

Stagecoach Inn

WILLIAMSON COUNTY

Jarrell · Forbes Exit (Yankee Road)

BELL COUNTY

IH-35

Leander Road

Georgetown

FM 2243

Inner Space Cavern

Kouri Ranch

Nash Ranch

Round Rock

TRAVIS COUNTY

0   5   10
MILES

Williamson County interests upset U.S. Bureau of Roads policy limiting access to new interstate highways. Within the county, four private efforts to alter or add interchanges—greatly enhancing the value of property—succeeded. *Drawn by Kristen Tucker Pierce.*

would get an "additional point of access." The Old Stage Coach Inn was saved.[21]

Eight miles south of Old Stage Coach Inn, in northern Williamson County, Georgetown's Charles A. Forbes, Humble Oil Company's Williamson County distributor, borrowed money, purchased land and built a truck stop where he reckoned an exit would dump traffic off Interstate 35 north of Jarrell. A friend at the Texas Highway Department had told him an exit would probably be built there, but warned that the Bureau of Public Roads in Washington had the last word. Forbes's widow, Mary Forbes, recalled her husband's hopes. "When Charles bought that property, he was so excited," she remembered. "He said, 'If this works out the way I think it's going to work out, this will be both boys' college educations.' But it didn't work out."[22]

Forbes cut an elegant figure. Quick on his feet, comfortable with national politicians as well as with the farmers and ranchers he traded with, he had recently served as Georgetown's mayor. But in 1964, his dreams were turning sour. Five times, the Texas Highway Department had asked the U.S. Bureau of Public Roads for an exit at an approved grade separation (overpass) over I-35 in north Williamson County so that two county roads would not be blocked. Five times the Bureau had refused the requests. But Forbes had bet heavily on the exit being built. In June 1963 he wrote Congressman Homer Thornberry asking for help. Thornberry queried the Bureau and learned that the exit Forbes sought had actually been partly constructed, but Jarrell citizens had stopped it with a protest. The Bureau's top man informed Thornberry:

> The State . . . upon being prompted by protests by the citizens of Jarrell, proposed that the exit ramp in the vicinity of Station 70 be eliminated and relocated to the south at a point near the Jarrell interchange structure. This proposal was approved . . . because it would provide more direct service to the community of Jarrell. . . . The addition now of the ramp requested by

[21] Letter, Johnson to Dion VanBibber, May 22, 1958, ibid.
[22] Interview, Mary Forbes, Mar. 26, 2000, Georgetown.

Mr. Forbes to provide access to his business establishment would result in a duplication of southbound exit ramps serving the Jarrell interchange.[23]

On January 14, 1964, the new interstate opened in north Williamson County—and business at the Forbes Truck Stop and Cafe began to evaporate. Before I-35 opened, from January 1 through January 14, Forbes had delivered 17,339 gallons of Humble gasoline to his truck stop. After the interstate opened, from January 15 through January 30, he delivered just 1,183 gallons—a 93 percent decrease. In the two weeks after the new highway opened, the cafe's business fell 80 percent.[24] Forbes was desperate. State Representative Gene Fondren of Taylor arranged a Texas Highway Commission hearing, at which state engineers agreed to try again.[25] They did, but for the sixth time Washington rejected the request for an exit ramp at Forbes's truck stop. The Bureau's chief wrote Congressman Jake Pickle, who had replaced Thornberry, "unless a policy is followed consistently, there results similar requests from other property owners which in fairness cannot be denied."[26]

Forbes was wretchedly disappointed. He felt that his government was killing "the American way of life."[27] He accepted "absolute failure" of his effort to convince the Bureau of Public Roads to allow the exit. "They hold the whip handle and are certainly using it," he wrote Pickle. "Everyone in Texas approved our request." At this point Forbes was seeking a loan from the Small Business Administration to "pay off the commercial bank for the debt for this property. . . . I am attempting to avoid a very embarrassing situation and prevent a default,

[23] Letter, Rex M. Whitton to Homer Thornberry, July 1, 1963, Box 17, Pickle Papers (Center for American History, University of Texas at Austin).

[24] Letter, Charles A. Forbes to J. J. "Jake" Pickle, Feb. 4, 1964, ibid. Forbes sent copies of the letter to Senators Ralph Yarborough and John Tower of Texas, both of whom contacted the Bureau of Public Roads on his behalf. Also see "Fondren Asks Highway Assistance for Jarrell," *Taylor Daily Press*, Feb. 6, 1964, p. 2.

[25] "Fondren Asks Highway Assistance for Jarrell," *Taylor Daily Press*, Feb. 6, 1964, p. 2.

[26] Letter, Rex M. Whitton to J. J. "Jake" Pickle, Mar. 31, 1964, Box 17, Pickle Papers (Center for American History, University of Texas at Austin).

[27] Letter, Charles A. Forbes to Homer Thornberry, June 13, 1963, ibid.

which I have no intention of committing." He likened his plight to "paying the funeral director on the installment plan."[28]

Pickle kept probing at the interstices of the Bureau's policy.[29] On May 14 he received a missive from Rex Whitton, the Bureau's director, denying the latest request for the exit Forbes sought. The bottom line, Whitton wrote, was that accepting this proposal would constitute "undue preferential treatment to some."[30] But that same morning, Pickle had coffee with one of Whitton's bureaucrats, Walt Osborne, chief of Special Procedures Branch, Project Coordination Division. After his meeting with Osborne, Pickle wrote Forbes that "there is a possibility of some relief." Osborne had indicated, Pickle wrote, "the Federal boys here would not object if the State would agree to an off-ramp, which I believe would be fairly near your station.... You understand that you will have to sell this to the State officials, but I do not anticipate that you will have a great deal of trouble."[31] That afternoon Pickle dictated a short note to Osborne's boss. "Dear Mr. Whitton," it read, "Thank you for your letter of May 14 regarding additional ramp connections with Interstate Route 35 north of Jarrell. Mr. Osborne came by the office today for a personal visit about this matter, which I appreciate very much."[32]

Another of Pickle's remarkable "understandings" had been achieved. The exit was built.[33] Unfortunately for Forbes, it came too late to save Forbes's truck stop. "That was Charles's biggest producer, and he lost it all. It was a disaster for him. By the time we got the deal done his business was gone," Gene Fondren recalls.[34] Forbes sold the property. Thirty-five years later, the land holds a ramshackle collection of trailers. Men in T-shirts lounge around, drinking beer on a Sunday afternoon, flicking spent cigarettes into tangled heaps of trash. It is a

---

[28] Letter, Charles A. Forbes to J. J. "Jake" Pickle, Apr. 14, 1964, ibid.

[29] Letter, J. J. "Jake" Pickle to Forbes, Apr. 21, 1964, ibid.

[30] Letter, Rex M. Whitton to J. J. "Jake" Pickle, May 14, 1964, ibid.

[31] Letter, J. J. "Jake" Pickle to Charles A. Forbes, May 14, 1964, ibid.

[32] Letter, J. J. "Jake" Pickle to Rex M. Whitton, May 15, 1964, ibid.

[33] The U.S. Bureau of Public Roads approved the exit on June 5, 1964. See memo to "jp" (Jake Pickle), June 5, 1964, ibid.

[34] Interview, Gene Fondren, Sept. 1, 2000, Austin.

portrait of the poverty rural interstates sometimes attract, a photo opportunity for a born-again Walker Evans.

The only remnant of Charles Forbes's fight for his business is an exit on Interstate 35's southbound track. "The boys at the Texas Highway Department were so irritated at the Yankees up in Washington who had been so difficult that they named it 'Yankee Exit,' " Mary Forbes said with a tight smile.[35]

Moses J. Kouri and his younger brother Tom traded in Austin real estate. At one time they owned 150 rent houses. They bought, improved, and sold. Sometimes they just bought and sold. They had prospered after moving to Austin from Manor when they were boys. Their father, James "Jim" Kouri, and his three brothers had immigrated to Texas from Lebanon in the late 1880s. "They had a hard fight," said Tom Kouri. "They couldn't speak English, and they became peddlers. They carried piece goods in backpacks all over Texas. They were ambushed in West Texas, but they liked Central Texas. The Germans and Swedes were all so nice."[36] The Kouri brothers settled in Manor, where Jim acquired several farms and a big mercantile store.

In the twenties Jim Kouri started buying property in Austin. He dispatched Moses, then barely seventeen, to Austin to oversee renovation of an old rooming house on Guadalupe Street. Moses invited his eight-year-old brother, Tom, to live with him in Austin and "learn the business." Tom learned about construction—and life—from Moses, whom he worshipped. In Tom's eyes, Moses was a "brilliant, brilliant man," the "genius of the family" who could foresee the future. "He gambled mainly on real estate. He saw what was coming and he dealt with it," Tom would say decades later. "If Moses were living today, I think Michael Dell would have to take a back seat to him." [37]

[35] Interview, Mary Forbes, Mar. 26, 2000, Georgetown.

[36] Interview, Tom Kouri, Sept. 22, 2000, Austin. I am indebted to Mr. Kouri for supplying generous details of his family's background, of his and Moses Kouri's purchases along U.S. Highway 81, later Interstate 35, and their efforts to influence highway builders to create an interchange and, years later, a new highway from scratch (State Highway 45). He is a delightful and remarkable gentleman.

[37] Interview, Tom Kouri, Sept. 22, 2000, Austin.

Each fall Tom and Moses drove up U.S. Highway 81 to Dallas to watch the University of Texas's annual grudge match against Oklahoma University. "Every year, we would drive by this ranch in Williamson County, and Moses would look at it and say, 'I *need* to own that land.' He talked to Owen Sherrill in Georgetown—Sherrill was the ranch king of Williamson County—and one day Sherrill called to say the ranch owner's widow wanted to sell." The ranch sat west of the highway halfway between Round Rock and Georgetown. On a cold misty Sunday afternoon in April 1962, the Kouri brothers drove up to Williamson County, dickered with Opal Overby, and bought her 1,020 acres.[38]

Moses Kouri had always believed Highway 81 would become Texas's chief north-south artery—not Highway 281, as some thought, or a more easterly route. Already Highway 81 was the main road from Mexico to Dallas. Now the interstate was being built, moving steadily toward Williamson County. If the Kouris could somehow cause a full interchange to be built at their property, its value would multiply many times. The physical setup at their ranch mirrored what Charles Forbes faced in north Williamson County, though at first glance their situation seemed less promising. A dirt road from the east (Chandler's Road) terminated at the soon-to-be interstate highway. There was no road heading west from Highway 81 near the Kouri property. There was no reason for one: hardly anyone lived out there. Three hundred people clustered in Leander; Cedar Park's post office served about fifty "cedar choppers." An overpass would link the existing Chandler Road to I-35 across from the Kouri place, and that was that.

The Kouri brothers went to work. Within seven months of buying the Overby ranch, they had hired attorney Gene Fondren, Williamson County's state representative, to represent their interest in creating a full interchange where their property touched I-35.[39] The Kouris teamed up with John H. Nash Jr., who owned about fifteen hundred acres across Highway 81 from their place and another sixteen hundred

---

[38] Ibid.; Warranty Deed Vol. 450, p. 683, Apr. 7, 1963, Williamson County Deed Records (County Clerk, Williamson County, Georgetown). The Kouris added another forty acres to the spread in a separate purchase.

[39] Interview, Gene Fondren, Sept. 1, 2000, Austin.

Tom Kouri, Moses Kouri, and Moses's wife Evangeline attend the Houston Live-stock Show, February 1962. *Courtesy Tom Kouri, Austin.*

acres north of them.[40] Nash, who also hired an attorney, owned Austin's Capitol Chevrolet car dealership and a name that commanded respect, particularly among politicians and the road lobby.[41]

Williamson County Judge Sam Stone wrote Ed Bluestein, the highway department's district engineer in charge of the interstate,

[40] Map: South Half of Williamson County, 1969, Tobin Surveys Inc. (San Antonio, Tex.). John H. Nash Jr. hired Robert Sneed of Sneed & Vine, Austin, to represent him with the Texas Highway Department and U.S. Bureau of Public Roads.

[41] "John H. Nash Jr. dies in Hong Kong," *Austin American-Statesman*, Oct. 1, 1979. Nash sat on the board of the Lower Colorado River Authority and was well known in Washington circles.

The Golden Egg

spelling out the county's interest in developing a full-bore inter-change halfway between Georgetown and Round Rock. "As county officials," Stone wrote, "we try to anticipate future county road system requirements to meet the needs and demands of our developing communities. . . . There are presently no east-west roads crossing the area west of the proposed U.S. Interregional 35 between the Brushy Creek Road near Round Rock and RM 2243 near Georgetown.[42] Although this area in the past has been owned by a few rather large landowners, current trends give every indication that we can expect rather rapid development with many additional people and residences in the foreseeable future."[43]

In other words, Judge Stone proposed an interstate link with a cross-county highway that did not exist. The road Stone envisioned curved southwesterly to connect with Brushy Creek Road west of Round Rock, rather than heading straight west toward Cedar Park, as it eventually would do. The Round Rock Independent School District threw its weight behind the interchange proposal.[44] Fondren lobbied Bluestein for the Texas Highway Department's blessings.[45] His central argument was that by 1975 Round Rock and Georgetown would become part of Austin, with "almost continuous development along Interstate 35."[46] It was a planner's argument, one highway builders should appreciate. The argument centered on the long-range public good derived from adding the interchange, on the government's nickel. The fact that the Kouri and Nash properties would escalate in value was not alluded to.

Convincing the U.S. Bureau of Public Roads was no snap. In the winter of 1964, while Charles Forbes was watching his truck-stop

---

[42] Ranch to Market 2243 was locally known as Leander Road; the Brushy Creek Road was called Harry Mann Road, the latter name interpreted by generations of Williamson County teenagers as "Hairy Man Road." Brushy Creek Road was one of several early cattle trails called the Chisholm Trail.

[43] Letter, Sam Stone to Ed Bluestein, Nov. 11, 1963, Box 17, Pickle Papers (Center for American History, University of Texas at Austin).

[44] Letter, Noel Grisham to Ed Bluestein, Nov. 12, 1963, ibid. Grisham was superintendent of the Round Rock Independent School District.

[45] He was joined by Nash's attorney Robert Sneed.

[46] Letter, Gene Fondren and Robert Sneed to Ed Bluestein, Dec. 3, 1963, Box 17, Pickle Papers (Center for American History, University of Texas at Austin).

business dry up, Fondren shuttled from Austin to Washington "seven or eight times" to argue the Kouris' case. "There was no exit ramp anywhere between Georgetown and Round Rock," he would recall later.[47] "We argued that [the nonexistent] 1431 would become a major highway, and eventually it did. It's *the* east-west connection in Williamson County."[48] The chief engineer at the Bureau of Roads was a Texas Highway Department alumnus: "He was our highway connection," said Fondren. "Jake, of course, was my entrée. And then, of course, we had LBJ in the background. That didn't hurt." It also didn't hurt that "the Kouris spent some serious money."

Back in Williamson County "people thought we were crazy," Tom Kouri remembered forty years later. "They thought we were spending way too much money. But Moses had it figured out." On June 5, 1964, Jake Pickle got a call from Walt Osborne at the Bureau of Public Roads. The Kouris and Nash would get a full interchange across their property.[49] Anticipating success, Tom and Moses Kouri had just sold the property to Houston car dealer Raymond Pearson. "We made a very handsome profit," said Kouri. "A very, very, very handsome profit."[50]

That summer the Kouri brothers sunk that profit into another piece of Williamson County, purchasing a dairy farm at the northwestern intersection of what would become Interstate 35 and Texas Farm-to-Market 1325, then a two-lane country road south of Round Rock and well north of Austin's northern physical limits. "It was so simple," Tom would remember. "Moses went out there and he would

[47] Actually, the original plan provided for an exit at Texas Crushed Stone, just south of Georgetown, and one just north of Round Rock, where I-35 veered west of U.S. 81, locally called Mays Street.

[48] Fondren was referring to Texas Ranch-to-Market 1431, which in Williamson County did not exist until 1986, when it connected Interstate 35 and U.S. Highway 183 at Cedar Park.

[49] Memo to "jp" (Jake Pickle), June 5, 1964, Box 17, Pickle Papers (Center for American History, University of Texas at Austin). The memo included the information that two exit ramps north of Jarrell had also received approval, including the one leading to Forbes Truck Stop.

[50] Interview, Tom Kouri, Sept. 22, 2000, Austin; General Warranty Deed, May 27, 1964, Williamson County Deed Records, Vol. 468, p. 654 (County Clerk, Williamson County, Georgetown).

The Golden Egg

drive by there and look at it. He called his realtor. 'See if you can buy that farm,' Moses said. Mr. Ernest Anderson said he was willing to sell 121 acres, but he said, 'I want to keep the house.' Then Mr. Anderson quoted a ridiculous price. Moses agreed to the price but he wanted the acreage *and* the house, regardless of the price. He told me, 'See what he wants—I *need* that. We've got to have it.' I say, 'Mose, that's a little high! It's a ridiculous price!' He says, 'We got the money, we're going to buy it!' "

Shortly after selling the Overby ranch in 1964, Moses and Tom Kouri bought Ernest Anderson's 121 acres, and house, at the corner of FM 1325 and U.S. Highway 81, just north of the Travis County line. In 1998 Tom sold it.[51] By that time Dell Computer sprawled directly across from his place on Louis Henna Boulevard just east of I-35 and the State Farm Insurance headquarters faced Kouri's land on the other side of FM 1325. The Kouri farm was an undeveloped island in a sea of commercial and retail property. "In 1977 I retired and started ranching," Tom said. "I went out there every day. I got an old tractor. I horned those cattle, mowed that hay. All the kids and grandkids had a great time. I miss it even today."[52]

Back at the old Overby ranch, others were profiting from Moses Kouri's instincts. Less than a year after the Kouris sold the place, Pearson resold it to the Lamoc Corporation. Six years later the Westinghouse Corporation, the first national corporation to move to Williamson County, bought the land, at least partly because it commanded an interchange with the expressway—the one the Kouris created.[53] In 1986, twenty-three years after Judge Stone predicted it, a new five-lane highway, Ranch to Market 1431, carved open the

---

[51] Moses Kouri died in January 1974.

[52] Interview, Tom Kouri, Sept. 22, 2000, Austin. Tom Kouri owned the Handy-Andy grocery franchise in Austin, which became the Humpty Dumpty grocery chain.

[53] Interview, Claude Hays, Aug. 20, 2001, Georgetown. Mr. Hays is senior partner of Georgetown Title Company, which provided invaluable research assistance on the history of the Kouri land purchase, as well as histories of land deals near Lake Georgetown and Granger Lake, including Ralph Moore's. See Williamson County Deed Records, Vol. 475, p. 163 for Pearson's sale to Lamoc Corporation (Walter S. Higgins acting for Pearson); and Vol. 479, p. 13 for details of Lamoc's July 13, 1971, sale to Westinghouse Corporation.

untouched heart of Williamson County ranch land. The speck of humanity that once had been Cedar Park stood ready to capture the next new wave of development off I-35.

———

In May 1963 Texas Highway Department engineers were hunting for solid rock into which to drive pilings for an overpass that would allow Interstate 35 to soar over the Georgetown Railroad tracks a mile south of Georgetown. This is not an easy task anywhere along the cave-riddled Balcones Escarpment. It was especially difficult on this portion of the Edwards Aquifer Formation, where the soil is like a kitchen sieve. Still, the engineers were astonished as they drilled hole after hole—nineteen in all—and broke through into thin air eleven times.

They decided to investigate. They dropped a highway construction worker into a core hole to explore. The man, whose name is lost to history, rode a drill bit down through forty feet of solid rock, armed with a flashlight, until he landed on a wet cave floor. Water dripping from the ceiling and tons of bat guano made the floor slippery, and the explorer tripped and dropped his flashlight. In total darkness, except for a pinprick of light from the core hole above, he scrabbled about, his panic growing. Eventually he found the flashlight.

He had landed in what became Inner Space Caverns' "Discovery Room," a theatrically huge space laced with frozen fountains forty thousand years old, waves of limestone "curtains" dripping from the ceiling, and stalagmite totems thrusting up from still pools and a thick mud floor. Interstate 35 had revealed one of Texas's most interesting caverns.[54] From the point of view of spelunkers and academic geolo-

———

[54] For this account I have relied heavily on a published account by Tommie Pinkard, "Under the Interstate," *Texas Highways,* 13 (May 1966), 15; and on interviews with George Norsworthy Jr., president of Inner Space, July 14, 1998, Dripping Springs; Doyle Clawson, July 14, 1998, Georgetown; and Ramsey Clinton, July 1, 1998, Georgetown. The Georgetown Corporation was created to develop the caves into a commercial attraction. Partners were Donald Duncan, a Burnet rancher and banker; Burnet mayor Ramsey Clinton, cave manager Doyle Clawson, and Dallas advertising men George and Tom Norsworthy, who had owned Longhorn Caverns since 1932. Also see "Inner Space Cavern," Walter Prescott Webb, H. Bailey Carroll, and Eldon Stephen Branda (eds.), *The Handbook of Texas* (3 vols.; Austin: Texas State Historical Association, 1952, 1976), III, 430.

gists and paleontologists, it was a riveting find.[55] From the Texas Highway Department's perspective, it was a major nuisance.

The road engineers wanted to seal the cavern's entrance, which lay directly under the planned expressway, and pretend they had never found it. That might have happened if news of the caverns had been kept quiet. But word leaked out. University of Texas student Bill Russell of Georgetown led the University of Texas Speleological Society to the cavern, and the cave buffs started mapping.[56] The superhighway project ground to a halt.

In December, six months after I-35's core-sampling crew discovered the caverns, Georgetown's *Sun* learned about the cave when *Dallas Times Herald* columnist Frank Tolbert called to inquire about it. "Big Cave Discovered South of Town, Parts of Camel & Elephant are Found," a *Sun* headline trumpeted.[57] Dr. William W. Laubach, the retired Lutheran minister who owned the property, had invited Southern Methodist University scientists to tour the caverns, and their excitement had found its way into the Dallas press. The professors found fossilized bones of a prehistoric form of camel and elephant (actually a mammoth), as well as those of a peccary, jaguar, dire wolf, and a small horse—all extinct.[58]

Meanwhile, highway department officials champed at the bit to seal "Laubach Cavern" and move on.[59] Greer and Bluestein agreed to close the cavern's entrance on February 10, 1964, so they could complete the unfinished I-35 bridge over Georgetown Railroad's tracks, but Georgetown mayor Thatcher Atkin and Chamber of Commerce president Robert F. B. "Skip" Morse pleaded for a few days' delay,

---

[55] Adventurous Georgetown teenagers had explored the caves on Dr. W. W. Laubach's place for years. They called it Bat Cave.

[56] They mapped seven thousand feet of caves and tunnels, which wound under the interstate somewhat in the shape of a sinuous dragon. Eventually four miles of cavern were explored. See Pinkard, "Under the Interstate," *Texas Highways,* 13 (May 1966), 15; "Roadside Attractions Steer New Course," *Wall Street Journal,* July 15, 1998.

[57] Don Scarbrough, "Big Cave Discovered" in "Passing Glance," *Williamson County Sun* (Georgetown), Dec. 19, 1963, p. 1.

[58] Bob H. Slaughter, "Downward, Ho! Georgetown Cave Hailed by 'Dallas Times Herald'," *Williamson County Sun* (Georgetown), Feb. 27, 1964.

[59] Memo for File, Bob [no last name], Pickle office, Feb. 7, 1964, Box 17, Pickle Papers (Center for American History, University of Texas at Austin).

Left: A state highway worker rides a core drill into a freshly discovered cavern underneath Interstate 35 south of Georgetown. The find stalled highway work for more than a year. *Courtesy Texas Department of Transportation, Austin.*

Below: Early visitors to Inner Space Caverns gaze in awe at fantastic geological formations lit in the French manner. *Dewey G. Mears Photo, Courtesy Williamson County Sun, Georgetown.*

smelling the prospect of a deal that could result in development of the caves as a commercial attraction. Atkin called his brother-in-law, Congressman Pickle, for help. State Representative Fondren, working furiously to save Charles Forbes's truck stop and to create an interchange on the Nash and Kouri land, tackled this problem as well.

Somehow the state was held off long enough to work out the conflicting interests of Dr. Laubach, the caverns' prospective operators, and the State Highway Department, which ended up redesigning an access road so that Inner Space visitors could enter the tourist attraction without difficulty. The cavern's progress from raw cave to commercial enterprise was managed by Doyle Clawson, who had run Burnet's well-known Longhorn Caverns for years.

"If nobody had wanted to develop it, the highway department would have sealed it. To them it was just trouble," Clawson would recall. "[But] in my years managing Longhorn Caverns, I had gotten to be friends with lots of spelunkers from the University of Texas Speleological Society . . . my friends came running to me and told me about these beautiful caverns. What sealed the deal for us was the fact that Dr. Laubach's son was college roommates with a young man named Michael Lorfin, who worked as a guide at Longhorn Caverns. That was the key to the whole deal."[60]

The cave's location was its key to success. "We were so excited to have that location right there on the interstate," said Ramsey Clinton, a partner in the enterprise.[61] Other partners were Tom and George Norsworthy, who owned a Dallas advertising agency and Longhorn Caverns. They were close friends with the developers of Six Flags Over Texas in Arlington. As a result, the cavern's promoters seriously considered turning their geological curiosity into a theme park along the lines of Six Flags.

"Partly we didn't do this because it seemed to run counter to the feeling of the natural caverns," said George Norsworthy Jr. many years later. "And partly it was a lack of sewer and water service. We couldn't do a big restaurant without access to sewer and water."[62]

Georgetown never ran a water or sewer line down the western face

[60] Interview, Doyle Clawson, July 14, 1998, Georgetown.
[61] Interview, Ramsey Clinton, July 1, 1998, Georgetown.
[62] Telephone interview, George Norsworthy Jr., July 14, 1998, Dripping Springs.

of Interstate 35, and no restaurant ever opened. But the caverns were a big hit. The Georgetown Corporation partnership invested 175,000 dollars in developing them. Among other expenses, they hired a French lighting firm to illuminate the "Lake of the Moon" room. After a name-the-cavern contest failed to produce the perfect name, the owners came up with "Inner Space"—a subliminal play on Americans' fascination with the outer space program in the mid-1960s.

Inner Space opened July 22, 1966. A brief trolley ride took the first visitors into the bowels of a two-mile trace of caverns winding under Interstate 35.[63] In a room as big as a football field, they gazed at the fossilized bones of elephant-sized giant sloths, Zimmerman deer (with 150-point antlers), and the armadillo-like *glyptodont*, which looks rather like a Volkswagen Beetle with a spiked tail. Visitors learned that most of these animals died ten to twenty thousand years ago, presumably after pausing for a drink and plunging through thin soil into the jumble of sink holes that became Inner Space Caverns.

Thirty-five years later Inner Space is still a successful operation. Some three million people have strolled through its corridors, goggling at geological formations that look like frozen waterfalls, Italianate baroque cathedrals, ten-foot-tall ice cream cones, Martha Graham's flowing veil—eighty thousand to one hundred thousand visitors a year.

The tone of the place has shifted from exotic to educational. People don't ramble up and down clogged Interstate 35 just for fun, as they once did, nipping in here for a Stuckey's snack or there to gawk at San Marcos's Snake Farm. Now they hurtle to work and back, or thunder up from Mexico in eighteen-wheeler trucks, with no time to spare. Today, school field trips comprise a big part of Inner Space's business. Kids and their teachers motor from Austin, San Antonio, Houston, Dallas, Fort Worth, and points between. For an extra fee, children can work at a mock dig, learning how to be paleontologists for a day while learning the structure of Central Texas geology.[64]

[63] Actually, the trolley was not ready on opening day, but it was installed shortly thereafter and has been one of Inner Space's chief attractions over the years. See "Inner-space Open Friday," *Williamson County Sun* (Georgetown), July 21, 1966. Much of the caves' extensive structure has never been opened to the general public.

[64] "Roadside Attractions Steer New Course," *Wall Street Journal*, July 15, 1998.

THE UNEARNED INCREMENT

Now, the labours of the nation at large do add daily and yearly to the value of the land, whether the landlord plays the part of an improver or not.

JOHN STUART MILL, *The Right of Property in Land*

In the early sixties, most of Williamson County's thirty-five thousand residents viewed the coming of Interstate 35 with distaste. They knew it would be a gigantic project—a sliver of the biggest public works project the United States had ever launched—and they knew it would change things. But they didn't know precisely how. In rural Williamson County the interregional highway was viewed as a foreign object, an amalgam of designs drawn by engineers and internationalists with questionable intentions. It did not seem calculated to serve local needs, and indeed, it didn't, certainly not at first. Public works water projects appeared potentially useful, but zipping easily from Mexico to Canada seemed an outlandish goal to farmers and ranchers of Williamson County. The county's concerns were local. Its politics were local. Its mix of German, Czech, Mexican, African-American, Swedish, Wend, and Anglo-American communities was exceedingly local, though springing from outlander roots.

The first leg of Interstate 35 opened in north Williamson County on January 15, 1964, a stormy winter's day. It swept past Charles Forbes's instantly inaccessible truck stop and barreled right through the middle of Jarrell, an unincorporated village of four hundred. Formed in 1912 when the Bartlett Western Railroad went through,

Jarrell was the county's youngest town.[1] When the rail line failed in 1935, Jarrell ceased to have a strong reason for existence. U.S. Highway 81 had helped keep it alive, but the interstate was a different creature. During construction a Georgetown newspaper editor described the highway's impact on Jarrell: "You might say that the east end of Jarrell's business district has been wiped clean by the highway department, which is taking the new road directly through that community. . . . [It was] necessary to destroy or move a number of business firms."[2]

The interstate eliminated Jarrell Motor Company, two service stations, a garage, a liquor store, and a cafe. "They done what they were going to do," said Jarrell's Emil Danek, owner of Danek Hardware and Lumber, of the highway's engineers. "They just took out the east side." The opening of the interstate was nonetheless a happy day for Danek. "The day they opened it was the day my son was born. It was snowing. I was driving my wife to the hospital [at Georgetown] and I missed the turnoff. I had to back up on the interstate to get off. It was just snowing sheets at the time."[3]

The Georgetown segment came next. The county seat's citizens were scared. "At any rate, we will just have to do the best with what we have left," wrote a glum *Sun* columnist. "Apparently there is no way out of the predicament. . . . It was either run the broad double lane through town or west of town."[4] Rancher Jay Wolf rode across his land west of Georgetown on horseback one day with a young cowhand named Larry Hausenfluke. "This is where Interstate 35 will come—it will come right through here," Wolf told Hausenfluke, indicating its path. Hausenfluke was shocked. He mumbled that the people of Georgetown surely wouldn't want *that*. "It doesn't matter what they want. It's what they'll get," Wolf snapped.[5] Not long afterwards

[1] For a delightful discussion of this little railroad that ran along the northern border of Williamson County from Bartlett to Florence, see Clara Scarbrough's *Land of Good Water,* 334–344. The railroad was known as the "Four Gospels Line."

[2] Don Scarbrough, "Passing Glance," *Williamson County Sun* (Georgetown), Jan. 3, 1963, p. A-1.

[3] Telephone interview, Emil Danek, June 9, 1998, Jarrell, Tex.

[4] Don Scarbrough, "Passing Glance," *Williamson County Sun* (Georgetown), Jan. 3, 1963, p. A-1.

[5] Telephone interview, Larry Hausenfluke, June 5, 1998. For many years, Hausenfluke was superintendent of the Jarrell Independent School District.

the rancher opened a Gulf service station at the intersection of Highway 29 and Interstate 35, the first Georgetown man to take that step. (Humble, Texaco, Phillips 66, and Mobil quickly bought sites along the expressway; forty years later Wolf's family would sell 103 acres at the intersection of I-35 and Highway 29 to a mall developer for a reported twenty-seven million dollars.)[6]

One of the few people in Georgetown not worried about the coming interstate was Owen Sherrill, who owned about two hundred acres of land on the interstate himself. He clearly understood the symbiotic relationship between new sources of water, the new interstate, and growth. He shepherded to fruition Georgetown's first significant Interstate 35 land sale. Early in 1963 Tom Joseph sold the Weir Estate, consisting of 475 acres and a magnificent Victorian house with a view of the San Gabriel's South Fork, for a quarter of a million dollars; the property would dominate the interchange at Interstate 35 and Leander Road for decades. The Weirs' beloved polo field, where Georgetown's crack polo team had practiced before winning the U.S. Southwestern Region title, was erased by bulldozers—I-35 gouged a deep path right through it.[7] The *Sun* reported, "Owen W. Sherrill, flamboyant realtor of Georgetown and the Southwest, brought the deal to a close. He predicts that the development of industrial, shopping center and residential buildings on the Weir tract will mean great things for Georgetown."[8] Despite its attractions, the development failed to "make."[9] Eventually the land would become San Gabriel

---

[6] Rebecca F. Sellers, "Park Land at Site Becomes an Issue," *Sunday Sun* (Georgetown), June 8, 2003, p. A-1.

[7] Georgetown had a terrific polo team in the early thirties led by one-armed O. W. Cardwell, who played polo by reining with his teeth. Two of his best players were "Doc" (Afton) and "Duddy" (Howard) Weir, whose father, H. M. "Greely" Weir, set up the polo field east of his house off Leander Road. Details can be found in Esther Messick Weir, "Polo at Georgetown," in *Georgetown's Yesteryears: Reaching for the Gold Ring,* ed. Martha Mitten Allen (4 vols.; Georgetown: Georgetown Heritage Society, 1985), I, 68–82.

[8] "'Development' Talk Started by Land Sale," *Williamson County Sun* (Georgetown), July 11, 1963, p. A-1; letter, Owen W. Sherrill to J. J. "Jake" Pickle, Dec. 3, 1964, Pickle Papers (Center for American History, University of Texas at Austin). The buyers were J. Mit Lee and a group of investors from Bryan, Texas.

[9] Interview, Jay Sloan, Sept. 11, 2000, Georgetown.

Heights, a pleasant, upper-middle-class residential subdivision where the river runs through a number of residents' backyards.

The interstate opened through Georgetown in September 1965. Local citizens steeled themselves for loss. The *Sun* headline read, "Highway 35 Bypass Opens and All's Quiet, Very" and then reported:

> State Highway Department officials move the barricades on Interregional Hwy 35 Thursday morning, and by this seemingly mundane act create a moment of history in Georgetown. . . . For the present at least, the immediate results will be simply quietness. For the first time in over 100 years, Georgetown will be sidetracked from one of the throbbing arteries of Western Hemisphere commerce. The Chisholm Trail is detoured.[10]

At its next meeting, the Georgetown City Council annexed the interstate.[11]

———

Round Rock felt that same ominous quietness in 1968 when the third and last segment of Williamson County's portion of Interstate 35 was completed. The entire 20.7-mile county stretch, from north of Jarrell to south of Round Rock, was budgeted at seven and a half million dollars. Probably it cost considerably more—twenty-one million dollars at least.[12] Whatever its cost, since its completion I-35 has operated like a gigantic magnet in Williamson County, sucking investment money, creative energy, and development into its ever-expanding force field. In America the interstate highway has generally thrown off money, making it a postmodern example of John Stuart Mill's "unearned increment" through which the state itself, through its

---

[10] "Highway 35 Bypass Opens and All's Quiet, Very," *Williamson County Sun* (Georgetown), Sept. 23, 1965, p. A-1. I-35 roughly followed one of the Chisholm Trail's paths through Williamson County.

[11] "City Annexes Highway 35!" *Williamson County Sun* (Georgetown), Oct. 7, 1965, p. A-1.

[12] "7.5 Million Interstate 35 Highway Construction Is Scheduled in Williamson County," ibid., Feb. 23, 1961, p. 1; "Highway 35 Bypass Opens and All's Quiet . . ." ibid. In "All's Quiet," the *Sun* reported that the county's 9.7-mile central segment of I-35, through Georgetown, cost one million dollars a mile. I was not able to confirm an official cost.

actions and investments, vastly increases the value of private property, however well or poorly the property is managed.

But in 1968 few suspected that would be the case. As in Georgetown, Round Rock businessmen were heartsick when I-35 "bypassed" the business district by about six blocks. Mays Street, or U.S. Highway 81—the city's former heartbeat—sat deserted. "Most everybody in town thought the interstate would kill Round Rock," N. G. "Bunky" Whitlow, longtime Round Rock banker and civic leader, would remember. Shortly after I-35 shunted traffic away from the Mays Street "strip," Whitlow said,

> I was working behind the counter at Farmers State Bank
> when Carlo Carlson came in. Carlo had the Texaco station on
> Old Highway 81, right in the middle of town. He was a
> Swede and real opinionated. "I'll be damned," he yelled at me.
> "There's a god-damned dog asleep in the middle of the street
> in front of my station! I want you to come out here and see
> him." So I followed Carlo down Main Street, and sure
> enough, a big old red dog was fast asleep in the middle of
> Highway 81. There were no cars in sight. Once in a while, one
> would come along and drive around him.[13]

A fluke brought Westinghouse Corporation to Round Rock. The groundwork had been laid with the Georgetown reservoir and Interstate 35, but still, the acquisition of a major national industrial player like Westinghouse broke the county's social and economic patterns. Tom E. Nelson Jr. grew up in one of Round Rock's leading families. Nelson's father, Tom E. Nelson, ran Farmers State Bank and farmed in Palm Valley, where *his* father had settled and founded Palm Valley Lutheran Church, which became the nucleus for Round Rock's Swedish farming community. The youngest Nelson moved to Austin as a teenager, and moved into privileged circles.[14] With substantial

[13] Interview, N. G. "Bunky" Whitlow, July 1, 1998, Round Rock.

[14] Tom E. Nelson Jr. attended elementary school in Round Rock, transferred to the Texas Military Institute in San Antonio, and graduated from Austin High School. He earned a business degree at the University of Texas, married Carol Corley of Austin, and settled down to raise a family in Tarrytown. In 1972, when Round Rock's revered Gus

land holdings in Williamson County and considerable financial training, he prospered. As the 1970s dawned he helped the Austin Chamber of Commerce snag International Business Machines from Kentucky by securing land for a new headquarters in North Austin.[15] Through his Austin Chamber connections Nelson heard that Westinghouse Corporation, then an industrial giant, was looking for a new plant location. Nelson's role in Round Rock's maiden courtship with an industrial powerhouse was pivotal. Nelson "talked up" Round Rock with Austin leaders and ended up, with Austin's blessings, negotiating deals for land west of Interstate 35 within Round Rock's extraterritorial jurisdiction and close to Georgetown. He negotiated contracts with Raymond Pearson, who owned a Ford-Mercury dealership in Houston, for 607 acres; rancher Leon E. Behrens, for seven hundred acres; Lamoc Corporation, 1,040 acres (Moses Kouri's first Williamson County investment, the old Overby ranch); and sheep trader James Garland Walsh, who optioned fourteen hundred acres for roughly a million dollars.[16] Altogether, Westinghouse bought outright and optioned thirty-five hundred acres. By Williamson County standards, it was a huge land deal.

Westinghouse wanted to build a "new town," a planned community through which private investors could "get into" HUD funds for up to 50 percent of the development's cost, according to one observer.[17] The anticipated population was thirty-five thousand—the

Lundelius stepped down as honorary chairman of the board of Farmers State Bank, Nelson bought the bank and appointed N. G. "Bunky" Whitlow as its president. Nelson was appointed a director of the American Bank of Austin in 1972. See "American Bank Adds Director," *Austin American-Statesman,* Jan. 19, 1972; *Round Rock Leader,* Jan. 27, 1972.

[15] Telephone interview, Tom E. Nelson Jr., Oct. 23, 2001, Austin. Nelson died June 6, 2005.

[16] Ibid. Ultimately, Westinghouse did not exercise its option on Walsh's land. Pearson's father was Col. T. M. Pearson, who had sold land in Friendswood near Houston that became headquarters for the National Aeronautics and Space Administration. It is interesting that the Nash, Henna, and Pearson families—all owners of Texas auto empires—owned ranches or homes within a few miles of each other in Williamson County.

[17] Telephone interview, Jim Hislop, Oct. 22, 2001, Round Rock; telephone interview, Tom E. Nelson Jr., Oct. 23, 2001, Austin. According to Hislop, most of the "New Towns" were created as municipal utility districts, because through them private developers legally "could recover most of what they put in." Hislop and Nelson agreed that

Westinghouse's giant turbine manufacturing facility in 1972. *Courtesy* Williamson County Sun, *Georgetown*.

size of Williamson County.[18] The model was Reston, Virginia.[19] "They had ambitions for a model city," Nelson remembered many years later. "They spent six or seven hundred thousand dollars on planning. Of course, that was just a book figure since they owned the planning company." Just the prospect of doubling the size of Williamson County boggled many minds, both within and outside the county. "It would have made such an impact on the communities out there," Nelson noted, with considerable understatement.

---

the only New Town that ever really "worked" was Reston, Virginia, because, according to Hislop, "Reston got a bunch of industry in there first, so the people who moved there could work there." An investment group started a "New Town" near San Antonio called The Ranch, but it did not succeed because the development could offer no jobs, according to Hislop.

[18] Telephone interview, Don J. Leonard, Aug. 1974, Round Rock. In 1970 the U.S. Census put the county's population at 37,305.

[19] Telephone interview, Jim Hislop, Oct. 22, 2001, Round Rock. Another federally funded "New Town" was The Woodlands, north of Houston.

Nelson encouraged his Round Rock friends to throw a party for Westinghouse executives. The party should be pure Williamson County, Nelson advised—barbecue and beer on someone's patio and a country-and-western band. The Westinghouse officials, accustomed to Pittsburgh, Pennsylvania, were unsure about what to expect. Would drunken cowboys molest their wives? Could they bring their children? "They just want to get to know you," Nelson reassured them. "It will be perfect for your kids and wives." He was right. "These people couldn't *believe* these folks from Round Rock were treating them so well. They had a great time. That party really sold the place."

In the spring of 1971 Round Rock officials offered a sweetheart deal to Westinghouse, promising nearly the world in return for the prospect of acquiring a projected fifteen hundred local jobs and a huge residential development to which Round Rock would provide water and sewer services. Round Rock made two major concessions: it promised not to annex Westinghouse for seven years; and Westinghouse would not pay a penny for the massive water and sewer system Round Rock would build to service the plant and its outlying "new town."[20] Had Westinghouse built its own utilities, it would have cost an estimated quarter of a million dollars.[21] Round Rock offered to build the utilities, so long as Westinghouse "would permit sales of 300,000 gallons of water to customers other than Westinghouse."[22] Cost to Westinghouse: nothing. Estimated cost to Round Rock: 160,000 dollars.[23] Westinghouse accepted. By January 1972 company

[20] Letter, Dale Hester to George Chapman, Apr. 23, 1971, Administration Department, City of Round Rock archives (City Hall, Round Rock). Hester was Round Rock's mayor; Chapman was Westinghouse's top man. Also see contract, concluded Mar. 8, 1974, by the City of Round Rock and Westinghouse Corporation.

[21] Letter, Tye Collins to Dale Hester, Mar. 31, 1971, ibid. Collins was an engineer with Knowlton-Ratliff-English-Collins, consulting engineers of Fort Worth and Austin, which Round Rock had hired.

[22] Letter, Dale Hester to George Chapman, Apr. 23, 1971, ibid.

[23] "Water Supply—Westinghouse Plant," plus enclosure, ibid. Five options were studied; Plan 5, with the lowest cost to Westinghouse and the highest cost to Round Rock, was Round Rock's choice. As it turned out, the cost ran much higher because the size of the pipeline required by Westinghouse was larger than originally planned. I have not attempted to determine an accurate "final cost," but according to Round Rock City Council Minutes attached to an agreement between the City of Round Rock and

managers were interviewing candidates for 750 jobs at their brand-new sixty-thousand-square-foot, rust-colored metal facility on Interstate 35.[24]

Viewed narrowly, Round Rock fared poorly in its relationship with Westinghouse. For one thing, city officials woefully underestimated the operation and maintenance costs of producing and distributing water to the manufacturing facility. For another, Westinghouse consumed less water than Round Rock had projected. (Round Rock had expected its water to cost ten cents per thousand gallons, and billed accordingly. In fact, the actual cost was $1.08 per thousand—ten times as much.)[25] The decision to leave Westinghouse outside the corporate city limits meant that Round Rock could not garner tax money from Westinghouse. "That was a mistake," said a former city official. "We should have annexed the plant."[26] When the 1974 oil crisis hit, the market for the plant's product—gargantuan gas turbines used by power plants to generate electricity—dried up overnight. Westinghouse shut down to retool and laid off four hundred employees, over half its work force.[27] The "new town" was stillborn. "The

---

Westinghouse Electric Corporation dated March 9, 1972, Westinghouse would front the cost of construction of the water and sewer utility by no more than $561,000; when Round Rock completed construction it would repay Westinghouse all the advanced monies; and then, pending an Internal Revenue Service ruling, Westinghouse would give Round Rock a "gift" of $241,000.

[24] Kathleen Sullivan, "Shutdown Caps Westinghouse Struggle," *Austin American-Statesman*, Oct. 31, 1986, p. G-1.

[25] Letter, Stephan L. Sheets to Westinghouse Electric Corporation, Oct. 12, 1978, with attached study, "Cost to Produce and Distribute Water for Westinghouse." Administration Department, City of Round Rock (City Hall, Round Rock). Sheets was Round Rock's city attorney, and the letter started negotiations for a second seven-year contract with Westinghouse.

[26] Telephone interview, Jim Hislop, Oct. 22, 2001, Round Rock.

[27] Kathleen Sullivan, "Shutdown Caps Westinghouse Struggle," *Austin American-Statesman,* Oct. 31, 1986, p. G-1. Eventually Westinghouse built back up to 850 workers. But in 1986 corporate officials announced Westinghouse would halt manufacturing operations in Round Rock. On January 1, 1988, 300 of 450 employees were laid off and production was moved to Taiwan. A joint venture, TECO Electric and Machinery Corporation, continues to utilize the old plant. Also see Mark Mitchell, "Westinghouse Looks to Year 101," *Williamson County Sun* (Georgetown), Jan. 8, 1986, p. A-3; Mark Mitchell, "Westinghouse Destiny Chills Area Leaders," ibid., Nov. 21, 1986, p. A-1; Mark Mitchell, "Down and Out: Westinghouse's Loyal Corps," ibid., Jan. 10, 1988, p. A-1.

Westinghouse development deal turned on profits they were going to make selling those turbines," said an observer. "The idea of a 'new town' was peaches on top of the ice cream."[28] Instead, Westinghouse had to go on a diet.

But the broader lessons of Westinghouse taught Round Rock many valuable things. It established the village as a major industrial recruitment player. It taught Round Rock how to capture industrial prospects years before "economic development" became a Texas watchword. It motivated Round Rock to become a big-time industrial center: the city owned water and sewer utilities halfway to Georgetown, and it needed to sell the utility's products. And most importantly Westinghouse made clear that Round Rock *needed* industrial and commercial development.

The city's first city manager, Jim Hislop, said that after he was hired and started examining the city's finances, "I realized that single family residential [development] couldn't pay for itself. You could have three kinds of development—multifamily residential, commercial, or industrial. The importance of Westinghouse wasn't so much that our skirts were tied to Westinghouse. It was that we realized what we had to have. Then we went after McNeil Laboratories. When we got them, we said, 'We can do this!' Up to that time, we thought Westinghouse was a fluke."[29]

Out of what Round Rock learned from Westinghouse came an impressive series of industrial relocations to Round Rock. In the next decade McNeil Laboratories moved there. So did Farmers Insurance Group, Hughes Tool Company, Texas Tool and Fastener, Cypress Semiconductor Corporation, AMP Packaging Systems, Inc., Tellabs Texas, Inc., Weed Instrument, Inc., Du Pont Tau Laboratories, Texas Nuclear Corporation, Applied Information Memories, Inc., Carroll Touch Industries, and many others. Some of these ventures moved, were sold, or failed, but others always seemed to be waiting in the wings to come to Round Rock.[30] (Eventually Dell, among the most

[28] Telephone interview, Jim Hislop, Oct. 22, 2001, Round Rock.

[29] Ibid.

[30] Terry Goodrich, " . . . to Tour Sites," *Austin American-Statesman,* May 13, 1984; Mark Mitchell, "Round Rock's Industry Success Glitters," *Williamson County Sun* (Georgetown), Apr. 24, 1985, p. A-5; Mark Mitchell, "Round Rock Recruits a New

successful computer companies in the world in 2005, relocated there from Austin.) Thousands of jobs were created between 1972, when Westinghouse opened, and 1986, when the collapse of oil prices and financial institutions contracted the Texas economy. Round Rock's population grew from 2,811 in 1970 to 12,740 in 1980 to 30,923 in 1990 to 61,136 in 2000 to 78,000 in 2004.[31] Overall Round Rock's growth rate averaged a steady 500 percent over a forty-year span—virtually unprecedented anywhere.[32]

Westinghouse gave Round Rock leaders confidence and a national reputation for kindness to big corporations, especially high-tech ones. When California's Cypress Semiconductor met with Austin and Round Rock recruiters, Round Rock mayor Mike Robinson "reached into his pocket, pulled out a building permit and handed it to Cypress."[33] The Cypress search team canceled its plane reservations to California, drove to Round Rock, and chose a site. "When did we see the light?" asked banker Whitlow. "It was in 1972, when Westinghouse came to town. That was when we realized what Round Rock could become, though we never imagined the size of it."[34]

Westinghouse figured large in Georgetown's subsequent develop-ment, too, but in an entirely different way. Many of Westinghouse's managers and engineers moved to Georgetown because it had more housing choices than Round Rock, a serene university setting, and a courthouse square dripping with potential charm. The Westinghouse families' influence on the community led directly to a more open,

---

Computer Plant," *Williamson County Sun* (Georgetown), Oct. 12, 1986; Paul Schnitt, "Round Rock Basks in Recruiting Victories," *Austin American-Statesman,* Dec. 4, 1986, p. A-1.

[31] U.S. Census, 1970, 1980, 1990, 2000; City of Round Rock Planning and Devel-opment Department, Round Rock.

[32] Telephone interview, Jim Hislop, Oct. 22, 2001, Round Rock.

[33] Paul Schnitt, "Round Rock Basks in Recruiting Victories," *Austin American-Statesman,* Dec. 4, 1986, p. A-1.

[34] Interview, N. G. "Bunky" Whitlow, July 1, 1998. Whitlow was appointed president of Farmers State Bank on January 27, 1972, after Tom Nelson Jr. bought the bank from Tom Joseph. To many observers, it appeared that Whitlow was Round Rock's behind-the-scenes leader throughout much of the 1970s and 1980s. Under Whitlow's leadership, the bank "boomed" Round Rock, but when the economy faltered in 1986, the bank was hurt badly.

professional city government emphasizing citizen oversight and planning—a movement that tended to cast a wary eye on unbridled development, as opposed to Round Rock, which embraced growth.[35] Leo Wood, Georgetown's city manager from 1968 to 1985, vividly recalls the "Westinghouse effect": "When Westinghouse came, what a change!" he exclaimed. "Their people had different personalities from what we were used to, different thoughts on the planning process. Most of them came from Pittsburgh or Minneapolis-St. Paul, Minnesota. They had seen growth and didn't necessarily like what they saw. They got very involved, helped set the tone and policy for our comprehensive plan. People like Frank and Helen Hubbard and MaryEllen Kersch came out of that."[36]

In 1972, the year Westinghouse opened its I-35 plant, Bobby Stanton arrived in Georgetown. After his graduation from the University of Texas at Austin, Stanton had gone to work at Tim's Air Park, a small private airport off I-35 near Pflugerville. Soon, with his wife's help, he was managing the airport—teaching flying, repairing airplanes, and selling planes and aviation fuel—"working our tails off," in his words. He got to know Walter Yates, an aviation buff who spent summers prospecting for gold in Alaska. Yates owned three hundred acres of land two miles west of I-35 on Andice Road northwest of Georgetown. Stanton wanted to buy and develop Yates's land, and five years at Tim's gave him the stake he needed. Georgetown Savings &

---

[35] This could qualify as a chapter on its own. Georgetown's ongoing desire for "transparent," well-planned development repeatedly collided with its baser needs (such as a commercial and industrial tax base), resulting in occasional rejections of mayors and city managers. Among these were Leo Wood, Bob Hart, and Mary Ellen Kersch. The trend that Westinghouse perhaps started (though some might argue that Southwestern University played this role earlier) accelerated in the 1990s when the Del Webb Corporation built Sun City, and Georgetown took that potentially giant community into its corporate limits.

[36] Interview, Leo Wood, June 30, 1998, Georgetown. Actually, MaryEllen Kersch arrived in 1980 with IBM, but Wood's point was still valid. Some years after Wood left the city manager's position, he was elected mayor of Georgetown, a position he held during this interview. In the spring of 1999 he was defeated by Kersh, the "newcomer." In an ironic twist, in early 2002, Mayor Kersh was recalled in Georgetown's first recall election, largely by votes of "newcomers" from Del Webb Sun City, an age-restricted planned community.

The Unearned Increment

Loan wanted to get into the development business and backed him.[37]
By the end of 1972 Stanton had carved winding roads, set up utility service, and opened the rambling Serenada Country Estates, the first major development "across the interstate" from Georgetown—and the county's largest. It was an immediate success. Employees of three north Austin and Round Rock blue chip technological companies—I.B.M., Texas Instruments, and Westinghouse—made up roughly 50 percent of Serenada's homeowners. The development lay outside Georgetown's corporate limits, but inside Georgetown's pub-

---

[37] Interview, Bob Stanton, July 8, 1998, Georgetown. Also see "Stanton the Man: How He Developed," *Sunday Sun* (Georgetown), Nov. 20, 1983, p. A-1. Georgetown Savings & Loan was owned by Grogan Lord, who brought in "Skip" Morse and Jay Sloan to "grow" his Georgetown financial institutions. Their interest in Stanton's land may have been triggered when Austin Savings & Loan "took over" San Gabriel Heights in southwest Georgetown. "The local S&L people thought if outsiders were coming in to do a development, maybe they should, too," Stanton said.

Road, River, and Ol' Boy Politics

lic school district. "I-35 obviously played a major role in making it work," Stanton mused years later. When Serenada Country Estates opened, Austin was twenty-five minutes away, with little traffic. "What interested me about the land was that beautiful treed stuff," he said. "It was close to Georgetown, but not too close—no taxes. And I guess everybody had heard about the lake."[38]

Stanton next acquired land from I. M. Hausenfluke, whose family had ranched west of Georgetown for generations. Hausenfluke's land became Sanaloma Estates, Brangus Ranch, and a portion of Serenada East, all just north of Georgetown Municipal Airport.[39] Between 1972 and 1986 Stanton acquired roughly 3,500 acres northwest of Georgetown, conveniently close (but not too close) to I-35. In addition to Serenada Country Estates and Sanaloma, he personally created Serenada West, Serenada East, Brangus Ranch, Reata Trails, Logan Ranch, and Berry Creek subdivisions. Ultimately these subdivisions physically dominated Georgetown.[40] On a map their bulk looks larger than Del Webb's Sun City, the age-restricted planned community which in May 1995 broke ground and almost overnight changed the city's demographic and political complexion.[41] Before Sun City, Stanton had developed most of Georgetown west of I-35, which

[38] Interview, Bob Stanton, July 8, 1998, Georgetown.

[39] Ibid. Stanton said that for most of the land he purchased northwest of Georgetown he paid in the neighborhood of $1,000 an acre.

[40] Map: "Welcome to Georgetown Texas," 1992, Mosher-Adams, Inc., Oak Creek, Wisconsin (on which Stanton marked his developments), Tobin Surveys Inc. (San Antonio, Tex.); map: "North Half and South Half Williamson County, Texas," 1969, ibid.; map: "Williamson County, Texas," 1999, by John V. Cotter, Map Ventures (Pflugerville, Tex.); "Stanton the Man: How He Developed," *Sunday Sun* (Georgetown), Nov. 20, 1983, p. A-1.

[41] By 2002 Sun City's voters had killed a school bond floated by Jarrell Independent School District, which lay inside part of the development which was to contain 9,000 "units," or houses; vetoed a Georgetown City Council bond proposal for library and parks additions; and kicked out Mayor Kersh in a recall election that split Georgetown down the middle, between "Old Town" and west of Interstate 35, particularly Sun City and Berry Creek. In August 2001, Del Webb Corp., developer of Sun City Georgetown, sold out to Pulte Homes, Inc., one of the nation's largest home-builders for middle class families. Since then, Pulte has made plans to build 5,500 non-age restricted homes—radically changing the mission of the original Sun City, which was aimed solely at retirees.

The Unearned Increment

today dwarfs Old Town. Stanton's genius flowered during a time when commuting in Texas was seen as a civilized pastime for upper-middle-class professionals. His developments west of Interstate 35 reflect what he believed commuters wanted. He lived the life himself, except that he worked in Georgetown, right in the middle of his development. While developing Serenada Country Estates, Stanton built a house for his family in the heart of Serenada. He still lives there. "I lived right in the center of my little world," he said. If somebody's toilet didn't work, it was, 'Call Stanton.' If there was a hole in the road, 'Call Stanton.' It never entered my mind to make a bunch of money and skip the country."[42]

His strategy was to "focus on where I lived, within a three-mile radius. I did have this idea about Andice Road [which now leads to Sun City]. I knew it would develop. Eventually, I owned all the land on Andice Road starting at the strip center next to The Pit to three miles out, except for two eighty-acre tracts. Once I paid $12,000 an acre for a small piece of land where the bank and strip center are. Even I wondered whether I paid too much. But that land sold the other day for seven dollars a foot, $280,000 an acre."[43]

"Bobby Stanton was probably the first person to envision modern Georgetown," said Bill Snead, owner of Texas Crushed Stone, one of Georgetown's most important industries. "I think Stanton was the unrecognized pioneer of modern Georgetown. Just think about what he did—Serenada, Berry Creek, McDonald's."[44]

---

[42] Interview, Bob Stanton, July 10, 1998, Georgetown. Like many Texas developers, Stanton nearly lost everything during the 1986 real estate bust. He had just finished developing Berry Creek, with a beautiful eighteen-hole golf course and a deluxe country club when Texas's financial bubble burst, and he had to give up ownership of his pet project. Since then he has recovered, completing several smaller projects for low- to middle-income homeowners, including Crystal Knoll Terrace and Raintree east of Old Georgetown.

[43] Interview, Bob Stanton, July 8, 1998, Georgetown.

[44] Interview, Bill Snead, July 2, 1998, Georgetown. Aside from Southwestern University and government, Texas Crushed Stone has long been Georgetown's largest employer. It moved to Georgetown from Austin in 1958, before Interstate 35 was built, because of the excellent rail connections it acquired on the site. But after I-35 arrived, the company handsomely capitalized on its enhanced capacity for trucking crushed stone to Austin and points around the state.

McDonald's? "It was this thing I had about Andice Road," Stanton explained. "As an investment, I bought a couple of acres where I-35 met Andice Road. It didn't take a rocket scientist to figure out it was going to be a major corner." Old-timers shook their heads. "It was said he paid one hundred thousand dollars for that old house and property," said one. "Everybody thought he was crazy. We thought people were taking advantage of this young stranger."[45] Then McDonald's bought the land. The corner proved a gold mine for Stanton and for the world's leading hamburger chain.

Beyond the stories of enterprising and farsighted individuals, what happened when the interstate showed up in Williamson County? Economically, how did I-35 change the lives of farmers, ranchers, investors, and townspeople? Did I-35 trigger the "unearned increment" vilified by the seventeenth-century British philosopher, John Stuart Mill? To unlock the landed gentry's stranglehold on Great Britain's land, Mill suggested that the state tax any "unearned increase" in the value of land when the owner made no "exertion or sacrifice" to that end. Mill argued that "The growth of towns, the extension of manufactures, the increase of population consequent on increased employment, create a constantly increasing demand for land. . . . By this increase of demand the landed proprietors largely profit, without in any way contributing to it."[46]

But Mill was trying to reform land laws in a country where a tiny proportion of its citizens owned virtually all the land. Mill's European peasant proprietors, "in the ungrudging and assiduous application of their own labour and care," were an entirely different matter from Black Waxy farmers who descended from European proprietor peasants, and Balcones ranchers, who descended from starving tenant farmers of Scotland and Ireland via the Ozarks and Appalachia. Williamson County's landowners did not remotely resemble Mill's landed gentry, most of whom idled while tenant workers toiled

---

[45] Interview, Sam Brady, July 7, 1998, Georgetown.

[46] John Stuart Mill, "The Right of Property in Land," *London Examiner,* July 19, 1873, in *Collected Works of John Stuart Mill,* Vol. III, *Newspaper Writings,* ed. Ann P. Robson and John M. Robson (Toronto: University of Toronto Press, 1986).

The Unearned Increment

on their land, paying rent "that runs into men's mouths as they sleep."[47]

Of course, it is true that the American taxpayer footed the bill for Interstate 35, not to speak of the Laneport and North Fork dams. And it is true that, for the most part, landowners along Interstate 35's route prospered hugely from the national investment in Williamson County. But as Charles Forbes discovered to his sorrow and Jarrell to its municipal misery, the interstate's impact on rural property values was not always a good thing. Much has been written about the impact of America's interstate system on urban property values, where engineers typically laid down six or eight divided lanes of concrete, plus access roads, through the hearts of dense neighborhoods that would never recover from the shock. For the rural areas of Williamson County the pattern was more complex. Some rural towns boomed while others died; some people benefited, some suffered.

[47] John Stuart Mill, "Should Public Bodies Be Required to Sell Their Lands?" *London Examiner*, Jan. 11, 1873, ibid.

Below: Looking east toward "Northtown" (north Georgetown) in 1965, with Interstate 35's San Gabriel North Fork bridge under construction. San Gabriel Park curls to the left of the river, whose regular flooding periodically stripped brush and trees from the flood plain. *Courtesy Texas Department of Transportation, Austin.*

Aerial view of same scene in 2005, shot from a higher altitude and further away. The two complexes at the bottom of the frame are SuperWal-Mart and Home Depot; Georgetown's central business district is above and to the right of I-35, while new commercial development lines the interstate. *Clark Thurmond Photo, Courtesy* Williamson County Sun, *Georgetown.*

Among those who benefited from I-35 were a number of ranchers who had struggled to survive the fifties. As the highway builders approached them with right-of-way offers, most of these ranchers immediately cashed out. Some felt nostalgia, even grief. But now they had money in the bank, some for the first time in their lives. Former Round Rock councilman S. C. Inman Jr. inherited ninety acres of land west of U.S. Highway 81 on the north side of town, across the highway from the Texas Baptist Children's Home. Inman farmed that land for a decade. "One year it got so dry nothing came up, so I had to go to work for someone else," he said. He and his wife ran the Clay Pot, a downtown eatery, to make a little cash.[48]

"There never was anybody who really wants a highway to go

---

[48] Interview, S. C. Inman Jr., July 1, 1998, Round Rock. The Inmans eventually sold the restaurant to a Taylor man for $500.

Aerial view of Round Rock in 1968, right after Interstate Highway 35 opened. Austin White Lime's pits, west of the interstate, were one reason the Texas Highway Department avoided changing the interstate's route. The road crossing the interstate at the bottom of the frame is FM 1325. *Courtesy City of Round Rock.*

Aerial of Round Rock in 2005, same view. At the bottom of the photo La Frontera, whose developers bought Tom Kouri's ranch, dominates the left side of I-35
at the FM 1325 intersection, while Dell Computers lies to the right of I-35. *Courtesy Ercel Brashear, Georgetown.*

through their place," he said, "and yet, you're sort of *hoping* it *will*. Dad had tried to sell the place for years and years. Nobody would look at it. Then I got this call from some fellow representing the highway department. He said, 'I'll give you one hundred dollars an acre for the land and the house, which came to eight thousand, eight hundred dollars. I knew I couldn't do better. I wanted more, but it was okay."[49]

Of course, if Inman had waited, his land would have fetched much more. Still, he was relieved to be shed of his land. He professes no regrets. Other sellers seem to have felt similar relief at the money they made, "unearned increment" or not. Even if it couldn't begin to match the standards of what came later.

Williamson County experienced Interstate 35 as an onslaught of federal power. The road was created in Washington and Austin; it would change everything it touched, depending on how local city officials, citizens groups, and entrepreneurs reacted to it. As a public works project, the interstate can be seen as a collage of forces—utopian, bureaucratic, democratic-populist, small-town boosterism, and capitalism writ large, represented in Williamson County by people like Forbes, Kouri, Nelson, and Stanton. Overall, the local interests provided a surprisingly sensible balance (with significant help from the Texas Highway Department) against a federal enterprise that was theoretically apolitical and rigidly egalitarian, but which failed to measure up to either of these goals. The nation's top highway engineers, so the theory went, would build the best highways for the greatest good, without political interference. Since Congress mandated the interstate system as a national concept, rather than requiring legislation for each individual road (as they did for dams), members of Congress did not tend to challenge the interstate builders. When Congress did challenge the program, it generally targeted waste and graft, not poor planning. With few exceptions, until the interstates were well under way Congress did not question the astonishing lack of regard that the interstate's builders seemed to have for the poor and powerless who happened to live in the project's way—whether the black and Hispanic residents of East Austin, Jarrell's struggling business owners, or a clutch of retired colonels and genteel preservationists in

[49] Ibid.

Round Rock—a lack of regard familiar to farmers whose farms were lost to accommodate dams on the San Gabriel River. Ironically, as the highway builders rolled their ribbon of asphalt across Williamson County, a handful of locals—County Judge Stone and businessmen like Henna, Stanton, Kouri, and Sherrill—discerned the course of the population surges the highway would bring to the county far more acutely than the interstate's creators.

A PLACE REMADE

It is fair to say that most people who anguish over the
population problem are trying to find a way to avoid the evils
of overpopulation without relinquishing any of the privileges
they now enjoy.

GARRETT HARDIN, *The Tragedy of the Commons*

It is September 1, 2003. My husband is flying our family back to
Georgetown from Brownsville, where we spent Labor Day on the
beach at Padre Island. We are bumping through patches of dirty
weather, remnants of a weak hurricane that flooded Houston and
drenched most of Texas all weekend. Below us nothing but rangeland
stretches to the horizon, essentially unchanged since Spanish *ranchero*
owners colonized South Texas two centuries ago. Corpus Christi
appears ahead, a city of miniscule Monopoly pieces dwarfed by anvil
clouds swollen with thunderbolts. After veering east over Corpus
Christi Bay to avoid getting struck by lightning, we enter a cathedral
of clouds. The airplane is enveloped in a fuzzy gray netherworld.
I pray that pilot and airplane instruments are humming along in perfect harmony.

When we break through the bottom layer of our cumulus womb
east of San Antonio, we see Blackland Prairie farms squared off on the
earth below. Water is everywhere; the ground fairly groans with wetness, the stock tanks are full. As I marvel at this wetness in a normally
dry land, it occurs to me that the weather pattern we have trailed
from Brownsville meteorologically resembles the killer storm brewed
in the Gulf of Mexico eighty-two years ago, almost to the day. A

smudged line of camouflage green signals the Balcones Escarpment, curving across the horizon as if we were seeing it from space instead of six thousand feet above sea level. This agricultural world looks not much different, I think, from its 1921 precursor. But as we near Williamson County, the landscape begins to change.

East of Austin, the Black Waxy never "played out," as Texas's once-copius oil fields did, but was rendered obsolete by a global and corporate farm economy that has made the small American family farm close to worthless, save as hobby or real estate investment. Here orchards and row crops barely register against their neighbors: gigantic trailer-home developments where mobile homes swirl in concentric circles like platters of raw oysters for partygoers, densely packed neighborhoods of "starter" homes, and leisurely sprawls of "Tara" mansions and faux ranches owned by Austin's supremely wealthy class, the "Dellionaires" who got rich on the strength of college dropout Michael Dell's computer marketing wizardry.

As we fly over Williamson County, agricultural land becomes the rare green daub surrounded by a sea of residential development. Spilling off the Balcones Escarpment and pushing Austin's flanks north and east, tumbling down from Owen W. Sherrill's "highlands" of Williamson County at the wellsprings of the San Gabriel River—from Cedar Park, Leander, and Liberty Hill, from the "rural" suburbs of Anderson Mill and Brushy Creek—gush great rivers of subdivisions, shopping centers, malls, and a bewildering crosshatching of highways. Spilling down from those heights and across Interstate Highway 35 is a flood of human ambition, energy, and engineering, pushing inexorably to enfold Hutto, Georgetown, and Taylor. Distant farm villages like Walburg, Andice, Florence, and Granger await the same fate with excitement and fear. The airplane descends and the earth rearranges itself, ever so slightly, from a form more concave than convex. As we drop from the sky, the Balcones Escarpment rises and blots out the sinking sun, touching off a firestorm along its rim; the farm bowl below us flattens and disappears into the gloaming. Our wheels touch the runway at Georgetown Airport and we are home.

Round Rock led Williamson County's transformation. In 1973, when Jim Hislop arrived for an interview as Round Rock's first city

Road, River, and Ol' Boy Politics

manager, it was a speck of a town. Hislop was fresh out of college with a planning degree and, in his words, "dumber than a boxload of rocks, but I thought I was smart." Until being recruited for the job, he had never heard of Round Rock. This was not surprising since it was indistinguishable from most small Texas towns of that era, except for the thick coat of limestone dust expelled by its chief industry, a white lime plant. After a good hard rain, the lime would wash off the trees. Then it would build up again until the town took on a ghostlike aspect. The population sign on the edge of town read 2,811. Having just snagged the big Westinghouse plant, the city's leaders decided they needed a city manager.

Moments after he arrived for his job interview, Hislop scanned the city's financial books. He remembers nearly choking when he saw the city barely had three thousand dollars in the bank.[1] The job he was being offered paid seven hundred dollars a month.[2] "I was amazed," he confessed, thinking this was absolutely the wrong place for him. "I decided right then I would go to graduate school. I walked across the street to the bank, where I was supposed to get introduced around, intending to tell them I couldn't take the job. But when I came away from that bunch, I had changed my mind. I thought if any group of people wanted it that bad, I would go for the deal."[3] "It" was growth. Hislop's planning-school friends thought he had lost his mind.[4]

Round Rock has been "going for the deal" ever since. The men who convinced Hislop he could lead Round Rock into a new age wanted growth—star quality growth—at an embryonic stage of the town's development. But the little city already had huge problems as a result of its ambition. Westinghouse, the feather in Round Rock's cap just two years earlier, now threatened financial disaster. To get Westinghouse, the city had built an extensive waterworks system, but few customers had hooked on. The thirty thousand people Round Rock had counted on to populate Westinghouse's "new town" had turned

[1] Interview, Jim Hislop, July 2, 1998, Georgetown.

[2] Interview, Bob Bennett, Sept. 14, 2000, Round Rock.

[3] Interview, Jim Hislop, July 2, 1998, Georgetown. The bank was Farmers State Bank, owned by Tom E. Nelson Jr. and run by N. G. "Bunky" Whitlow, from which most of Round Rock's major economic moves were orchestrated.

[4] Interview, Bob Bennett, Sept. 14, 2000, Round Rock.

into a fairy tale. Round Rock was growing, all right, as Anglos fled a new Austin public school policy of busing minority students to "white" schools, but this residential boomlet did little to subdue Round Rock's economic headaches.[5] Interstate 35 made it easy for people to commute from Round Rock to Austin, but the school district was in crisis, overloaded with students and strapped for cash.[6]

Round Rock's leaders wanted to capture industry that could pay for the city's expensive new utility plant, boost the city's tax base, and improve its schools. As soon as North Fork Dam's gates closed on the San Gabriel River, the city would control enough water to serve any number of new industries and subdivisions, but that would not happen for six years. "The deciding factor was the acquisition of water from Lake Georgetown," Hislop remembers. "Without that water, we never could have gotten McNeil Laboratories, Cypress Semiconductor or even Dell."[7] Another factor played a critical part in Round Rock's industrial development: the "people's desire to grow," as Hislop put it. "It was incredible. Usually twenty percent of a community wants growth. But in Round Rock, *everybody* was in on the deal." Probably that stemmed from Round Rock's small size and stability through the sixties and seventies, as well as the outsized influence of Louis Henna.

If a town could have been genetically encoded, Round Rock's genes would have measured strongly pragmatic. Seventy years earlier it had rebuilt itself in a new location after a Brushy Creek flood washed away its business district, the original "Brushy Creek" that lay dormant until the "army colony" gathered there after World War II.[8] The little town clung to life through the Depression and war years by the grace of Highway 81. Compared to Georgetown and Taylor, Round Rock had little going for it. Compared to many small towns in Texas, it seems to have been remarkably egalitarian. Few African

---

[5] Mark Mitchell, "Boom. Bust. Comeback." *Williamson County Sun* (Georgetown), Sept. 30, 1998, p. E-8.

[6] Interview, Noel Grisham, Round Rock school superintendent, July 1974, Round Rock.

[7] Interview, Jim Hislop, July 2, 1998, Georgetown.

[8] This was the "Old Town" that Helen Irvin tried in 1956 to save from destruction by I-35.

James ("Jim") Hislop, early 1970s.
Courtesy Jim Hislop, McNeil, Texas.

N. G. ("Bunky") Whitlow. Courtesy N.
G. ("Bunky") Whitlow, Round Rock.

Americans lived there, but Mexican Americans owned half a dozen
businesses and were politically active, though they were overwhelmingly outnumbered by Anglos. While Hislop managed the city's
affairs, African-American Garfield W. McConico, a Henna protégé
whose influence would be felt for decades, and Lorenzo Rubio, a
Mexican American, were elected to the city council. In the early seventies this was far from normal in Williamson County or, for that
matter, in Texas.[9]

During Hislop's four years as city manager, Round Rock grew
from a "real" population of thirty-five hundred to fifteen thousand—
300 percent. In 2004 seventy-eight thousand people lived within its
city limits.[10] Between 1975 and 2000, the City of Round Rock grew

[9] McConico served on the Round Rock Council for twelve years, part of that time
as mayor pro tem. Today Round Rock's Planning and Development Department is
housed in the McConico Building. In the early 1970s, Rubio served on the council and
Chris Perez was a director of the Chamber of Commerce, while Isaac Lopez Jr. was a
member of the Round Rock Independent School Board from 1971 to 1980.

[10] Round Rock Planning and Development Department, Feb. 24, 2004, City of
Round Rock, Round Rock.

1,600 percent, an extraordinary growth rate to maintain for such a long period of time. "The overriding force was that the people wanted growth. They never varied from the course," Hislop said. Shaping and solidifying the people's support were four important leaders: Chevrolet dealer Louis Henna and three principals at Farmers State Bank—Gus Lundelius, Tom E. Nelson Jr., and N. G. "Bunky" Whitlow.[11] All were key players in Round Rock's development, but in Hislop's view, Whitlow was most important. "Round Rock would not have grown as fast as it did and with the quality it did if Whitlow had not been there," Hislop said. "If Nolan Ryan had come to Round Rock with a baseball team in 1973, he would have gotten a stadium in weeks instead of years. 'Bunky' would have declared one of his work days and everybody would have pitched in. Anybody who owned a bulldozer or road grader would have been out there clearing ditches."[12]

In the summer of 1976 disaster struck. A drought dried up Brushy Creek and the wells that, at that time, produced all of Round Rock's water. Though Round Rock owned rights to half the water from the reservoir at Georgetown, the dam was still years from completion. Industrial prospects evaporated. The city slapped a moratorium on development. Here was "go-go" Round Rock, the envy of small-town Texas's industrial recruiters, dying of thirst. Hislop telephoned an old planning-school buddy who was working in Houston's planning department. "The stock tank's dry and the fish are dead in the creek," he told his friend. "Bobbo, I need you."[13] Thus Round Rock acquired Bob Bennett as its first city planner. A year later, when Hislop stepped down under fire, Bennett took over as city manager. He directed the city until 2002, when he retired, the grand old man of Texas's city managers. Bennett capped his career by bringing Dell Computer Corporation, which would become the world's leading computer supplier, from Austin to Round Rock's undeveloped southern perimeter, infus-

[11] Whitlow's predecessor at Farmers State Bank, Gus Lundelius, another Henna man, may have been equally important as a city builder, though his quiet style was entirely different from Whitlow's.

[12] Interview, Jim Hislop, July 2, 1998, Georgetown. Hislop was referring to Round Rock's hesitation in the late 1990s to build a multimillion-dollar stadium for the Round Rock Express team owned by ball player Nolan Ryan.

[13] Interview, Bob Bennett, Sept. 14, 2000, Round Rock.

ing the city's coffers with a stunning mix of tax monies garnered from Dell's real property and its sales operation.[14]

Bennett believes credit for Round Rock's rocketship growth trajectory must include Austin's aversion to growth, the results of which pushed massive housing developments and significant industries into Williamson County. Lake Georgetown's water (to be augmented by a pipeline from Lake Stillhouse Hollow in Bell County), Interstate 35, and local backing for growth were critical as well. But Bennett thinks Hislop, the unbureaucratic bureaucrat, created the future. "We'd be riding around at night in Hislop's old truck, drinking beer, and he'd be seeing it all, pointing it out as if it were real: 'Here's where fifteen blue chips will be lined up,' he'd say. There was nothing there, just old ranch country. He could drive you around in that truck and almost convince you of anything. At first I thought, he's nuts. Then, you start thinking, he's right! This is just waiting to happen. He had the vision."[15]

It is hard to escape the conclusion that one of Round Rock's greatest assets in reinventing itself as a budding Edge City was an unusual cluster of talented visionaries working in concert. Henna, Nelson, Lundelius, Whitlow, Hislop, and Bennett, among others, kept the city driving hard toward a single goal: avoiding the typical bedroom-community economic trap by aggressively recruiting high-technology industry and big-box retail centers, and figuring the rest would fall into place. By doing so, Round Rock took the stage and rewrote the play in Williamson County.

————————

As in other Dry Sun Belt hot spots, with the advent of the interstate and control of an ample new water supply, Williamson County experienced profound change over a relatively short time. Despite global economic woes of the early twenty-first century, it remains one of the nation's fastest-growing counties, enjoying prosperity of a sort unimaginable in pre-interstate, pre-dam days.[16] The county no

[14] Bennett briefly retired to work for a development company that went broke after the economic collapse of 1986. Round Rock welcomed him back with open arms.

[15] Interview, Bob Bennett, Sept. 14, 2000, Round Rock.

[16] In 2000 the median household income in Williamson County was $60,642, according to the Bureau of Labor Statistics, compared to its 1990 median household income of $33,695, a Capital Area Planning Council figure.

longer depends on agriculture, which in 2001 contributed only 67 million dollars of the county's 28 billion dollars of annually produced wealth.[17] Rather its economy was driven by Dell Computer's Round Rock headquarters, whose sixteen thousand employees in 2002 led the U.S. in home-computer sales and the world in business- and consumer-computer sales.[18] The county's other leading economic forces included computer-component and other high-technology companies, the real estate and construction industries, and retail sales from new malls and "power centers"—the new term for huge constellations of retail stores like La Frontera in Round Rock. As Austin's economy bellyflopped in 2002, causing its population to shrink, Williamson County's suburbs continued to grow.[19]

In thirty years—actually less—the fortunes of east and west Williamson County completely reversed. Before Interstate 35 and the dams, Taylor's position on the rich Black Waxy enabled it not only to dominate the county economically and politically but to play a minor role in state and national politics. But when Granger Dam permanently flooded the county's rich farm belt, eastern Williamson County started a downward spiral that has manifested itself as widespread poverty. (To be fair, the trend mirrored what was happening to agriculture throughout the nation, but east Williamson County's experience was particularly severe.) Meanwhile Lake Georgetown and Interstate 35 vastly increased the ability of city leaders and entrepreneurs to "grow money" in west Williamson County, where once a handful of ranchers had barely scratched out livings. The effect of this shift was seismic. When the engineers finished their public works projects, Williamson County's land contours remained stable, but everything else changed. Agriculture, the county's bedrock, was swept aside. It was as if the hand that, seventy million years earlier, had torn a seam through the land and tipped the richest soil in North America

---

[17] Interview, Ronald Leps, Feb. 20, 2002, Georgetown. Leps served as the county extension agent for 20 years.

[18] Dell was also the world's leading seller of computers to businesses and consumers combined in 2002, according to an IDC research study. John Pletz, "Dell Takes Over Top Spot in Home PC Sales," *Austin American-Statesman*, Mar. 6, 2002, p. A-1; Amy Schatz, "The Dell Curve," ibid., June 22, 2003, pp. A-1, A-16.

[19] Robert W. Gee, "Austin was Lone Big Texas City to Shrink in '02," *Austin American-Statesman*, July 10, 2003, p. A-1.

Road, River, and Ol' Boy Politics

into east Williamson County had reversed the gesture. The flow of possibility headed west.

In the space of two decades, the weight of the county's wealth shifted to the west. In 2001 the market value of property within Round Rock's school district was 13.7 billion dollars. Georgetown's school district was worth 3.5 billion dollars, while Taylor's was assessed at 723 million dollars. This was not just a brute reflection of population; Round Rock was startlingly wealthy compared to the rest of the county. By 1990 the three cities with the highest median family incomes—Round Rock, Cedar Park, and Leander—had merged into a Southern Californiaesque carpet of subdivisions in the county's southwestern quadrant.[20] In 2000 Williamson County's median income of 60,642 dollars represented a ten-fold increase in personal wealth over its 1970 median income of 6,413 dollars.[21]

Most of the new wealth came from new emigrés, people who differed greatly from the Williamson County natives they were numerically overwhelming and replacing. Most migrated from urban centers, often from other parts of the country. They came for jobs in Austin's hot high-tech sector, though after 2002 that cooled considerably; they came to raise families; and they came to retire. In 2003 more than 70 percent of the people in Williamson County lived along the Interstate 35 corridor or west of it. Between 1990 and 2000 the county population grew by 80 percent, but east Williamson County barely held steady.[22] In 2003 the county's population reached 301,000.[23]

[20] The *2000 Census of the Population* showed median household populations of $54,98 in Georgetown, $60,354 in Round Rock, and $38,549 in Taylor. Mel Pendland, the Georgetown Chamber of Commerce director, estimated in early 2004 that those figures had risen to $61,000 in Georgetown, $71,000 in Round Rock, and $$41,000 in Taylor.

[21] *1970 Census of Population*, U.S. Department of Commerce, Vol. 1 Characteristics of the Population, Part 45, Texas sec. 1, General Social and Economic Characteristics, Table 44, May 1973; Capital Area Planning Council, Austin, data from 2000 Census, Feb. 11, 2002; Camille Wheeler, "Growth in Business, Homes Mirrored Dell Success," *Austin American-Statesman*, June 22, 2003, p. A-18.

[22] The 2000 Census listed 249,967 people living in Williamson County, which grew faster than any other county in Central Texas. The Greater Austin–San Antonio Corridor Council projected that Williamson County would have half a million residents in 2015, 634,000 in 2020, and 783,000 in 2025.

[23] Telephone interview, County Judge John Doerfler, Dec. 3, 2003, Georgetown.

The new residents came largely in two varieties: young middle-class couples seeking the American dream of leafy suburbs with safe, high-quality schools for their children; and retired couples bent on enjoying their golden years in an environment relatively safe from crime and high taxes. Unfortunately these two streams of emigrants increasingly clashed as the growth they engendered led to rising crime rates, traffic jams, and overcrowded schools, the costs of which could be borne only by higher property taxes.[24] Generally, parents with school children supported school bond proposals and modest tax hikes while the retirees, lured to Del Webb's Texas Sun City in Georgetown (an age-restricted community with a capped population projected at twenty thousand) and similar communities around the county, opposed them. In Round Rock and Georgetown, traditional town cohesiveness cracked along the Balcones Fault line: residents living east of I-35 tended to be poorer and more liberal, while west of I-35 residents tended to be whiter, richer, more conservative, more likely to commute, and less "local" in their world view. Rather suddenly, local politics got expensive and nasty, despite the fact that most newcomers identified themselves as Republicans.[25] As Round Rock took on an Edge City cast, the notion of "community" seemed increasingly quaint.

Throughout the Dry Sun Belt, as in Williamson County, powerful Democratic Party politicians largely created the new dams and inter-state highways that created ebullient new economies and growth patterns throughout the region.[26] Ironically, the great public works projects gave birth to "safe" Republican districts by replacing the tra-ditional agrarian demographic with a suburban counterpart. In the Williamson County of 1973, before water could fuel new growth,

[24] Texas depends on property taxes to fund local school districts, as the state does not levy an income tax and has substantially cut its outlays to public schools over the past decade. Thus, a property owner's tax may compare unfavorably to a comparable prop-erty in another state, while no Texan pays an income tax.

[25] Jeff Dorsch, "Mayor Recalled; May Elections Start," *Williamson County Sun* (Georgetown), Feb. 6, 2002, p. A-1; Tony Plohetsky, "Winds of Change in Round Rock," *Austin American-Statesman,* Feb. 28, 2002.

[26] President Eisenhower was a notable exception, but the interregional highway concept had been supported by Democratic presidents since Franklin D. Roosevelt.

Road, River, and Ol' Boy Politics

Wilson Fox (center), the powerful Williamson County Democratic leader, switched to the Republican Party to support Richard Nixon, who was elected president. In September 1972 Fox is flanked by Pat and Richard Nixon who are flanked by former governor John Connally and his wife Nellie at their South Texas ranch. *Courtesy Dr. James ("Jim") Fox, Austin.*

there were so few local Republicans that the entire county political party could—and did—hold its annual convention while driving around on back roads in a Pontiac Catalina, drinking beer. Today Williamson County is a Republican stronghold.[27]

"It would have been unthinkable in the early seventies to have a Republican *candidate* for county or precinct office, much less a Republican officeholder," said District Judge Billy Ray Stubblefield, who, like most county officials, started his political life as a Democrat. Now he is a staunch Republican.[28] No Democratic official remains in Williamson County government. For the most part, this political

---

[27] Interview, Stephen Benold, June 1998, Georgetown.

[28] Billy Ray Stubblefield, "In Quarter Century, Political Life Takes Revolutionary Turn," *Williamson County Sun* (Georgetown), Sept. 30, 1998.

revolution came not because new Republican candidates ousted incumbent Democrats in elections, but because Democratic Party incumbents defected to the GOP. They had no choice. It was a question of demographics, plain and simple. In 1973 the county had forty thousand residents. In 1998 Williamson County had 130,000 registered voters, almost all of them members of the Republican Party. Roughly 80 percent of Sun City Texas's four thousand residents considered themselves Republicans, not a huge number in itself, but Sun City voters quickly established a reputation for political discipline, even activism, to become a bloc feared and wooed by every candidate, from local school trustee to congressman.[29]

The county's agrarian heartbeat faded. As farmers sold or subdivided their land, their children scattered. Where conversations in German or Czech were once heard on bustling streets, boarded-up storefronts now prevailed. Today Granger's Davilla Street appears deserted, except for clots of activity around the Red and White Grocery and Granger Medical Clinic. The languages, customs, religions, and stories that gave meaning to the county's farm hamlets are mostly gone. In 1974, 66 percent of Williamson County's 722,560 acres were devoted to row crop farming, almost all of it east of I-35. The rest was "waste" land good only for grazing goats and cattle. In 2001, subdivisions and cities covered a third of the county's land surface, most of it west of I-35, while farmers—most of them part-timers—cultivated just 22 percent of the land.[30]

While there are still enclaves of Germans in Walburg, Czechs in Granger and Taylor, and Swedes in Hutto and Round Rock, they no longer form impenetrable clans or dominate local voting or religious and cultural institutions. Instead they have been swallowed, rendered invisible by waves of newcomers. Demographically the county is "white" and homogeneous. While African Americans, Mexican Americans, and Asian Americans have moved into Williamson County in substantial numbers over the last two decades, the far greater tide of Anglo emigrants has blotted out their "otherness" and diluted the

[29] Interview, "Gaz" Green, Feb. 19, 2002, Georgetown; interview, Ricia Gittins, Feb. 23, 2002, Georgetown.

[30] Telephone interview, Ronald Leps, Feb. 20, 2002, Georgetown.

county's old ethnic patterns.[31] Of course, there are positive things to be said for this state of affairs. Most people, even the poor, are better off financially, and overt racism is not tolerated by mainstream society. Marrying across ethnic or racial lines, which once required exile to Austin, Houston, or New York City, has become commonplace.

The staggering influx of new people, predicted to reach eight hundred thousand by 2025, has created huge transportation and environmental challenges. For the last decade, school administrators in Georgetown, Round Rock, Cedar Park, Leander, Liberty Hill, and Hutto have struggled to keep up with new arrivals. Road builders strive unsuccessfully to unclog traffic. The problem was not just the influx of new residents; Lakeline Mall in Cedar Park, La Frontera in Round Rock, and "big box" retail stores like H-E-B, Super Wal-Mart, and Home Depot have trailed this exciting new customer base, creating their own traffic gridlocks. By 2000, Interstate 35 had become the national interstate system's problem child.

In the maiden years of I-35, driving on the interstate could be likened to a Sunday afternoon drive in the country. In Georgetown, department-store owner Harry Gold advertised on television, "Gold's is just a conversation away," and grew prosperous on customers from Austin. Before I-35 was built, Texas Highway Department engineers confidently predicted that the interstate would stop congestion forever. Thirty years later, traffic crawled. In the mid-1990s, Governor Ann Richards declared I-35 a "disaster zone" through Williamson County. The stretch between Austin and Round Rock was Texas's most congested for the number of lanes. In Austin the centrally located Colorado River Bridge, which in 1959 had handled 6,287 vehicles a day, tallied a daily average traffic count of 201,000 in 2002. In North Austin 220,000 vehicles plied the interstate daily. In Round Rock 168,000 vehicles a day inched under the La Frontera-Dell Computer exchange at Farm to Market 1325, where Tom Kouri once raised cattle, and 154,000 vehicles crossed under the State Highway 620 overpass, where Colonel Todd and highway builder Travis Long downed drinks on the patio. Georgetown had not yet encountered regular

[31] U.S. Census, 1980–2000.

gridlock; there, only eighty-six thousand vehicles a day used I-35.[32] The average speed between Georgetown and Austin dropped to forty-five miles per hour—slower than before the interstate was built.[33]

No one knows how bad traffic will get before it frightens industrial and residential prospects away, but fear of such a future has driven a determined effort to build a "virtual" interstate, a 1.5-billion-dollar State Highway 130, which will roughly parallel I-35 along its busiest stretch.[34] A driver taking the new toll road south will peel off I-35 north of Georgetown, veer east around Georgetown, Round Rock, and Austin, and rejoin the free interstate in southern Travis County. Highway 130 will erase any remaining gaps between Hutto and Round Rock and will boost Taylor's population, much as I-35 might have had Dewitt Greer's first sketch of the interstate been built. Theoretically, 130 Toll, as the new highway has been dubbed, will pull some commercial truck traffic away from I-35.[35] But no one really knows how much. Clearly, the new highway will appeal to truck drivers making the Dallas–San Antonio run, but whether Williamson County commuters will use it enough to make a dent in I-35 traffic is an open question. Most of the people who commute to Austin live west of I-35; they will have to cross Georgetown or Round Rock on city streets to reach the toll road. Potential for gridlock is great, leading highway planners to propose widening Georgetown's only real east-west artery, University Avenue (Texas Highway 29) through the heart of the city's elegant Old Town—a plan that has been beaten back, so far, by Georgetown's negative response, but which could be resurrected.[36]

[32] Interview, John Hurt, Jan. 24, 2002, Austin.

[33] Michelle M. Martinez, "The Roads Test," *Austin American-Statesman*, Oct. 12, 2000.

[34] In a novel scheme designed to speed up the process of getting the highway online, the Texas Department of Transportation contracted with a private consortium called Lone Star Infrastructure to design, construct, maintain, and partly finance 130 Toll. Lone Star is a joint venture between Fluor-Daniel, Balfour Beatty Construction, and T. J. Lambrecht Construction.

[35] Stephen J. O'Brien, "One Sure Thing: 130 Will Slice County into New Pie," *Williamson County Sun* (Georgetown) Sesquicentennial Edition, Nov. 8, 1998, p. F-3. Toll roads are often shunned by truckers because of their expense; no one knows what to expect on this route. Truck traffic increased dramatically after the 1993 North America Free Trade Agreement came into being.

[36] Kelly Daniel, "Buckle Up: Texas 130 is Finally a Go," *Austin American-Statesman;*

Most property owners along its path fought 130 Toll, seeing it as a Berlin Wall that will block east-west traffic flow except at a few points, and because it cuts through bucolic farm country. Most government leaders in Williamson County, however, including Round Rock, Hutto, Taylor, and Georgetown, fought hard for it. Round Rock wanted the toll road to give more flexibility to Dell's employees and shipping operations.[37] In 2000, county voters—sick of gridlock on I-35 and on U.S. 183, which connects Austin to Cedar Park and Leander—strongly backed 130 Toll in a 350-million-dollar bond election, with approximately fifty-four million dollars directly bankrolling the county's share of the project.[38] No doubt the virtual interstate will launch industrial and residential development. "It's virgin territory," said Leo Wood, a former Georgetown mayor and roadsgrant consultant.[39] State Representative Mike Krusee added, "State Highway 130 will be an engine for growth."[40] At the road's October 2003 groundbreaking ceremony, public officials waxed estatic about 130 Toll's impact on the Central Texas economy while the Eagles' road anthem, "Take It to the Limit," blasted from loudspeakers. But the officials made no promises to frustrated I-35 drivers. The toll road would "postpone gridlock"—not eliminate it.[41]

A dwindling supply of water is the county's greatest long-term

---

interview, Quevarra Moten, Texas Department of Transportation, Feb. 25, 2002, Austin. The SH 130 toll road is the largest part of a $3.6 billion project which includes two other major highways: a short but critical State Highway 45 in south Round Rock which will link U.S. 183 to SH 130 and an even shorter extension of Farm-to-Market 734 (Parmer Lane) to link State Highway 45 in south Williamson County and Highway 195 in northwest Georgetown, near Sun City Georgetown.

[37] Dell Computer Properties, 2001 Certified Values, Williamson County Appraisal District, Georgetown.

[38] Several other road projects were included in the bond proposal, notably State Highway 45, which will improve east-west transportation south of Round Rock, especially for Dell's vast operation. Travis County and the City of Austin also shared in the cost of the project. The entire 130 Toll road is supposed to cost $1.5 billion and open in December 2007.

[39] Interview, Leo Wood, June 30, 1998, Georgetown.

[40] Interview, Mike Krusee, June 1999, Austin. Krusee represents Williamson County in the state capital.

[41] Ben Wear, "Long, Rocky Path Full of Close Calls Leads to Toll Road," *Austin American-Statesman,* Sept. 28, 2003, pp. A-1, A-11.

environmental challenge.[42] In 2002 water from the Granger and Georgetown reservoirs, once believed to be more than ample for Williamson County's needs, could not slack the county's growing thirst. Round Rock and Georgetown had sucked up all of Lake Georgetown and depended on Lake Stillhouse Hollow, forty miles away in Bell County, to hold them until 2020.[43] Taylor had more water than it needed locked up in Granger Lake, but faced financial hardships in maintaining a deteriorating treatment plant and pipelines. In 2004 the Brazos River Authority purchased Taylor's plant and pipelines for 5.1 million dollars, promised to invest fourteen million dollars in the facility, and guaranteed to cap Taylor's water rates for fifty years. The deal will free up to 40 percent of Granger Lake's water for sale elsewhere, giving the BRA more clout within its own river basin.[44] Long-range planners predicted that Liberty Hill, Cedar Park, Florence, Granger, Hutto, and Thrall would run out of water by 2020.[45] Leander ran dry in the nineties, forcing its supplier (the BRA) and the Lower Colorado River Authority to agree in 1998 that the city could pump water from the Colorado River Basin rather than the Brazos River watershed, an emergency move requiring state legislation.[46] West Williamson County purchased most of its water supply from the privately owned Chisholm Trail Water Supply Corporation, which, along

[42] County politicians, fearing that conservative Republican voters might link them with the dreaded "tree-hugging" liberals, preferred to refer to environmental concerns as "quality of life" issues. Interestingly, the county's solid GOP voting bloc routinely favored "quality of life" issues at the polls.

[43] Georgetown City Manager Bob Hart, quoted in "Fifty Years Out," *Williamson County Sun* (Georgetown), Sesquicentennial Edition, Nov. 8, 1998, p. F-1.

[44] Telephone interview, Judy Pearce, Brazos River Authority, Feb. 25, 2004, Waco; Kate Alexander, "Taylor Deal Would Give River Authority a Boost," *Austin American-Statesman*, Aug. 11, 2003, pp. B-1, B-3; Walter Howerton Jr., "Taylor Voters Nix Selling Water Plan," *Sunday Sun* (Georgetown), Nov. 9, 2003, p. A-4; Laura Heinauer, "Taylor to Vote Again on Water Plant Sale," *Austin American-Statesman*, Dec. 4, 2003, p. B-1; Dick Stanley, "Taylor Voters OK Water Plant Sale," ibid., Feb. 8, 2004.

[45] Brazos G Regional Water Plan, HDR Engineering, July 2000, Table 4-72, 4-155 (Brazos G Regional Water Planning Group, Waco); "Water for Texas," Vol. II, pp. 3-150, 3-155, Aug. 1997 (Texas Water Development Board, Austin); Alan Lindsey, "Alliance Works on Water Plan for Williamson," *Sunday Sun* (Georgetown), Dec. 17, 2000, p. A-1.

[46] Erin J. Walter, "Leander Will Give Away Water Rights," *Austin American-Statesman*, Jan. 4, 2001, p. B-2.

with Georgetown and Round Rock, contracted for a major new source of water that was to be funneled to Lake Georgetown through a 36.5-million-dollar, twenty-eight-mile pipeline from Lake Stillhouse Hollow. The Bell County water will be stored in Lake Georgetown and pumped to its customers, theoretically solving Williamson County's major water needs for twenty years. In December 2001, use of the new water spigot was delayed when engineers turned the system on and the pipes exploded. At first vandalism was suspected, but eventually investigators concluded that an "inappropriate grade of fill" caused the pipes to be "smushed flat like a pancake."[47] The mistake cost twenty million dollars, almost all of which was covered by insurance. The reconstructed pipeline was scheduled to start delivering water to Williamson County in May 2004. But even with that extra source of water in play, water experts say that by 2020 Williamson County will need the equivalent of another Lake Georgetown to make up the expected water deficit.[48] By the middle of the twenty-first century, twice as much will be needed.

How this much water supply can be developed no one really knows. Owen Sherrill's desired dam on the San Gabriel's South Fork, though authorized by Congress, is considered dead because the amount of water it would provide cannot justify its cost, and the environmental impact of such a dam would be difficult, if not impossible, to mitigate. The Brazos River Authority has proposed building a dam in Milam County near Cameron (which, if built, would bury the farm land that the Granger Dam was built to protect), but opposition from Milam has been vociferous and most experts agree it will never be built.[49] The most promising approach may be to utilize existing water supplies by pumping them long distances, sometimes against

---

[47] Telephone interview, Judy Pearce, Brazos River Authority, Feb. 25, 2004, Waco ; Jeff Dorsch, "36 Million Water Pipeline Springs a Big Leak," *Williamson County Sun* (Georgetown), Feb. 6, 2002; Tony Plohetsky, "Lake Stillhouse Pipeline Leaking, Must be Repaired," *Austin American-Statesman*, Feb. 12, 2002.

[48] Brazos G Regional Water Plan, HDR Engineering, July 2000, Table 4-72 (Brazos G Regional Water Planning Group, Waco), 4-151–4-155.

[49] Michelle M. Martinez, "Williamson Reservoir Plan Draws Opposition," *Austin American-Statesman*, Aug. 29, 2000, p. B-1; Michelle M. Martinez, "Fighting to Keep Above Water," ibid., Dec. 28, 2000; interview, Horace Grace, Aug. 7, 2000, Georgetown.

gravity, to where they are needed. Thus, tapping the Carrizo-Wilcox Aquifer in Milam and Lee Counties and pumping water uphill to Williamson County, or transferring water from one river basin to another, though still not legal, are options that look better all the time. Water conservation receives little official attention.[50]

Today, a few souls still canoe the San Gabriel River, though white-water runs are gone. During dry periods, the Brazos River Authority allows a trickle of water—the river's "low-flow average"—to be released from the Georgetown and Granger reservoirs.[51] Thus, though the river never dries up completely, as it sometimes did before the Corps controlled its flow, the brief bursts of boiling river after heavy rainfalls that so delighted canoeists and tube floaters in pre-dam days are gone. During "low-flow" months, few county residents swim or eat fish caught from the river—common practices before the dams. Hundreds of underground springs have dried up. Persistent complaints about Georgetown's and Taylor's water treatment plants leaking suspect contaminants have made people wary about drinking city-treated water.[52] Georgetown plans to build a new wastewater treatment plant, this one west of town on the river's South Fork, and residents downstream are protesting. The Tonkawas' *takachue pouetsue*, land of good water, has vanished.

At Lake Georgetown, North Fork Dam has affected the San Gabriel River's flood plain by killing hundreds of ancient down-stream cottonwoods, cypress, pecan, and burr oak that had grown to prodigious heights and circumferences. Before the dams altered the river's natural rhythms, seasonal floods gradually built up bottomlands

[50] Alan Lindsey, "Water Agreement Will Meet Needs for Next 20 Years," *Williamson County Sun* (Georgetown), Aug. 23, 2000; Alan Lindsey, "G'town should expand water intake . . . " ibid., Sept. 20, 2000; Alan Lindsay, "Alliance Works on Water Plan for Williamson," ibid., Sept. 17, 2000.

[51] The Corps of Engineers operated the dams under the control of the Brazos River Authority, which "owned" the water on behalf of the State of Texas. The BRA negotiated contracts with potential water customers and controlled the release of water from all of its dams, unless a flood threatened, at which point the Corps took over.

[52] At the time of this writing, Georgetown's Dove Springs treatment plant was once again being inspected by the Texas Natural Resource Conservation Council. Carter Nelson, "State Agency Reviewing Dove Springs Treatment Plant," *Austin American-Statesman*, Feb. 28, 2002.

Road, River, and Ol' Boy Politics

with ever-thickening layers of silt, adding to the fertility of the Black Waxy.[53] But after the dams interfered with the river's flow, this changed. Now when heavy rains threaten flooding on the San Gabriel, Corps employees shut the gates at North Fork Dam, causing Lake Georgetown to rise. The goal is to reduce the river's downstream flow while its tributaries are dumping flood waters into the main stem. After the flood threat subsides, the Corps releases massive volumes of water from the swollen lake, forcing the San Gabriel to run at high velocity for days. It is as if the dam operators are aiming a powerful garden hose at a pile of soft dirt: Black Waxy riverbanks melt under the water's scouring action, and buckle into the river, taking trees with them. Every year, river-front property owners see their acreage shrink.

Granger Lake has not produced wealth for Granger or Taylor in the form of real estate development, as Ralph Moore, Wilson Fox, and Congressmen Poage and Pickle expected. The lake attracts fishermen and recreational users, but compared to other Central Texas lakes, it is lightly used. Its broad surface, surrounded by a flat prairie, turns into a dangerous high chop on windy days, drowning unwary boaters.[54] Henry Fox predicted Granger Lake would become a giant mud hole. As it nears its twenty-seventh birthday, the lake is experiencing what its managers call "heavy silting." Some doubt it will last fifty years.

The most obvious environmental change wrought by the dams and Interstate 35 is obvious to the naked eye: the landscape. The county's former pastoral beauty, once as picturesque, in its way, as the hills of Tuscany, is mostly gone. Happily, a few remnants of the county's agrarian past are being salvaged. The county has designated several "Heritage Roads" and hundreds of "Heritage Trees" for protection, though how this will actually be achieved remains elusive. The county has also purchased hundreds of acres of land for two county parks, one (between Round Rock and Cedar Park) primarily recreational, the other (near Georgetown) a lovely pecan orchard where trees meet in a natural arch over County Road 152 where it

---

[53] However, the big floods of 1921 and 1957 did cause severe erosion.

[54] Interview, Dan Thomison, Oct. 4, 2000, Granger Lake. For many years, the number of drownings at Granger Lake was quite high compared to other Corps lakes.

crosses Berry Creek. The site of a nineteenth-century water mill is on the property, which is being developed as a nature preserve and historic center.[55] Round Rock and Cedar Park are building a hike-and-bike trail along Brushy Creek, and Georgetown is extending its river hike-and-bike trail to connect with the Army Engineers' primitive trails around Lake Georgetown.[56] These efforts have cheered almost everyone, but, even if successful, they will result in pockets of preservation within an overwhelming pattern of suburban sprawl, which 130 Toll will aggravate.[57] Increasingly Williamson County is becoming a place where, as Gertrude Stein said about Oakland, California, "there is no there there."

County political boss Wilson H. Fox "made" Lake Granger in an effort to sustain Taylor's long-held position as a small city "on the rise."[58] Thirty years after his death, Fox's belief that Taylor would become an industrial magnet if it could but control a dependable source of water has not yet materialized. Taylor has grown from 10,619 in 1980, when Lake Granger's water first became available, to 13,575 in 2000. In the same span of time, Granger gained sixty-three people, from 1,236 to 1,299.[59] Both cities have become poor relative to the rest of the county and have lost scores of businesses. Today, school systems that once sent the children of Taylor and Granger to elite universities struggle for funds and students. Both cities look sad and tattered. But Taylor leaders are eyeing the approach of 130 Toll with cautious hope: perhaps some of the wealth the toll road is expected to generate will flow east, anointing Taylor with new riches.

Wilson Fox died of a heart attack on an icy day in 1974, while

[55] Aimee Michels, "Berry Springs: A Real Beauty," *Williamson County Sun* (Georgetown), Aug. 4, 2002, p. A-1. County Commissioner Mike Heiligenstein of Round Rock pushed hard for the recreational park, while Commissioner David Hays provided the impetus for the purchase and development of Berry Springs Park.

[56] The grant for this project was landed through the work of Georgetown congressman Chet Edwards of Waco, a Democrat of the Jake Pickle stripe.

[57] Linda Scarbrough, "Fifty Years Out," *Williamson County Sun* (Georgetown), Sesquicentennial Edition, Nov. 8, 1998, p. F-2.

[58] Fox may also have been influenced by personal financial incentives, as well as by the Corps' goal of efficiently developing the entire Brazos River watershed.

[59] U.S. Census, 1970 and 2000.

delivering a tax return to a client. His estate was appraised at 129,000 dollars, the bulk of this being his Riverside Ranch.[60] His widow, Hilda, quickly sold the 313 acres the Corps wanted for Laneport Dam for 213,000 dollars. She kept 148 acres. Ten years later she died in a Taylor nursing home.[61] The Fox's only child, James ("Jim") Wilson Fox, is an Austin plastic surgeon gifted with the family aptitude for storytelling. After all this time, he tends to sympathize with his Uncle Henry's negative position on Laneport Dam, "though I sort of like the concept of having a lake."[62] With his son, Josh, Jim enjoys visiting his old home at Riverside Ranch. Several years ago the Corps allowed Josh to throw a party for his high school friends at Hilda's Bottom, now government property. Eleven hundred toga-clad teenagers frolicked that night under the great burr oaks; the ghosts of Wilson Fox's barbecue-blowout guests would have felt right at home.[63]

More than any other person, Owen W. Sherrill brought into being Georgetown's North Fork Dam, one of the two most transformative human deeds in Williamson County's history. (The other being the construction of Interstate 35.) But Sherrill left few traces. One of his two sons, nicknamed "Bo," drowned in 1960 in a boating accident on Lake Travis, where Sherrill owned a cabin.[64] His other son, Owen W. ("Jack") Sherrill Jr., who walked with difficulty supported by crutches, worked at his Dad's real estate office in the early sixties, but moved to Taylor and was living there when his father died in 1976. Owen W. Sherrill's will dictated that Jack receive one thousand dollars in cash. Sherrill's wife (Jack's stepmother), Kay Deaver Sherrill, got

---

[60] "Wilson Fox Died Saturday, Buried in Taylor," *Williamson County Sun* (Georgetown), Feb. 14, 1974, p. B-5; "Last Will and Testament," Wilson H. Fox, Feb. 13, 1974 (Williamson County Clerk's Office, Georgetown); interview, Dr. James "Jim" Fox, Sept. 6, 2000, Austin.

[61] Obituary, Hilda Fox, *Sunday Sun* (Georgetown), Apr. 22, 1984, p. A-5.

[62] Interview, Dr. James "Jim" Fox, Sept. 6, 2000, Austin.

[63] Ibid.

[64] Letter, Jack Sherrill to J. E. Owen, Jan. 11, 1961, Owen W. Sherrill Papers, Cushing Memorial Library (Texas A&M University, College Station); telephone interview, Martha Burt Nelson, Oct. 2, 2003, Frederick, Md. Nelson, a distant relative of Kay Sherrill's, lived with the Sherrills while attending Southwestern University in Georgetown and continued to visit them frequently, especially during Kay Sherrill's last years.

everything else—an estate appraised at four hundred thousand dollars. Like Fox's, Sherrill's estate was comprised mostly of real property, a tad over eight acres across from Inner Space Caverns, south of Georgetown off Interstate 35.[65]

Sherrill's first purchase of land at that prime location was 164 acres in 1931, when he worked as a banker. Over the years, he added to the property until he owned 216 acres fronting U.S. Highway 81. When I-35 arrived, this piece of property sat across from the interstate's first exit to Georgetown from the south. In 1974, his health failing, Sherrill sold almost all of his land to National Housing Industries of Round Rock. Georgetown Railroad Company bought it next and sold some to developer Gregory Hall. When Kay Sherrill died in 1992, she still owned 8.299 acres, including the 1951 limestone ranch house she and Owen had designed and built, several deteriorating frame houses, and a beautiful stone-faced barn. Her estate, appraised at 210,000 dollars, went entirely to Martha Burt Nelson, a distant relative who lived with the Sherrills while attending Southwestern University, becoming the "daughter" Kay and Owen never had.[66] In 1996 Nelson sold the Sherrill property to Hall, who tried to build a water park on it, setting off a controversy that embroiled Georgetown in a series of philosophical and political fights.[67] Today about half of Sherrill's amassed land package, tucked behind a business called Discount Hobbies and a tourist attraction called The Candle Factory, awaits a

[65] "Last Will and Testament" and "Inventory and Appraisement of the Estate," Owen W. Sherrill, Dec. 14, 1976, No. 8677 (Williamson County Clerk's Office, Georgetown). Under Texas's community property laws, half of that $400,000 belonged to Kay Sherrill. I drew some of my information from interviews with Martha Burt Nelson, Oct. 2, 2003, Frederick, Md.; J. D. Thomas Jr., Sept. 8, 2000, Georgetown; retired District Judge William Lott, Sept. 25, 2003, Georgetown; and "Abstractor's Certificate," a title search I commissioned from Georgetown Title Company, Dec. 2, 2003.

[66] "Last Will and Testament," Kay Deaver Sherrill, Nov. 13, 1991, No. 13904 (Williamson County Clerk's Office, Georgetown). According to numerous sources, including Martha Burt Nelson, Kay Sherrill and Martha Burt were extremely close.

[67] Telephone interview, Martha Burt Nelson, Oct. 2, 2003, Frederick, Md. At the time, the Hall episode seemed a turning point for Georgetown's policy on growth. City leaders had unofficially approved a financial package for Hall's water park that enraged much of the local electorate; most of these leaders were turned out in a subsequent election. But several years later the "reformers" were turned out themselves, paving the way for a cozier relationship between the city and developers.

buyer. Even in its slightly dilapidated state the ranch house, enclosed by a serpentine stone fence, thick oak groves, and a line of towering crepe myrtle, still telegraphs its showplace origins. A refrigerator propped on the front porch signals despair. Like Sherrill, developer Hall has used the land as if it were a bank, leveraging it for cash when needed. Under both men's stewardship, those nearly two hundred acres—sandwiched between I-35, Georgetown's southern entrance, and Inner Space Caverns—have been a reservoir for ambition.

Sherrill's imprint on Williamson County is not entirely forgotten. His portrait greets visitors to Lake Georgetown's Corps headquarters, along with those of other Georgetown men who lobbied for the reservoir. The Georgetown Chamber of Commerce occasionally bestows an Owen W. Sherrill Award for exemplary business and community leadership. Sixty boxes of Sherrill's papers, which in 1975 he personally delivered to his alma mater, Texas A&M, can be found in Cushing Memorial Library, stored in one of its four warehouses. A few years ago, during a space crunch, the Cushing's chief archivist considered destroying Sherrill's papers. But in October 2003 they still lay deep in the library's vaults, awaiting scholarly examination.[68]

While Charles Forbes, Moses and Tom Kouri, and John H. Nash Jr. were lobbying politicians in Washington to create golden exchanges that would increase the value of their land, one Georgetown rancher took a more direct route to secure an interchange at Leander Road. Whenever A. C. ("Doc") Weir drove into town from his ranch to drink coffee at the L&M Cafe, visit the post office, and purchase supplies, he would pull over and stop where a thick black rubber traffic counter stretched across Leander Road, roughly where it would one day intersect with Interstate 35. Weir would get out of his truck, walk

---

[68] Interviews, David Chapman, Apr. 14, 2000, Oct. 6, 2003, College Station. Until I showed up to examine Sherrill's 75 linear yards of uncatalogued files in April 2000, Chapman, chief of Texas A&M's archives at the Cushing Memorial Library, could remember no other request to look at the collection. He knew it well; as a young librarian he had helped Owen Sherrill unload scores of boxes in 1975, shortly before Sherrill died. What I saw (and I didn't see nearly all of it) was fascinating stuff; it is a rich opportunity for the right researcher. In addition to the materials he gave to the library, Sherrill donated $25,000 to establish a Texas A&M Presidential Scholarship.

around to the truck-bed, and grab an old hammer. Then he would walk over to the traffic counter and give it a few whacks, as if he were killing a snake. "What on earth are you doing, Daddy?" his daughter asked him during one such performance. "I'm just holding up Georgetown," Weir said.[69] His efforts seem to have paid off. Georgetown got an overpass that connected it to Leander Road and points west.[70] People living out Weir's way, all the way to Leander, could drive straight into town rather than having to circle miles out of their way. The Interstate 35–Leander Road interchange encouraged banks and developers to build subdivisions southwest of Old Town. Doc's widow, Esther, and daughter, Laura, invested early and heavily in Georgetown's Courthouse Square renovation, inspiring other preservationists to follow.[71] Eventually Williamson County will build a highway loop west around Georgetown. The loop will funnel most of Georgetown's suburban population, including Sun City Texas and Serenada Estates west of Georgetown, straight onto I-35 south of town. Its course will take all that traffic right through the front yard of Doc Weir's ranch.[72]

As a young man, Jay Wolf barely had two nickels to rub together. As a World War II veteran, he entered Southwestern University on the GI Bill, arriving in Georgetown in 1949 with his wife, Bettie, and two

[69] Interview, Laura Weir-Clarke, Nov. 18, 2003, Hamden, Conn.

[70] It is possible that the Highway Department was not entirely taken in by traffic-counter finagling, on Leander Road or elsewhere, but the fact remains that a full exchange was added where none had been planned.

[71] "Doc" Weir's son, Mike Weir, an Army physician stationed in the Northwest while his mother and sister were leading the Georgetown preservation movement, supported their work, which was bankrolled largely by the ranch. Laura Weir-Clarke, trained as an architect, carried out the restoration of the Steele-Makemson Building and the Masonic Lodge, two of downtown Georgetown's most imposing structures.

[72] Interview, David Hays, June 2002. Hays was a county commissioner with strong interests in long-range planning and the environment. This was not the first time that a highway has collided with the Weir family. The first Weir ranch in Georgetown, owned by "Doc's" parents, lay directly in the path of Interstate 35, which wiped out the polo field in front of the two-story Victorian home, which can be seen today west of I-35 across from Georgetown Hospital. Since the Weir family sold it, the house has been an antique shop, tearoom, bed and breakfast, and dinner theater.

Road, River, and Ol' Boy Politics

Left: A. C. ("Doc") Weir. *Courtesy Esther Weir, Georgetown.*

Below: Jay Wolf, owner of Wolf's Wool and Mohair, beams at Al Kauffman, winner of a 1965 Williamson County Livestock Show prize with his French Rambouillet merino sheep. Kay Braun looks on. *Evans Studio Photo, Courtesy Jay Wolf family, Georgetown.*

small children. Early every morning he milked cows for a dairy farm north of campus and peddled milk in San José, Georgetown's Mexican-American *barrio*. He borrowed money to open Wolf's Wool and Mohair, and when it became a gathering place for ranchers, he learned about ranch properties. Soon he started investing, and acquired a string of ranches that he traded, his wife said, "the way some people trade horses." Bettie coveted one Georgetown ranch owned by Red Messer because of its towering oak trees, so in the early fifties Wolf bought it. When Interstate 35 came through, he put a filling station there, at the intersection of I-35 and Highway 29, the first local man to stake his claim on the interstate.[73]

Then he hung onto it. Jay Wolf died in 1996. The Wolf family rejected numerous offers for the land, which had become Georgetown's prime undeveloped I-35 intersection. In 2003 Bettie and her five children sold Red Messer's old place—103 acres fronting I-35, State Highway 29, and the South San Gabriel River—to Simon Property Group, the nation's largest mall developer, for a reported 27 million dollars.[74] Signs announce a 2005 opening of "Wolf Ranch," described as an outdoor "power center" valued at 110 million dollars. A Target store will anchor the complex, and Simon plans fifty-five stores, fifteen restaurants, and outdoor plazas and trails along the San Gabriel River.[75] Wolf Ranch's economic impact on Georgetown will be huge: the city's projections show the retail center employing roughly twelve hundred people, generating a million dollars annually in school-district property tax revenues, and adding more than two million dollars annually to the city's general revenue fund. However, the deal

[73] Interview, Bettie Wolf, Nov. 7, 2003, Georgetown; interview, Judy Wolf Hindelang, Nov. 7, 2003, Georgetown; and "Family Honors Local Businessman," advertisement in the *Williamson County Sun* (Georgetown), Sesquicentennial Edition, Nov. 8, 1998, p. F-3.

[74] The family has not confirmed a sale price, but City of Georgetown officials reported the land went for at least six dollars per square foot, or $261,360 per acre, including 17 acres of flood plain, which in Williamson County was typically appraised at $1,500 per acre. See Rebekah F. Sellers, "Park Land at Site Becomes an Issue," *Sunday Sun* (Georgetown), June 8, 2003, p. A-1.

[75] "Wolf Ranch Coming in Summer 2005," *Georgetown City Reporter*, Summer 2003; R. Michelle Breyer, "Masterminds of the Mall," *Austin American-Statesman*, Aug. 17, 2003.

Road, River, and Ol' Boy Politics

enveloped the city in a controversy over growth policy: incentives worth as much as fifty million dollars angered many taxpayers who saw the subsidies as setting a dangerous precedent and eroding the city's financial health; other citizens believe the development will be such an asset to Georgetown that the incentives will be worth it.[76]

One Georgetown old-timer credits Jay Wolf with having brilliantly orchestrated his land purchases along Georgetown's major arteries in order to ensure ownership of I-35-fronting property.[77] The Wolf family doubts their patriarch was *that* prescient, but they honor his business acumen. Bettie summed it up. "Jay liked his money better than he liked his places," she said. His eldest child, Judy Wolf Hindelang, imagines her father's reaction to the family's deal with Simon: "I think he would be proud of us."[78]

Williamson County's metamorphosis from a mythological pastoral paradise to pattern-book Dry Sun Belt Superburb illuminates a form of democratic pressure politics I will call "gilded democratic action" and its impact on the development of America in the twentieth century.[79] The gilded democratic action is led by an unusually forceful

---

[76] Simon Property negotiated a sweet deal with the City of Georgetown through which Simon will receive 53 percent of the 1 percent city tax on sales generated by Wolf Ranch for up to twenty years, plus interest. Also, the city will pay $15 million for roads inside the retail center, plus another $10.5 million for off-site highway improvements—a deal that infuriated other developers who had been required to pay all of the cost of transportation enhancement within their developments, and sometimes outside them as well. See Rebekah F. Sellers, "Park Land at Site Becomes an Issue," *Sunday Sun* (Georgetown), June 8, 2003, p. A-1. Rebekah F. Sellers, "Landscape, Parking Lot Part of Fee," *Williamson County Sun* (Georgetown), June 11, 2003, p. A-1; Rebekah F. Sellers, "City Council Takes its First Public Look at Simon Contract," ibid., July 9, 2003, p. A-1; and Sam Pfiester, "Dialogue Between Two Sides," ibid., p. A-10. Many of the Simon deal's strongest supporters lived in Georgetown's Sun City Texas.

[77] Interview, William S. Lott, Sept. 25, 2003, Georgetown. Lott is a retired district judge who practiced law in Georgetown during the 1960s and 1970s.

[78] Interview, Judy Wolf Hindelang, Nov. 7, 2003, Georgetown.

[79] "Gilded democrat" is my phrase denoting intense democratic activism that is essentially marshaled by a single individual, as opposed to more familiar forms of democracy such as populism, interest-group politicking, or, in the last decade of the twentieth century, focus groups. It does not connote identification with any political party; Republicans are just as prone to be gilded democrats as Democrats.

A Place Remade   *359*

individual endowed with native political talent, hard-won connections and unmatchable perseverance. Superficially working on behalf of or in partnership with powers that be, the gilded democrat remaps the future in ways that ultimately confound entrenched political and economic powers. Sometimes gilded democrats work for the public; Robert Moses, New York City's master builder, was one such creature. Mostly they operate outside the purview of public oversight. Two well-known gilded democrats were Alvin J. Wirtz, the Texas attorney who engineered the rise of the Lower Colorado River Authority and Lyndon Baines Johnson's political career, and newspaper magnate Harry Chandler, whose *Los Angeles Times* cheered the city's takeover of the Owens Valley watershed while he was privately developing the San Fernando Valley on the promise of that water.[80]

As I see them, Wilson Fox and Owen Sherrill were "gilded democrats," operating behind the scenes to eclipse or stall the efforts of powerful government officials. Fox and Sherrill wrapped themselves in the cloaks of organizations they controlled, but each man passionately pursued his own vision of a bridled San Gabriel River for personal as well as public reasons. Sherrill's dream of dams guarding Georgetown was driven by his bravura salesmanship of Georgetown ("highest point between Dallas and San Antonio," his brochures proclaimed) and a genuine desire to protect the county seat and agricultural "lowlands" to the east from flooding, after having seen the wholesale devastation wrought by the 1921 flood. On Wilson Fox's side, much earlier than his contemporaries he came to accept that agriculture was a lost cause on Williamson County's Black Waxy. He believed Taylor needed a dependable source of water to attract industry to replace the vanishing agricultural economic base, and he saw the reservoir at Laneport as the answer. Never a wealthy man, he may also have been tempted to sell his money-losing Riverside Ranch in the only way allowed by his father's will—through a government "taking." In their idiosyncratic ways, Sherrill and Fox outwitted, outmaneuvered, and outlasted the opposition (including, at times, each other) for twenty years, ultimately winning the dams they sought. In

---

[80] Blake Gumprecht, *Los Angeles River: Its Life, Death, and Possible Rebirth* (Baltimore: Johns Hopkins University Press, 1999), 118.

doing so, they overturned the carefully wrought plans of congress-men, governors, water managers, and flood engineers officially charged with controlling the entire Brazos River Basin.

In a less visible and less democratic fashion, a handful of property owners and Williamson County elites twisted the intentions of the national and state road bureaucracies in charge of planning and build-ing the interstate highway system. In the early days of interstate plan-ning, road engineers typically decided, announced, and then, if neces-sary, defended the interstate's proposed path—as they did in Round Rock, when property owners objected to the Texas Highway Depart-ment's announced route.[81] In other instances, highway planners sim-ply decided, as Dewitt Greer did when he determined I-35's Williamson County route. Landowners Forbes, Kouri, and Nash "worked the system" until the U.S. Bureau of Roads and the Texas Highway Department retooled facets of their Great Road, contra-dicting bureaucratic policies designed to protect the public. And yet—and this is important—the changes goaded into being by these gilded democrats worked for the general good. The fact is, at least in Williamson County's case, local politicians and property owners almost always had a clearer "take" on the impact of projected public works projects on average people than did the professional planners and government experts.

The lesson: one highly motivated and canny individual, acting as a gilded democrat or a behind-the-scenes elite, can strikingly alter important public works projects. Throughout the development of the dams and the interstate highway in Williamson County, this was assuredly the case, a point often missed by academic critics who have tended to see the massive government public works projects of the twentieth century as being imposed from the top down upon a help-less population. Today, as our political leaders become increasingly remote from their constituencies and government agencies seem impenetrable to the average citizen, the notion of an unknown local hero rising up to smite (or bend) a public works project sounds more like movie script than reality. But this is not necessarily the case.

Gilded democratic action can still work. The route of Williamson

---

[81] "The Road to Preserving History," *New York Times,* editorial, Nov. 12, 2003.

County's 130 Toll is a prime example. The 1.5-billion-dollar reliever highway for I-35 was first proposed twenty years ago. At that time the highway's path through the county seemed obvious: the abandoned Missouri-Kansas-Texas Railroad bed running east of Georgetown and Round Rock would be perfect. The Texas Department of Transportation recommended it. Georgetown and Round Rock leaders, and with them county commissioners, believed in this route so strongly that they purchased most of the right-of-way. The cost was not great but enormous energy was invested in planning and in protecting the route.[82] But in the two decades that passed between the first proposal and final approval for the project, important realities changed.[83] For one, Round Rock's population surged east and lapped over the proposed route. If built, the highway as planned would pass less than a football field's length from several dense developments of "starter homes" for young families—politically the equivalent of yesteryear's Czech farmers. For a time, it didn't seem to matter, because in the economic downturn of the eighties State Highway 130 seemed doomed. In the nineties, however, the project was revived, the neighborhoods protested and highway planners studied alternative routes, all of which were judged inferior to the old Mo-Kan corridor. Again and again the highway men insisted that the original route was the most efficient and economical route.[84] In the end, though, Highway 130 planners decided to skirt angry homeowners by routing the toll road several miles further east, just west of Hutto, substantially decreasing the highway's effectiveness as an I-35 alternative and making it more expensive to build. The biggest cost was to county taxpayers, who voted to foot the bill for the new right-of-way through the county, amounting to about fifty-four million dollars.[85]

Why did the Department of Transportation change its position? "One man killed it—John Gordon," said one of the county's most

---

[82] Telephone interview, John Doerfler, Feb. 25, 2004, Georgetown.

[83] See Ben Wear, "Long, Rocky Path Full of Close Calls Leads to Toll Road," *Austin American-Statesman,* Sept. 28, 2003, pp. A-1, A-11.

[84] Alternative routes would attract fewer drivers and cost more to build, highway engineers said, and they were undoubtedly correct.

[85] Telephone interview, John Doerfler, Feb. 25, 2004, Georgetown.

Road, River, and Ol' Boy Politics

knowledgeable transportation experts.[86] Gordon, a Round Rock politician, whipped up emotions and inspired the residents of the threatened subdivisions to fight. Gordon is a giant of a man who sports lumberjack shirts and work boots. He booms his opinions and is hard to ignore. The transformation of Williamson County from Democratic to Republican can be credited in large part to his organizational brilliance as Republican Party chairman during the early 1980s. In mostly Anglo Round Rock, the "race card" does not exist, but the "class card" does. Round Rock's least affluent tend to live east of Interstate 35; richer residents live mostly to its west. Gordon argued that if highway commissioners ran 130 Toll along its "preferred" Mo-Kan route, they would be discriminating against Williamson County's underclass. He won his point. As a result of Gordon's intervention, fewer commuters will drive 130 Toll, raising its cost and lowering its utility.[87] Hutto's population will explode. Scores of family farms will disappear. On the other hand, several thousand first-time homeowners will not see their investments savaged and their lives made miserable by the highway's inevitable offspring: the constant roar of eighteen-wheelers, the bustle of new construction, and localized traffic nightmares.

Imagine the shape of Williamson County had certain people not flown the banner of their dreams. Had Williamson County's great federal projects been built according to the first impulses of the Army Corps of Engineers and Dewitt Greer's Texas Highway Department, there would be one dam on the San Gabriel River at Granger, creating a giant reservoir lapping at the edges of Granger, Jonah, and Circleville. There would be no dams on the North or South Fork of the San Gabriel River west of Georgetown. Interstate 35 would ply the Blacklands between Temple and Granger, veering slightly west on its southward journey toward Taylor. At Taylor it would splice through a maze of sprawling subdivisions at the old railroad crossing, Frame Switch, on State Highway 79, and then head straight into Austin, avoiding Round Rock and Hutto. Round Rock would be a commu-

---

[86] Interview, Charles Crossfield, June 2002, Georgetown.

[87] Ben Wear, "Long, Rocky Path Full of Close Calls Leads to Toll Road," *Austin American-Statesman,* Sept. 28, 2003, pp. A-1, A-11.

nity of twenty-five thousand and growing, partly due to Austin's bulge north along U.S. Highway 81. As a result of the "army colony's" inspired leadership in the fifties, Round Rock's biggest industry would be tourism sparked by the elaborate restoration and development of historic Brushy Creek, a project rivaling San Antonio's River Walk in importance. Taylor would control virtually all of the county's water reserves and sell water to Round Rock and Georgetown— pumping water twenty miles against gravity, an expensive proposition. Georgetown would not have changed much since 1968, when the interstate was completed through Taylor. Its biggest employers would be the university and the county government. Williamson County's south*eastern* sector—especially Taylor and Hutto—would lead the county's economic prosperity. A new Sun City Texas would bask in the neighborhood of Granger Lake; Granger and Bartlett would be retirement Meccas. The logical culmination of Ralph Moore's dreams would have been realized.

Of course, other scenarios might prevail. If Henry Fox's "little dams" program for the San Gabriel River had been built, as indeed a similar network of earthen dams was on Brushy Creek, Williamson County would have had no substantial supply of water to grow on. With fewer water resources, the county might have remained ranch and farm country, at least for a time. Or, if the U.S. Bureau of Reclamation had won its battle with the Army Corps of Engineers, two dams would plug the San Gabriel River's forks west of Georgetown. That would have intensified residential development on the ecologically fragile Edwards Plateau west of the Balcones Escarpment, while Stacy Labaj's "fabulous" Black Waxy would have been irrigated in the California fashion, producing—who knows?—grapes for a wine industry developed by Czechs and Germans, or boutique organic crops for Austin's Whole Foods market. It is a game of historical fiction, tracing alternative universes that might have been.

But what, after all, have we learned? It is clear that the law of unintended consequences often shakes intended patterns into something unimagined by planners. It is clear that throughout the arid Southwest in the twentieth century, the impulse to develop agricultural wealth led to gigantic federal dam projects that altered the landscape, providing huge pools of water that fed not just orange groves and cot-

Shortly before Interstate 35's 1968 Round Rock opening, this aerial was shot. The
view is to the north, with Round Rock in the distance. The Farm-to-Market
1325 and I-35 intersection cannot be seen because it is underneath the airplane.
Courtesy Ralph Rich, Austin.

Same view of the Interstate 35 corridor through Round Rock, March 2005, except that the Farm-to-Market 1325 intersection with I-35, beefed up by State Highway 45, is visible. Dell and La Frontera dominate Round Rock's south rim. *Clark Thurmond Photo, Courtesy* Williamson County Sun, *Georgetown.*

ton fields, but also astonishing population booms in places like Austin, Phoenix, Dallas–Fort Worth, El Paso, Salt Lake City, Denver, and Albuquerque. It is clear, too, that wherever the new interstate highway system converged with these new sources of water, the Dry Sun Belt blossomed, spreading suburban living into rural hinterlands and creating nascent Edge Cities like Plano, Texas; Fort Collins, Colorado; Mesa, Arizona; Provo, Utah; and now Round Rock.

But beneath the simple tools of road and water engineering that it took to spark development of the Dry Sun Belt, there was one thing more. Always, it seems, it took a combination of hubris and clear thinking combined in a single individual—from Senator Lyndon B. Johnson, who in 1953 proposed to turn Texas into a California-style industrial giant, to Owen W. Sherrill's dogged insistence that what the future inhabitants of the remote San Gabriel River Valley really needed was dams—to create new worlds, not just in Williamson County, Texas, but across the American Southwest.[88]

---

[88] Lyndon B. Johnson, "Water Supply and the Texas Economy," U.S. Senate, July 29, 1953, Senate Rivers and Harbors Files 1949-53 (LBJ Library, Austin).

SELECTED BIBLIOGRAPHY

Archival Collections

Austin Presbyterian Theological Seminary, Austin, Texas
   Church records, Diary of Rev. Walter S. Scott
Baylor University Collections of Political Materials, Waco
   Poage, W. R., Papers: Correspondence, San Gabriel Dam, Granger
   (Laneport) Dam, Flood Prevention, Brazos River Conservation, Belton
   Dam, Oral Memoirs
Brazos River Authority, Waco, Texas
   Assurances, Water Supply
   Biographical materials: Howard Fox, John Short Fox, R. F. Holubec
   Federal Reservoirs Water Revenue
   Minutes of board of directors meetings
   Poage, W. R., Congressman
   San Gabriel River Files
Center for American History, University of Texas at Austin
   Biographical Files
   Oral History
   Pickle, J. J. "Jake," Papers: Correspondence with Thatcher Atkin,
   Roman Bartosh, Col. Fickessen, Henry Fox, Wilson Fox, Harry
   Provence, Don Scarbrough, Owen Sherrill, Files on Circleville
   BBQ, Taylor BBQ, and Pickle Appreciation Party; Files on San
   Gabriel (Laneport), P.O. Georgetown, P.O. Round Rock, P.O. Taylor
   Texas Newspaper Collection
   Texas University *Bulletins*
   TXC Collection
   Williamson County Scrapbook
Clark, Edward A., Texana Collection, Southwestern University, George-
town
   San Gabriel River Crossings, movie, Texana Collection

Cushing Memorial Library, Texas A&M University, College Station
Owen W. Sherrill Papers
Department of the Army, Fort Worth District Corps of Engineers
Acquisition Records
Civil Works Project File, Brazos River
Condemnation Records
Maps
Harry Ransom Humanities Research Center, University of Texas at Austin
Jno. Trlica, Photographs, Photo Collection
Labaj, Stacy, Papers, lent by Dorothy "Dot" Daniel, Murfreesboro, Tennessee
Lyndon Baines Johnson Library, Austin
Correspondence with Thatcher Atkin, George Brown, Henry Fox, Wilson Fox, Owen Sherrill, William Lott, John McCall, Harris Melasky, J. J. "Jake" Pickle, W. R. Poage, Don Scarbrough, John Sharpe, Homer Thornberry
House Papers 1937–49
Highways, Rivers—San Gabriel Dam Project and Laneport, Taylor Bedding
LBJA Subject Files: Public Works Highway, San Gabriel River, Press, Case File, Confirmations, BBQ, Selected Names, Senate Papers (Brazos River Authority)
Oral Histories: George R. Brown, Sam Gidean, Frank W. Mayborn, Harris Melasky, W. R. Poage, Sam V. Stone, J. J. ("Jake") Pickle, Homer Thornberry, Bob Waldron
Senate Case and Project Files, 1957–1960
Highway Program, San Gabriel dams, Laneport dam, Georgetown, Granger, Taylor, Williamson County, Taylor (Bedding)
Senate Rivers & Harbors Files 1949–1953: Rivers general, Rivers–Laneport Dam, Rivers–San Gabriel
Thornberry, Homer, Papers
VP: Public Works San Gabriel
Merrick, Ben A., Manuscript, lent by the late Loretta Mikulencak, Granger, Texas; owned by Tessa Mikulencak Knox, Glen Rose, Texas
Moore, Ralph W., Papers, lent by Opal Wilks, Taylor, Texas
National Archives & Records Administration, Southwest Region, Fort Worth
U.S. Corps of Engineers Records, Correspondence, Memoranda, San Gabriel River Valley, U.S. Bureau of Reclamation

Round Rock, City of
    Contracts
    Maps
    Minutes
    Statistics
Scarbrough, Donald L., Papers, owned by author
Sharpe, John M., Papers, Georgetown Public Library, Georgetown, Texas
Stearns, Margaret, Papers, owned by author
Taylor Public Library
    Newspaper Archives
    Obituaries: Howard "Son" Bland, David Fontaine Forwood, Henry
    Fox, Wilson Fox
    Photographs
Texas Department of Transportation, Austin
    Commission Files by County, 1940s: Bell, Travis, Williamson
    Dewitt Greer Papers, 1920s–1960s
    General Correspondence
    Photographs
    Maps
    Texas Highways Archives
Texas State Library, Austin
    Dewitt Greer Papers, 1970s–
    Map Collection
    Texas Parade
Travis County Collection, Austin Public Library
    Newspaper Clippings
U.S. District Court for the Western District of Texas, Austin Testimony,
    Findings Court Case
Williamson County Tax Appraisal District
    Certified Appraisals
    Maps

Interviews

Beard, Truett, Granger, Tex., March 28, April 13, 2000
Bennett, Bob, Round Rock, Tex., September 14, 2000
Benold, Stephen, Georgetown, Tex., June 22, 1998
Brady, Sam, Georgetown, Tex., July 7, 1998
Brashear, Ercel, Georgetown, Tex., April 20, 2005
Broad, Morton, Austin, Tex., March 23, 2000
Brogren, Eleanor, Georgetown, Tex., May 1974

Bukala, Mike, Waco, Tex., October 31, 2000
Burleson, Bob, Temple, Tex., June 1974
Bullion, Tom, Taylor, Tex., May 12, 2000
Cannon, Billye "Bill" Fulcher, Rockdale, Tex., January 28, 2000; San Marcos, Tex., April 19, 2005
Cariker, Billie Sue Henna, Round Rock, Tex., September 14, 2000
Chapman, David, College Station, Tex., April 14, October 6, 2000
Clawson, Doyle, Georgetown, Tex., July 14, 1998
Clinton, Ramsey, Georgetown, Tex., July 1, 1998
Cooke, Bill, Rockdale, Tex., January 2000
Crossfield, Charles, Georgetown, Tex., June 2002; Round Rock, Tex., April 15, 2005
Culpepper, Charlie, Round Rock, Tex., October 5, 1998
Danek, Emil, Jarrell, Tex., July 9, 1998
Daniel, Dorothy "Dot" Labaj, Round Rock, Tex., June 12, 2001
Doerfler, John, Georgetown, Tex., June 15, October 5, 1998, February 25, 2004
Doering, Carl, Georgetown, Tex., September 10, 2003
DuBose, Glenda, Georgetown, Tex., September 19, 2001
Duncan, Donald, Jr., Houston, Tex., July 7, 1998
Evans, Edward Lee, Georgetown, Tex., March 25, 2000
Fairburn, Jeannine, Georgetown, Tex., November 2001
Fondren, Gene, Austin, Tex., March 9, September 1, 2000
Forbes, Mary, Georgetown, Tex., March 26, 2000
Forbes, Tom, Austin, Tex., May 23, 2000
Fox, Carol, Circleville, Tex., April 13, 2000, June 1, 2001, April 20, 2005
Fox, Geraldine, Granger, Tex., September 20, 2000
Fox, Henry B., Circleville, Tex., May 1974
Fox, Dr. James, Austin, Tex., September 6, 2000
Fox, Marie, Circleville, Tex., April 13, 2000
Fox, Paul, Austin, Tex., August 31, 2000
Fox, Susan, Taylor, Tex., April 13, 2000
Garry, Mahon "Buzz," Austin, Tex., April 5, 2000
Gittins, Ricia, Georgetown, Tex., February 23, 2002
Graves, Linda Crawford, Georgetown, Tex., May 1974
Graves, Tom, McKinney, Tex., November 15, 2001
Griffith, Edward, Taylor, Tex., August 17, 2001
Green, Gaz, Georgetown, Tex., February 19, 2002
Gunn, Sam, Kirrama, Queensland, Australia, February 1986; Minerva Hills, Queensland, Australia, September 5, 2001
Gunn, Tate, Kirrama, Queensland, Australia, February 1986; Rockport,

Road, River, and Ol' Boy Politics

Tex., September 4, 2001

Hajda, Betty, Granger, Tex., May 1974, July 6, 1998, September 23, 1999, June 1, 2001

Hammer, Jerry, Georgetown, Tex., July 1974

Hart, Bob, Georgetown, Tex., October 2, 1998

Hausenfluke, Larry, Georgetown, Tex., July 7, 1998

Hays, Claude, Georgetown, Tex., August 20, 2001

Hays, David, Georgetown, Tex., June 2002

Heiligenstein, Mike, Round Rock, Tex., June 2002

Higginson, Scott, Phoenix, Arizona, August 30, 2002

Hindelang, Judy Wolf, Georgetown, Tex., November 7, 2003

Hindelang, Paul E., Jr., Georgetown, Tex., July 2, 1998

Hislop, James, Georgetown, Tex., July 1974, July 2, 1998

Hull, Ken, Hilton Head, S.C., August 28, 2002

Hurt, John, Austin, Tex., March 22, 2000, January 24, 2002

Inman, S. C., Jr., Round Rock, Tex., July 1, 1998

Irvin, Helen, Austin, Tex., January 14, 2005

Kimbro, "Fat," Georgetown, Tex., July 6, 1998

Kouri, Tom, Austin, Tex., September 22, 2000

Krusee, Mike, Austin, Tex., June 1999

Lawrence, Virginia Forwood, Taylor, Tex., March 28, 2000

Leonard, Don J., Round Rock, Tex., August 1974

Leps, Ronnie, Georgetown, Tex., February 20, 2002

Lott, William S., Georgetown, Tex., September 25, 2003

McAdams, Jane, Cedar Park, Tex., February 23, 1999

McLachlan, Iva Wolf, Georgetown, Tex., February 2005

Mantor, Ruth, Taylor, Tex., October 29, 1999

Mehevec, Jerry, Taylor, Tex., January 27, 2002

Mikulencak, Loretta, Granger, Tex., May 1974, July 6, 1998, September 23, 1999, June 1, 2001

Miles, James, Taylor, Tex., September 21, 1999

Mills, Jim, Georgetown, Tex., December 15, 1998

Moore, Ethel M., Georgetown, Tex., September 22, 2003

Moore, Tommy, Granger, Tex., July 6, 1998

Nelson, Martha Burt, Frederick, Md., October 2, 2003

Nelson, Tom E., Jr., Austin, Tex., October 23, 2001

Norsworthy, George, Dripping Springs, Tex., July 14, 1998

Patterson, Charles, New York, N.Y., June 1974

Pearce, Judy, Waco, Tex., February 25, 2004

Pendland, Mel, Georgetown, Tex., February 25, 2004

Pickle, J. J. "Jake," Austin, Tex., July 3, 1998, April 18, 2000, May 19, 2001

Raesz, Jan, Taylor, Tex., January 2004
Rich, Ralph, Austin, Tex., March 31, 2000
Salvato, Frank, Taylor, Tex., September 23, 2003
Scarbrough, Clara Stearns, Georgetown, Tex. (many conversations)
Scarbrough, Don, Georgetown, Tex., May 1974, July 1998
Schwartz, "Babe," Galveston, Tex., November 2001
Shannon, Richard A., Austin, Tex., July 1974, November 3, 2001
Shell, "Wiggie," Georgetown, Tex., September 20, 2000
Sloan, Jay, Georgetown, Tex., November 8, 1999, September 11, 2000
Smith, David, Warrenton, Va., December 4, 2000
Smith, Helen Fox, Giddings, Tex., September 11, 2000
Snead, Bill, Georgetown, Tex., July 2, 1998
Spencer, Patsy Gunn, Georgetown, Tex., February 1986, September 2, 21, 2001, April 18, 2005
Stanton, Bobby, Georgetown, Tex., July 8, 10, 1998
Stearns, Margaret Tegge, Taylor, Tex., May, July 1974
Stubblefield, Billy Ray, Georgetown, Tex., June 22, 1998
Stump, William, Georgetown, Tex., November 8, 1999
Teague, Dave, Granger, Tex., July 1974
Todd, Betty, Georgetown, Tex., July 1998
Todd, William "Bill," Round Rock, Tex., September 14, 2000, February 2, 2002
Thomas, J. D., Georgetown, Tex., September 8, 2000
Thomison, Dan, Granger Lake, Granger, Tex., October 4, 2000
Weber, Carey, Lake Georgetown, Georgetown, Tex., December 14, 1999
Weir, Esther, Georgetown, Tex., January 19, 2001
Weir-Clarke, Laura, Georgetown, Tex., May 2004
Whitlow, N. G. "Bunky," Round Rock, Tex., July 1, 1998, September 14, 20, 2000
Wilks, Opal, Taylor, Tex., November 7, 2000
Wolf, Bettie, Georgetown, Tex., November 7, 2003
Wood, Leo, Georgetown, Tex., July 1974, June 30, 1998
Woodruff, C. M. ("Chock"), Jr., Austin, Tex., July 1974
Zimmerhanzel, Betty, Circleville, Tex., December 8, 2000
Yawn, Nancy, Round Rock, Tex., February 26, 2002

Periodicals

Austin American-Statesman, 1946–1980, 1998–2004
Fauquier [County] Times Democrat, Warrenton, Va.
Granger News, 1946–1964, 1973

Round Rock Leader, 1956–1972
San Antonio Express, 1921, 1956
The Sunday Sun (Georgetown), 1974–2004
Taylor Daily Press, 1921–1980
Taylor Times, 1946–1955
Temple Daily Telegram, 1935, 1968, 1972
Texas Highways, 1961, 1962, 1966
Texas Landsman (Austin), July 1964
Texas Parade, 1950–1956
Waco News-Tribune, 1966, 1972
Wall Street Journal, 1959, 1998
Williamson County Sun (Georgetown), 1921–2004

Books

Abbot, Patrick L. and C. M. Woodruff Jr. *Balcones Escarpment, Central Texas.* San Antonio: Geological Society of America, 1986.

Albion, Robert G. *The Rise of New York Port, 1815–1860.* New York: Charles Scribner's Sons, 1939.

Allen, Martha Mitten, editor. *Georgetown's Yesteryears: Reaching for the Gold Ring.* Georgetown, Tex.: Georgetown Heritage Society, 1985.

————, *Georgetown's Yesteryears: The People Remember.* Georgetown, Tex.: Georgetown Heritage Society, 1985.

Alperin, Lynn M. *Custodians of the Coast: History of the United States Army Engineers at Galveston.* Galveston: U.S. Army Corps of Engineers, Galveston District, 1977.

Anderson, Ken. *You Can't Do That, Dan Moody!* Austin: Eakin Press, 1998.

Boyle, Robert H., John Graves, and T. H. Watkins. *The Water Hustlers.* San Francisco: Sierra Club, 1971.

Branda, Eldon Stephen, editor. *The Handbook of Texas. Vol. III. Supplement.* 3 vols. Austin: Texas State Historical Association, 1952, 1976.

Brandt, E. N. *Growth Company: Dow Chemical's First Century.* East Lansing, Mich.: Michigan State University Press, 1997.

Buck, Solon Justus. *The Granger Movement: A Study of Agricultural Organization and Its Political, Economic, and Social Manifestations, 1870–1880.* Cambridge, Mass.: Harvard University Press, 1913.

Buffington, Jesse L. *Restudy of Changes in Land Value, Land Use and Business Along a Section of I-35.* College Station, Tex.: Texas Transportation Institute, 1964.

Caro, Robert A. *The Years of Lyndon Johnson: The Path to Power.* New York: Vintage Books, 1983.

———. *The Years of Lyndon Johnson: Means of Ascent*. New York: Alfred A. Knopf, 1990.

———. *The Power Broker: Robert Moses and the Fall of New York*. New York: Vintage Books, 1975.

Cather, Willa. *My Ántonia*. Boston: Houghton Mifflin Co., 1949.

Central Texas Business and Professional Directory. Austin: Centex Publications, 1950.

Clay, Grady. *Real Places: An Unconventional Guide to America's Generic Landscape*. Chicago: University of Chicago Press, 1994.

Colegrove, Bill. *Episodes: Texas Dow 1940–1976*. Houston: Larksdale, 1983.

Copelston, Frederick. *History of Philosophy, Vol. VII, Fichte to Nietzsche*. London: Burns and Oates Ltd., 1963.

Cronon, William. *Nature's Metropolis: Chicago and the Great West*. New York: W. W. Norton & Company, 1991.

Crosby, Alfred W. *Ecological Imperialism: The Biological Expansion of Europe, 900–1900*. Cambridge, UK: Cambridge University Press, 1986.

deBuys, William. *Enchantment and Exploitation: The Life and Hard Times of a New Mexico Mountain Range*. Albuquerque: University of New Mexico Press, 1985.

Denton, Sally and Roger Morris. *The Money and the Power: The Making of Las Vegas and Its Hold on America, 1947–2000*. New York: Alfred A. Knopf, 2001.

Douglas, William O. *Farewell to Texas: A Vanishing Wilderness*. New York: McGraw-Hill Book Co., 1967.

Dykstra, Robert R. *The Cattle Towns*. New York: Alfred A. Knopf, 1968.

Fehrenbach, T. R. *Lone Star: A History of Texas and the Texans*. New York: American Legacy Press, 1983.

Ferejohn, John A. *Pork Barrel Politics: Rivers and Harbors Legislation, 1947–1968*. Stanford, Calif.: Stanford University Press, 1974.

Fergusson, Erna. *New Mexico: A Pageant of Three Peoples*. New York: Alfred A. Knopf, 1951.

Field, Donald R., James C. Barron, and Burl F. Long. *Water and Community Development: Social and Economic Perspectives*. Ann Arbor, Mich.: Ann Arbor Science, 1974.

Foley, Neil. *The White Scourge: Mexicans, Blacks, and Poor Whites in Texas Cotton Culture*. Berkeley: University of California Press, 1997.

Fox, Henry B. *The 2000-Mile Turtle and Other Enjoyable Episodes from Harold Smith's Private Journal*. Austin: Madrona Press, 1975.

———. *Dirty Politics is Fun*. Seattle: Madrona Publishers, 1982.

Fradkin, Philip L. *The Seven States of California*. New York: Henry Holt and Co., 1995.

Garreau, Joel. *Edge City: Life on the New Frontier.* New York: Doubleday, 1991.

Geertz, Clifford. *The Interpretation of Cultures: Selected Essays.* New York: Basic Books, 1973.

George, Henry. *Progress and Poverty.* New York: Robert Schalkenback Foundation, 1960.

Gottlieb, Robert. *A Life of Its Own: The Politics and Power of Water.* San Diego: Harcourt Brace Jovanovich, 1988.

Goudie, Andrew. *The Human Impact on the Natural Environment.* Fourth Edition. Cambridge, Mass.: MIT Press, 1994.

Graham, Otis L., Jr. *Toward a Planned Society: From Roosevelt to Nixon.* New York: Oxford University Press, 1976.

Graves, John. *From a Limestone Ledge: Some Essays and Other Ruminations about Country Life in Texas.* New York: Alfred A. Knopf, 1980.

————. *Goodbye to a River.* New York: Alfred A. Knopf, 1983.

Grisham, Noel. *Round Rock, Texas, U.S.A.!!!* Round Rock, Tex.: Sweet Publishing Co., 1972.

Gumprecht, Blake. *The Los Angeles River: Its Life, Death, and Possible Rebirth.* Baltimore: The John Hopkins University Press, 1999.

Hendrickson, Kenneth E., Jr. *The Waters of the Brazos: A History of the Brazos River Authority 1929–1979.* Waco, Tex.: The Texian Press, 1981.

Hoelscher, Steven D. *Heritage on Stage: The Invention of Ethnic Place in America's Little Switzerland.* Madison: University of Wisconsin Press, 1998.

Holman, Alma Lee, editor. *Welcome to Taylor.* Taylor, Tex.: Taylor Daily Press, 1994.

Jackson, John B. *American Space: The Centennial Years, 1865–1876.* New York: Norton, 1972.

————. *Landscapes: Selected Writings of J. B. Jackson.* Edited by Ervin H. Zube. Amherst: University of Massachusetts Press, 1970.

————. *A Sense of Place, a Sense of Time.* New Haven: Yale University Press, 1994.

Jordan, Terry G. *Texas: A Geography.* Boulder, Colo.: Westview Press, 1984.

————. *Environment and Environmental Perceptions in Texas.* Boston: American Press, 1980.

Kelley, Ben A. *The Pavers and the Paved.* New York: D. W. Brown, 1971.

Kunstler, James Howard. *The Geography of Nowhere: The Rise and Decline of America's Man-Made Landscape.* New York: Simon & Schuster, 1993.

————. *Home from Nowhere: Remaking Our Everyday World for the Twenty-first Century.* New York: Touchstone Books, Simon & Schuster, 1998.

Labaj, Stacy. "Friendship, Texas." Unpublished manuscript, date unknown, lent to the author by Dorothy "Dot" Daniel, Murfreesboro, Tenn.

Landsberger, Henry A., editor. *Rural Protest: Peasant Movements and Social Change.* London: Macmillan Press, 1974.

Least Heat-Moon, William. *PrairyErth (A Deep Map).* Boston: Houghton Mifflin Co., 1991.

Leavitt, Helen. *Superhighway—Superhoax.* New York: Ballantine Books, 1970.

Lefebvre, Henri. *The Production of Space.* Oxford, UK: Blackwell, 1974.

Leffler, John J. *Historic Williamson County: An Illustrated History.* San Antonio, Tex.: Historical Publishing Network, 2000.

Lewis, Tom. *Divided Highways: Building the Interstate Highways, Transforming American Life.* New York: Viking, 1997.

Luckingham, Bradford. *Phoenix: The History of a Southwestern Metropolis.* Tucson: University of Arizona Press, 1989.

Maass, Arthur. *Muddy Waters: The Army Engineers and the Nation's Rivers.* Cambridge, Mass.: Harvard University Press, 1951.

McCandless, Barbara. *Equal Before the Lens: Jno. Trlica's Photographs of Granger, Texas.* College Station: Texas A&M Press, 1992.

McCarter, Robert. *Frank Lloyd Wright.* London: Phaidon Press, 1997.

McCullough, David. *The Jonestown Flood.* New York: Touchstone Books, Simon & Schuster, 1968.

———. *The Path Between the Seas: The Creation of the Panama Canal, 1870–1914.* New York: Simon and Schuster, 1977.

Machann, Clinton, and James W. Mendl. *Krásná Amerika: A Study of the Texas Czechs, 1851–1939.* Austin, Tex.: Eakin Press, 1983.

McHarg, Ian L. *Design with Nature.* New York: Garden City Press, 1969.

McPhee, John. *Encounters with the Archdruid.* New York: Noonday Press, Farrar, Straus and Giroux, 1971.

Mann, Roy. *Rivers in the City.* New York: Praeger Publishers, 1973.

Mantor, Ruth. *Our Town: Taylor.* Taylor, Tex.: First-Taylor National Bank, 1983.

Marx, Karl. *Capital.* Chicago: Encyclopaedia Britannica, Inc., University of Chicago, 1952.

Masterson, V. V. *The Katy Railroad and the Last Frontier.* Columbia: University of Missouri Press, 1988.

Meikle, Jeffrey L. *Twentieth Century Limited: Industrial Design in America, 1925–39.* Philadelphia: Temple University Press, 1979.

Meinig, D. W. *Imperial Texas: An Interpretive Essay in Cultural Geography.* Austin, Tex.: University of Texas Press, 1969.

Merrick, Ben A. "The Granger Farm, 1902–1908." Unpublished manuscript, 1990, lent to the author by the late Loretta Mikulencak, Granger, Tex.; owned by Tessa Mikulencak Knox, Glen Rose, Tex.

Milam County Heritage Preservation Society, editor. *Matchless Milam: History of Milam County, Texas.* Dallas, Tex.: Taylor Publishing, 1984.

Mill, John Stuart. *Collected Works of John Stuart Mill: Newspaper Writings.* Vol. 25. Edited by Ann P. Robson and John M. Robson. Toronto: University of Toronto Press, 1986.

————. *Principles of Political Economy with Some of Their Applications to Social Philosophy.* Vol. 3. Edited by J. M. Robson. London: University of Toronto Press, 1965.

Mitchell, W. J. T., editor. *Landscape and Power.* Chicago: University of Chicago Press, 1994.

Montejano, David. *Anglos and Mexicans in the Making of Texas, 1836–1986.* Austin, Tex.: University of Texas Press, 1987.

Morehead, Richard. *Dewitt C. Greer: King of the Highway Builders.* Austin: Eakin Press, 1984.

Morris, John Miller. *El Llano Estacado: Exploration and Imagination on the High Plains of Texas and New Mexico, 1536–1860.* Austin: Texas State Historical Association, 1997.

Paher, Stanley W. *Las Vegas, As It Began, As It Grew.* Las Vegas: Nevada Publications, 1971.

Palmer, Tim. *Lifelines: The Case for River Conservation.* Washington, D.C.: Island Press, 1994.

Pickle, Jake and Peggy Pickle. *Jake.* Austin: University of Texas Press, 1997.

Porter, Eliot. *The Place No One Knew: Glen Canyon on the Colorado.* San Francisco: Sierra Club, 1963.

Reisner, Marc. *Cadillac Desert: The American West and Its Disappearing Water.* New York: Penguin Books, 1986.

Robinson, John. *Highways and Our Environment.* New York: McGraw-Hill Book Co., 1971.

Rose, Mark R. *Interstate Express Highway Politics, 1941–1956.* Lawrence, Kan.: Regents Press of Kansas, 1979.

Roske, Ralph J. *Las Vegas: A Desert Paradise.* Tulsa: Continental Heritage Press, 1986.

Scarbrough, Clara Stearns. *Land of Good Water: Takachue Pouetsu, A Williamson County History.* Georgetown, Tex.: Williamson County Sun Publishers, 1973.

Scarpino, Philip V. *Great River: An Environmental History of the Upper Mississippi, 1890–1950.* Columbia: University of Missouri Press, 1985.

Schneiders, Robert Kelley. *Unruly River: Two Centuries of Change along the Missouri.* Lawrence: University Press of Kansas, 1999.

Shallat, Todd. *Structures in the Stream: Water, Science, and the Rise of the U.S. Corps of Engineers.* Austin: University of Texas Press, 1994.

Shepard, Paul. *Man in the Landscape: A Historic View of the Esthetics of Nature.* College Station: Texas A&M University Press, 1991.

Smythe, William E. *The Conquest of Arid America.* New York: Macmillan Co., 1907.

Stilgoe, John R. *Borderland: Origins of the American Suburb.* New Haven, Conn.: Yale University Press, 1988.

Sullivan, Robert. *The Meadowlands: Wilderness Adventures on the Edge of a City.* New York: Anchor Books, 1998.

Taylor, Alan. *William Cooper's Town: Power and Persuasion on the Frontier of the Early American Republic.* New York: Alfred A. Knopf, 1995.

Taylor and its Opportunities. Taylor, Tex.: Taylor Chamber of Commerce, 1940.

Texas Highway Fact Book. Austin: Texas Highway Department, State of Texas, 1959.

Vidich, Arthur J. and Joseph Bensman. *Small Town in Mass Society: Class, Power, and Religion in a Rural Community.* Princeton, N.J.: Princeton University Press, 1968.

Wallach, Bret. *At Odds with Progress: Americans and Conservation.* Tucson: University of Arizona Press, 1991.

Webb, Walter Prescott. *The Great Plains.* Boston: Ginn and Co., 1931.

Wiley, Peter and Robert Gottlieb. *Empires in the Sun: The Rise of the New American West.* New York: G. P. Putnam's Sons, 1982.

Wilson, Alexander. *The Culture of Nature: North American Landscape from Disney to the Exxon Valdez.* Cambridge, Mass.: Blackwell, 1992.

Worster, Donald. *Rivers of Empire: Water, Aridity, and the Growth of the American West.* New York: Oxford University Press, 1985.

————. *Dust Bowl: The Southern Plains in the 1930s.* New York: Oxford University Press, 1979.

Wright, Frank Lloyd. *The Living City.* New York: Meridian, 1958.

Zube, Ervin H. and Margaret J. Zube, editors. *Changing Rural Landscapes.* Amherst: University of Massachusetts Press, 1977.

Government Documents

"Acreage Acquired by the U.S. Government for Lake Granger Project," January 1, 1976. Granger Independent School District, Granger, Texas.

"An Order approving the feasibility of the North San Gabriel, South San Gabriel, and Laneport Reservoirs Project," Texas Water Commission, June 25, 1962. Brazos River Authority, Waco.

"Act to create county of San Gabriel," House Bill 512, House Journal, 2nd Leg. Session, State of Texas, February 3, 1848. Texas State Archives, Austin.

"Act to create county of San Gabriel," House Bill 625, House Journal, 2nd Leg. Session, State of Texas, February 11, 1848. Texas State Archives, Austin.

"Brazos River Conservation and Reclamation District of State of Texas. Brazos River, Texas Flood Control and Conservation Project," September 26, 1935. Baylor Collections of Political Materials, Waco.

"Brazos G Region Water Plan," July 2000. Brazos River Authority, Waco.

Brazos River Authority water agreement with SCB Development Corporation, January 23, 1985. Brazos River Authority, Waco.

"Brazos River Conservation and Reclamation District and Dow Chemical Company, A Water Supply Contract," December 30, 1942. Brazos River Authority, Waco.

C. C. Allison v. Robert Froehlke, Secretary of the Army, et al., No. A-71-CA-84, June 2, 1972. U.S. District Court for the Western District of Texas, Austin.

C. C. Allison v. Stanley R. Resor, Cival Action No. A-71-CA-84, Aug. 13, 1971, U.S. District Court for the Western District of Texas, Austin.

Capital Area Planning Council, "Regional Outdoor Recreation and Open Space Plan 1973–1990," 1973. Copy in author's possession.

City of Austin Resolution, February 9, 1956. Lyndon B. Johnson Library, Austin.

City of Georgetown Resolution, January 14, 1957. Brazos River Authority, Waco.

City of Round Rock, Engineering Study: "Water Supply—Westinghouse Plant," March 31, 1971. City of Round Rock.

———, Agreement, City of Round Rock and Westinghouse Electric Company, March 9, 1972, City of Round Rock.

———, Contract, City of Round Rock and Westinghouse, March 8, 1974. City of Round Rock.

———, "Cost to Produce and Deliver Water to Westinghouse," October 12, 1978. City of Round Rock.

———, Resolution, November 8, 1956. Texas Department of Transportation, Austin.

Conference, Army Corps of Engineers and U.S. Bureau of Reclamation, February 5, 1946, Austin, Tex. RG77, Records of the Corps of Engineers, Fort Worth District, Civil Works Project Files, 1934–1961, National Archives, Southwest Region, Fort Worth.

Deed Records, Williamson County Clerk, 1946–1948. Williamson County Clerk's Office, Georgetown, Texas.

"Dow Chemical Company's First Request for Production of Documents," Request RR-20, Fiscal Year 1989. Brazos River Authority, Waco.

"Facts . . . from the Brazos River Authority," Mineral Wells: Brazos River Authority, 1956. Lyndon B. Johnson Library, Austin.

"Federal-aide Highway Act of 1948," Public Law 834–80th Congress, *Public Roads,* Vol. 25, No. 5, September 1948. U.S. Printing Office, Washington, D.C., pp. 107–108.

"A Federal-Local Flood-Prevention Program," June 19, 1952, U.S. House of Representatives. Baylor Collections of Political Materials, Waco.

Flood Control Act, June 23, 1946. Baylor Collections of Political Materials, Waco.

Flood Control Act of 1954. Lyndon B. Johnson Library, Austin.

Final Environmental Impact Statement: Laneport, North Fork and South Fork Lakes, San Gabriel River, Texas, February 24, 1972. U.S. Army Engineers District, Fort Worth.

Final Environmental Statement, 1972. U.S. Corps of Engineers, Fort Worth.

"Granger Independent School District Lost Revenues from Lake Granger." Compiled by Granger Independent School District Tax Assessor; owned by author.

"Granger Lake, Texas," Audit #Ft W-2-008, Index, Tract Register, August 14, 1978. Department of the Army, Fort Worth District, U.S. Corps of Engineers, Fort Worth.

"Highways and the Nation's Economy," Joint Committee on the Economic Report sec. 5(a) of Public Law 304, 79th Congress. U.S. Government Printing Office, Washington, D.C., 1950, Perry-Castañeda Library, University of Texas at Austin.

"North San Gabriel Dam-Lake Georgetown," Audit #Ft W-2-0010, Real Property Title/Historical Files. Department of the Army, Fort Worth District, U.S. Corps of Engineers, Fort Worth.

Minutes, Brazos River Authority Board of Directors, April 18, 1966. Brazos River Authority, Waco.

Minutes, City of Round Rock, Nov. 13, 1956. Texas Department of Transportation, Austin.

Petition to Board of Engineers for Rivers and Harbors, December 1, 1961. Lyndon B. Johnson Library, Austin.

Petition to Texas Highway Commission, November 21, 1956. Texas Department of Transportation, Austin.

"Report on Survey of Brazos River and Tributaries, Texas," August 1947. Baylor Collections of Political Materials, Waco.

"Taylor, Texas, Water Supply," January 1959, Freese and Nichols, Civil Engineers. Brazos River Authority, Waco.

Texas Highway Commission Hearing, April 13, 1943. Texas Department of Transportation, Austin.

Texas Highway Commission Hearing, July 31, 1944. Texas Department of Transportation, Austin.

Texas Highway Commission Hearing, November 21, 1956. Texas Department of Transportation, Austin.

"Texas Roads and Highways," Texas Legislative Council, January 1953. Texas Department of Transportation, Austin.

Texas Water Commission, "Summary of Conference," January 7, 1965, Austin. Brazos River Authority, Waco.

U.S. Census Reports, 1940–2000 (*Census of the Population*).

"U.S. Highway System 1931–1932," General Files, Texas Highway Department. Texas Department of Transportation, Austin.

"Water for Texas. A Consensus-Based Update to the State Water Plan." Vol. II, 1997. Texas Water Development Board, Austin.

"Water Resource Development in Texas, 1995." U.S. Army Corps of Engineers, Southwestern Division, Dallas.

"Water Supply and the Texas Economy," Department of the Interior, Bureau of Reclamation. Center for American History, University of Texas at Austin.

Williamson County Commissioners Court Resolution, March 18, 1949, Georgetown. Lyndon B. Johnson Library, Austin.

Williamson County Deed Records, 1946–48. County Clerk's Office, Georgetown.

———, Vol. 450, April 7, 1963, ibid.

———, Vol. 468, May 27, 1964, ibid.

———, Vol. 479, July 13, 1971, ibid.

Williamson County Clerk's Office (Georgetown): Last Will and Testament, Wilson H. Fox, February 13, 1974; Last Will and Testament and Inventory and Appraisement of the Estate, Owen W. Sherrill, December 14, 1976, No. 8677; Last Will and Testament, Kay Deaver Sherrill, November 13, 1991, No. 13904.

Williamson County County Clerk's Office (Georgetown): Tax Roll 1964, Plat Maps 1974.

Williamson County Tax Appraisal District: Certified Appraisal Rolls for Granger Independent School District, Georgetown Independent School District, Taylor Independent School District, Round Rock Independent School District, Williamson County, and Williamson County Appraisal District, 1985, 1990, and 2001; Dell Computer, La Frontera, Del Webb Sun City, 2001. Williamson County Appraisal District, Georgetown.

Articles and Book Chapters

Atkinson, Jim, "Brave New 'Burbs," *Texas Monthly*, September 2003.

Alderman, William B., "The Truth About Scandal of the Interstate Highway Program," *Texas Parade*, June 1960.

Alexander, Kate, "Taylor Deal Would Give River Authority a Boost," *Austin American-Statesman*, August 11, 2003.

Banks, Jimmy, "Louis Henna: Hobby: Orphans," *Texas Parade*, Vol. 11, No. 9, February 1951, Texas State Archives, Liberty, Texas.

Biffle, Kent, "The Sly Old Fox," *Dallas Morning News*, April 28, 1984.

"Big Cave Discovered South of Town, Parts of Camel and Elephant are Found," *Williamson County Sun* (Georgetown), December 19, 1963.

Breyer, R. Michelle, "Masterminds of the Mall," *Austin American-Statesman*, August 17, 2003.

Brooks, David, "Exiles on Main Street," *New York Times Magazine*, April 9, 2000.

———, "Our Sprawling Supersize Utopia," *New York Times Magazine*, April 14, 2004, pp. 46–51.

Buchner, Rue and Anselm Strauss, "Professions in Process," *American Journal of Sociology*, Vol. 66, No. 4, January 1961.

Caro, Robert A., "The City-Shaper," *New Yorker*, January 5, 1998.

"Congress Allots $30 Million in Highway Funds for Texas," *Texas Parade*, Vol. II, 1950.

"Corn Becomes King for a Day," *Dallas Morning News*, September 29, 1938.

Daniel, Kelly, "Vintage Plan Shows What I-35 Might Have Been," *Austin American-Statesman*, November 28, 1999.

Detzer, Carl, "Our Great Highway Bungle," *Reader's Digest*, July 1960.

Dorsch, Jeff, "Mayor Recalled; May Elections Start," *Williamson County Sun* (Georgetown), February 6, 2002.

———, "36 Million Water Pipeline Springs a Big Leak," ibid., February 6, 2002.

Drew, Elizabeth B., "Dam Outrage: The Story of the Army Engineers," *Atlantic*, April 1970, Vol. 225, No. 4.

Eager, Timothy, "Las Vegas Stakes Claim in 90s Water War," *New York Times*, April 10, 1994.

Fairchild, Dan, "IH 35: Man's Effort to Join Nation by a Highway," *Austin American-Statesman*, January 10, 1971.

"Finance It Some Other Way," editorial in *Taylor Times*, April 28, 1955.

FitzGerald, Frances, "Sun City–1983," *Cities on a Hill: A Journey through Contemporary American Cultures*. New York: Simon and Schuster, 1986, pp. 203–245.

Fowler, Tom, "Geography Primer," *Williamson County Sun* (Georgetown), September 30, 1998.

Fox, Henry B., "IS THIS A GOOD THING?", advertisement, *Granger News,* January 1949.

———, "Two Super Highways thru County Leave Questions Unanswered," *Taylor Times,* April 28, 1955.

"Flood's Toll in San Gabriel Valley," *Austin Statesman,* September 14, 1921.

Friend, Tad, "River of Angels," *New Yorker,* January 26, 2004.

"From Salt Marsh to Chemical Center: The Texas Division," *Dow Diamond,* Vol. 18, No. 3, October 1955.

Fulton, William, "Operation Desert Sprawl," *Governing,* August 1999, pp. 16–21.

"Grade Crossing Tragedy at Round Rock Takes Ten Lives," *Williamson County Sun* (Georgetown), May 19, 1977, reprint from January 28, 1927.

Gee, Robert W., "Austin was Lone Big Texas City to Shrink in '02," *Austin American-Statesman,* July 10, 2003.

Godwin, Marshall R. and Jones, Lonnie L., editors, Preface, *The Southern Rice Industry,* College Station: Texas A&M University Press, 1970.

Griffin, Ronald C., "Water Use and Management in the Texas Rice Belt Region," June 1984, Department of Agricultural Economics, Texas A&M University.

Hardin, Garrett, "The Tragedy of the Commons," *Science,* Vol. 162, December 13, 1968.

"Highway Needs of the National Defense, a New Publication," *Public Roads,* August 1949.

"Highways," *New Handbook of Texas,* edited by Ron Tyler, et al., 6 volumes, Austin: Texas State Historical Society, 1996, III, 608.

Hill, Robert T., "Roads and Materials for their Construction in the Black Prairie Region of Texas," December 1889, No. 53, *Bulletin of the University of Texas,* Austin.

Howerton, Walter, Jr., "Taylor Voters Nix Selling Water Plan," *The Sunday Sun* (Georgetown), Nov. 9, 2003.

Jackson, John B., "In Search of the Proto-Landscape," *Landscape in America,* edited by George F. Thompson, Austin: University of Texas Press, 1995.

———, "The Discovery of the Street," pp. 61–66, "Landscape as Theater," p. 67, chapters in *The Necessity for Ruins and Other Topics,* Amherst: University of Massachusetts Press, 1980.

Janicke, Tim, "The I-35 Odyssey," *Austin American-Statesman,* November 26, 2000.

Jordan, Jay, "Circleville Philosopher Works 'Gentle Fraud,'" *Sherman Democrat*, July 31, 1988.

Kemp, Mrs. Jeff T., "Significance and Origin of the Names of the Rivers and Creeks of Milam County," Cameron, Texas, 1929.

Kittrege, William K., "Home Landscape," *Landscape in America*, edited by George F. Thompson, Austin: University of Texas Press, 1995.

Laas, William, "Why Your Road Leads to 1975," Ford Motor Company distribution, May 1955, Lyndon B. Johnson Library, Austin.

Lash, Kenneth, "Notes on Living with Landscape," *Landscape in America*, edited by George F. Thompson, Austin: University of Texas Press, 1995.

Lidell, Chuck, "LCRA Model Shows Devastation Central Texas Would Face in Flood," *Austin American-Statesman*, August 23, 2001.

———, "Growth Rates Slow Slightly," *Austin American-Statesman*, March 18, 1998.

Lindsey, Alan, "Allliance Works on Water Plan for Williamson," *The Sunday Sun* (Georgetown), December 17, 2000.

———, "Water Agreement Will Meet Needs for Next 20 Years," ibid., August 23, 2000.

———, "G'town Should Expand Water Intake," ibid., September 20, 2000.

Luberoff, David, "A Tale of Two Tables," *Governing*, Vol. 10, No. 8, May 1997.

Martinez, Michelle M., "The Roads Test," *Austin American-Statesman*, October 12, 2000.

———, "Williamson Reservoir Plan Draws Opposition," ibid., August 29, 2000.

———, "Fighting to Keep above Water," ibid., December 28, 2000.

Michels, Aimee, "Berry Springs: A Real Beauty," *Williamson County Sun* (Georgetown), August 4, 2002.

Mitchell, John G., "Urban Sprawl," *National Geographic*, July 2001, pp. 54–71.

Mitchell, Mark, "Boom. Bust. Comeback." *Williamson County Sun* (Georgetown), September 30, 1998.

———, "Sale Price Named for Berry Creek," ibid., September 11, 1988.

———, "Stanton Files MUDs That May Double City," ibid., November 20, 1983.

Moses, Greg, "The Farm Crisis," *Texas Business*, May 1982.

"The National System of Interstate Highways," *Public Roads*, Vol. 25, No. 1, September 1948, pp. 19–20.

Nelson, Carter, "State Agency Reviewing Dove Springs Treatment Plant," *Austin American-Statesman*, February 28, 2002.

Norton, Shana, "Water's for Fighting: The Edwards Aquifer Dilemma," Working Paper No. 80, 1994, Lyndon B. Johnson School of Public Affairs, University of Texas at Austin.

O'Brien, Stephen J., "One Sure Thing: 130 Will Slice County into New Pie," *Williamson County Sun* (Georgetown) Sesquicentennial Edition, November 8, 1998.

Osborne, David, "The Asphalt Bungle," *Inquiry,* October 1981.

"Owen W. Sherrill Claimed by Death," *Williamson County Sun* (Georgetown), May 6, 1976.

Peterson, Elmer T., "Big Dam Foolishness," *Country Gentleman,* May 1952.

Pfiester, Sam, "Dialogue Between Two Sides," *Williamson County Sun* (Georgetown), July 9, 2003.

Pinkard, "Under the Interstate," *Texas Highways,* May 1966.

Pletz, John, "Dell Takes Over Top Spot in Home PC Sales," *Austin American-Statesman,* March 6, 2002.

Plohetsky, Tony, "Winds of Change in Round Rock," *Austin American-Statesman,* February 28, 2002.

———, "Lake Stillhouse Pipeline Leaking Must Be Repaired," ibid., February 12, 2002.

Pollan, Michael, "The Triumph of Burbopolis," *New York Times Magazine,* April 9, 2000.

"Progress Along Interstate 35," *Texas Highways,* March 1965.

Ritter, Dale, "The Geological Perception of Landscape," *Landscape in America,* edited by George F. Thompson, Austin: University of Texas Press, 1995.

"Road to Preserving History," *New York Times* editorial, November 12, 2003.

"R. Rock Historical Ass'n Protests Expressway Survey," *Round Rock Leader,* November 8, 1956.

Roddy, Mary Agnes, "Blands of Taylor Wed 52 Years," *Austin Statesman,* November 13, 1932.

"Salvation from Seed," *Houston Chronicle Rotogravure Magazine,* September 22, 1957.

Sanger, David E., "Sometimes National Security Says It All," *New York Times,* May 7, 2000.

Scarbrough, Don, "Holy Mudballs, Super Pickle It Rained!" *Granger News,* September 8, 1966.

Scarbrough, Linda, "A Tale of the Gunns," *The Sunday Sun* (Georgetown), March 23, 1986.

———, "Fifty Years Out," *Williamson County Sun* (Georgetown), November 8, 1998.

Schatz, Amy, "The Dell Curve," *Austin American-Statesman,* July 10, 2003.

Seely, Bruce E., "Conclusion," *Building the American Highway System,* Philadelphia: Temple University Press, 1987.

Sellers, Rebekah, "Park Land at Site Becomes an Issue," *The Sunday Sun* (Georgetown), June 8, 2003.

————, "Landscape, Parking Lot Part of Fee," ibid., June 11, 2003.

————, "City Council Takes Its First Public Look at Simon Contract," ibid., July 9, 2003.

"She Heard 23 Die," *Rockdale Reporter,* January 22, 1998.

Sherrill, Owen W., "A Supplemental Report to the U.S. Corps of Engineers on Resurvey Dam Sites," circa 1954, unpublished, Brazos River Authority, Waco.

Sibley, George, "The Desert Empire," *Harper's,* October 1977.

Skrabanek, Robert L., "The Influence of Cultural Backgrounds on Farming Patterns in a Czech-American Rural Community," *Southwestern Social Science Quarterly,* Vol. 31, 1951.

Slaughter, Bob H., "Downward, Ho! Georgetown Cave Hailed by 'Dallas Times Herald'," *Williamson County Sun* (Georgetown), February 27, 1964.

Smith, Griffin, "The Highway Establishment and How It Grew," *Texas Monthly,* April 1974.

Smith, Rick, "Writer-Philosopher Took Shortcut to Beat Rat Race," *Austin American-Statesman,* March 5, 1979.

Stanley, Dick, "Taylor Voters OK Water Plant Sale," *Austin American-Statesman,* February 8, 2004.

Stubblefield, Billy Ray, "In Quarter Century, Political Life Takes Revolutionary Turn," *Williamson County Sun* (Georgetown), September 30, 1998.

Tanner, C. L., "The Enterprise of Granger," *The Texas Magazine,* Vol. 1, No. 6, April 1910.

"Taylor, the Biggest Little City," *Taylor Daily Democrat,* January 24, 1923.

Taylor, T.U., "County Roads," *Bulletin of the University of Texas,* No. 54, March 1890.

"Texas Highways: 2000 A.D.," *Texas Highways,* Vol. 9, No. 5, May 1962, Texas State Archives, Austin.

Turner, Thomas E., "Main Street of Texas," *Texas Parade,* Vol. 13, No. 11, April 1953, Baylor University Texas Collection, Waco.

Tyson, Kim, "Shutdown Caps Westinghouse Struggle," *Austin American-Statesman,* October 31, 1986.

Vargo, Joe, "State Highway Pioneer Dies," *Austin American-Statesman,* November 18, 1986.

Walton, Frenkel and Judy Walton, "Bavarian Leavenworth and the Symbolic Economy," *Geographical Review,* Vol. 90, No. 4, October 2000.

"Warning: Icy Road," *Wall Street Journal,* editorial, December 12, 1959.

Water, Erin J., "Leander Will Give Away Water Rights," *Austin American-Statesman,* January 4, 2001.

Wear, Ben, "Long, Rocky Path Full of Close Calls Leads to Toll Road," *Austin American Statesman,* September 28, 2003.

"Why Not Industry," *Taylor Daily Press,* advertisement, September 23, 1964.

Willson, Herbert G., "Taylor, Black Land and Cotton," *The Texas Magazine,* Vol. III, No. 4, February 1911.

Wood, Denis, "The Spell of the Land," *Landscape in America,* edited by George F. Thompson, Austin: University of Texas, 1995.

Woods, Tyler, "Southwestern Almost Says 'Finis'," *Williamson County Sun* (Georgetown), September 30, 1998.

Worth, Troy, "Evaluating the Interstate," *Texas Highways,* February 1962.

Film

Koenig, Diane, *San Gabriel River Crossings,* circa 1974–75, commissioned by Linda Graves, Georgetown, Tex.

Pamphlets and Brochures

"Good Water Trail," U.S. Army Corps of Engineers, Fort Worth District, 1994.

facts . . . for Agriculture Industry Municipalities Recreation-seekers, Mineral Wells, Tex.: Brazos River Authority, 1956.

"North Fork Lake," U.S. Army Corps of Engineers, Fort Worth District, undated.

"A Program for Texas Highways," Texas Research League, 1957, Center for American History, University of Texas at Austin.

"Round Rock, Texas, 1840–1965," City of Round Rock, July 1965, *Austin American-Statesman* morgue.

"Seeing Central Texas out of Austin," Tour No. 5 (northeast). Austin: Austin Centennial Committee, Austin Chamber of Commerce, 1936. TX-Z Collection, Center for American History, University of Texas at Austin.

State of the River, edited by Robert Cullick, Austin: LCRA Corporate Communications, 1993.

Texas and Louisiana Rice, Houston: Sunset Route Passenger Industrial

Department, Southern Pacific Co., circa 1911. Edward A. Clark Texana Collection, Southwestern University, Georgetown.

Texas Rice Book, Houston: Gulf, Western Texas & Pacific Railway, circa 1900. Edward A. Clark Texana Collection, Southwestern University, Georgetown.

Welcome to Brazoria County: The Receiving End of Wild River! Angleton: Brazoria County, 1957.

Texas Roads and Highways, Texas Legislative Council, No. 52–3, October 1952, Austin.

Speeches

Johnson, Lyndon B., "Water Supply and the Texas Economy," July 29, 1953, U.S. Senate. Lyndon B. Johnson Library, Austin.

Greer, Dewitt, "The Future of Highways and Highway Transportation," March 3, 1961, Pacific Regional Conference, International Road Federation, Sydney, Australia, *Texas Highways*, Vol. 8, No. 3, March 1961. Texas Department of Transportation, Austin.

Irvin, Helen, to Texas Highway Commission, November 21, 1956, Texas Department of Transportation, Austin.

Pickle, J. J., "Water Resources Developments," April 5, 1965, Proceedings: Seminar on Management of River Basins, Richard W. McKinney Engineering Library, University of Texas at Austin.

Wells, Walter J., "Practical Aspects of Water Resource Management—Brazos River Authority," April 5, 1965, Proceedings: Seminar on Management of River Basins, Richard W. McKinney Engineering Library, University of Texas at Austin.

Unpublished Theses and Dissertations

Cocovinis, Dimitri B., "Areal Map of Southeastern Williamson County, Texas," M.A. thesis, University of Texas at Austin, 1949.

Fox, Mary Elizabeth, "Road Systems in Texas," M.A. thesis, Southwestern University, Georgetown, 1931.

Helaakoski, Reijo, "Economic Effects of Highway Bypasses on Small Cities," M.A. thesis, University of Texas at Austin, 1991.

Langert, Margaret Eleanor, "The History of Milam County," M.A. thesis, University of Texas at Austin, 1949.

Michalik, Edward E., "Study of 32 Dropouts in the Granger High School," M.A. thesis, University of Texas at Austin, 1956.

Scanlon, Sister Francis Assisi, "The Rice Industry of Texas," M.A. thesis, University of Texas at Austin, 1954.

INDEX

Note: Illustrations are indicated with *italic* page numbers.

Collins, R. D., 55–56, 91, 94, 130, 140–41

Colorado River, 189, 199; *see also* Lower Colorado River Authority (LCRA)

Colorado, The (Waters), 35

Committee on Water Conservation, 76

commuting, 204, 260, 268, 289, 321–25; land development and, 204

"compromise" three dam package, map, *172*

Connally, John, 7, 134, 136, *343*

Conquest of Arid America, The (Smythe), 65

conservation. *See* environmental issues; soils; water conservation

construction costs: for 130 Toll, 362; bids awarded, 219; BRA assumes responsibility for San Gabriel dams, 198–99; cost-benefit ratios for dam construction, 216n24; for dams, 132, 151, 159–69, 198, 199n4, 216n24, 237; delays and increase in, 222; estimates for Laneport/North Fork dam package, 159–69, 198, 215–16n24; for Laneport Dam, 199n2, 237; for North Fork Dam, 199n2; Taylor loop I-35, 256

Cooke, M. C., 247

corn, 57, *58*

cotton, 20, *22*, *58*, 162, *182*

Cotton, James A., 159n1

Coupland, 19

Crawford, Joe E., 206, 208

crickets, 290

Critz, Richard, 7

Crockett Gardens, 76, 227–30, *228*, *229*

culture: automobiles and American culture, 9, 243, 249, 272–73; Czech communities and, 114, 179; decline of ethnic enclaves and cultural distinctions, 344; ethnicity and politics, 133; identity and community membership, 9; intercultural tension in Central Texas, 27–28; interstate highway culture, 9; landownership as Czech cultural value, 23–24, 177–78, 180; place and placelessness, 12; ranching as, 27–28, 195, 197; suburbanization and, 2, 279–80n25

Cypress Semiconductor Corporation, 319–20, 336

Czech-American farmers: cultural identity of, 179; landownership as cultural value, 23–24, 177–78, 180; Laneport Dam and displacement of, 68, 83, 110–15, 170, 175–76; Martinka family members, *24*; Melasky as advocate of, 86; ministers, *112*; *Nasinec* (Czech-language newspaper), *23, 25*; Poage and, 111, 114; settlement of Williamson County by, 21–23; small dams program and, 105

D

dams: agricultural land lost to production, 13, 53–56, 66–67, 111, 349; on Brazos River, 161–66; "compromise" three dam package, map, *172*; construction costs, 132, 151, 159–69, 198, 216n24, 237; debates over location of, 8, 53–56, 82, 143–51, 363; economic impacts of, 3, 11, 147, 159, 200–201, 203, 234–36, 239–40, 317–18, 336, 348; environmental opposition to, 171–72, 200–201, 215–26; for flood control, 12, 31n42, 45, 48–50, 53–56, 71–72, 89, 148–49, 198, 216; funding for projects, 52–53, 132, 139, 140, 160–61, 170–71; for hydropower, 50, 51–52, 80, 88, 91, 163, 199; local *vs.* regional planning, 56; municipal and industrial water supplied by, 148–49, 163–66, 364; Pickle as proponent of San Gabriel dams, 139, 147, 166–68, 233; plans for three dams, *172*; as politicized, 11, 68–69, 73–74, 126–27, 145, 200; as pork barrel projects, 73, 145, 200; as public works projects, 49; recreation and, 50, 90, 211–13, 350; rice producers as influential, 70; "simultaneous" dams project, 129–31, 140, 155; small dams program, 102–5, 115–16, 118, 122–23, 146, 153; water conservation and, 89, 146, 156, 217; *see also* Laneport Dam (Granger Dam); North Fork Dam

Danek, Emil, 311

Daniel, Price, 130–31

Death of a Salesman (Miller), 71

Defense Highway Act, 250

Dell Computer, 1, 12, 234, 304, 319–20, 333, 338, 340

Democratic Party: county-level organization and control, 6; Czech farmers as loyal to, 132–33; "liberal" and "moderate" factions of, 137n8; local political "machine," 69

demographics: age-restricted communities, 323; Williamson County and changing demographics, 341–42

Department of Agriculture, small dams program, 102–4

displacement, 225; of Czech farmers by Laneport Dam, 68, 83, 110–15, 170, 175–76, 177n7; of landowners by North Fork Dam, 176–77, 188–94, 196

Dobbs, Jim, 126, 131–33, 137

Doggett, Lloyd, 220–26

Double File Trail, 25

Double N Acres, *279*

Dow Chemical, 13, 70, 86–87; demand for

Fox, John Howard, 55–56, 93–94, 108, 115, 150, 153, 238

Fox, John Short, *106*, 106–7

Fox, Mary Elizabeth, 108

Fox, Robert Bryan, 55–56, 93, *106*, 107–8, 115, 187–88

Fox, Walter, *106*, 108

Fox, Wilson H., *106, 126, 343*; *Allison v. Froehlke* case and, 217–18, 220, 224–25; death of, 187–88, 233, 352–53; H. B. Fox and, 56, 153; Hilda's Bottom barbecues hosted by, 124–25; Laneport Dam and, 55, 60, 71, 93, 107, 115, 202, 233, 353; local politics and, 6, 132; Pickle and, 154, 157, 169, 188n36; population growth projected by, 204; price paid for condemned property, 187–88; sale of lands, 360; Sherrill and, 79; Thornberry and, 116, 128, 129, 134; Wells and, 149–50

Friendship, 24, *113,* 176; 1921 flood, 41; Laneport Dam and inundation of, 68

frontage roads (access roads), 292

Fulcher, Billye "Bill," 40; with siblings Katye, J.B., and Laura, *41*

funding: contingent upon sale of water, 142; of dam projects, 52–53, 132, 139–40, 142, 160–61, 170–71, 218; frozen during *Allison v. Froehlke* case, 218; for highway projects, 347; for interstate highway system, 272, 326; for Laneport/North Fork Dams package, 160–61, 170–71, 218; Pickle and dam funding, 139–40, 170–71; for pre-construction planning (dams), 147; for resurvey of San Gabriel River watershed, 97–98; for "simultaneous" dams project planning, 140; Texas Water Board and funding of Laneport/North Fork Dams package, 160–61; Thornberry and dam projects as political capital, 131–32

G

Garry, Mahon, 261–62

Garry, Mahon "Buzz" (son), 261–62

gas or service stations, 296–99, 312

Geertz, Clifford, 11

Geography of Nowhere (Kunstler), 289

geology: Austin Chalk, 17–18; Balcones Escarpment and, 4; Edwards Formation, 222, 223; faults and seismic risk to dams, 221–22, 223; geomorphology of region, 15; Inner Space Caverns, 222; suitability of dam sites, 222–223; as "weather maker," 31

Georgetown, 5–6; 1957 flood, 95–96, *98, 99,*

101; aerial view of Northtown (1965), *326*; aerial view of Northtown (2005), *327*; annexation of I-35, 313; claim on upper fork waters, 202; commuting and growth of, 321–22; dam proposed near, 94; dam resolution, 155; debates over dam locations and, 143–44; demand for water, 206–9; establishment of, 26n30; growth of, 236, 334, 354n67, 364; historic preservation in, 356; I-35 and, 8, 247, 251, 290–91, 311–13, 345–46; Inner Space Caverns and, 308–9; Johnson and dam projects near, 62–63; Lake Stillhouse Hollow and water supply, 348–49; local politics, 5–6; municipal planning and politics, 321; Round Rock as rival for water supply, 209, 238–39; San Gabriel River confluence, 30; small dams program supported by citizens, 123; subdivisions and residential development, 322–25; tax base, 341; Taylor as rival for county courthouse, 261; Taylor as rival for dam projects, 143–51; traffic volume, 345–46; treatment plant, 206–7, 209; water demand exceeds available supply, 348; "Westinghouse effect" on, 320–21

Gianelli, William, 74

"gilded democratic action" (democratic pressure politics), 13–14, 359–62; term defined and described, 359–60n79

Gonzales, Nicanor and family, 43

Gordon, John, 362–63

Graham, Otis L., Jr., 48, 91–92

Grand Canyon, Colorado River dam projects and, 171, 229

Granger, 19, 57, 247, 344; Czech presence in, 24; dam resolution, 155; economic impacts of Laneport (Granger) Dam, 234–35; I-35 route and, 251; Laneport Dam and, 68, 184; local political machine, 60; main street (1931), *22*; population of, 203, 353; small dams program supported by citizens, 123; water supply for, 348

Granger Dam. *See* Laneport Dam (Granger Dam)

Granger Lake (Laneport Dam reservoir), 234; claims on water by communities, 202–3, 237–38n16; demand exceeds supply, 348; recreational use, 220, 222, 351; reservoir size, 71, 83, 119–21, 201–2; water sales and water prices, 200, 206–8, 239n20

Granger News (newspaper), 69–70, 169, 179

Graves, Linda Crawford, 210–11, 231–32, *232*

Great Plains, The (Webb), 15, 198

35 and federal control, 330; Laneport Dam resolution, 88; lobbying for highway construction compared to dam projects, 267–68; local political clout and federal influence, 2, 4, 9, 11, 13–14, 21, 45, 69, 95, 267–68 (*see also* "gilded democratic action" *under this heading*); local political machinery, 24, 60, 133; pork barrel politics (log rolling) and dam construction, 73, 145, 200; redistricting, 218, 233; Republican Party, 126–27, 234, 342–44, 348n42, 363; shift in party loyalties, 342–44

pollution, water contamination, 350

Pool, Fred, 81

population: growth of Williamson County, 1, 19; I-35 linked to population growth, 146, 166, 203–4, 331; impacts of population increase, 333; Round Rock population growth, 8, 173, 315–16, 320; stress on infrastructure and services, 345; Westinghouse and population increase in Round Rock, 315–16; Williamson County, overall growth in, 236, 339–40

pork barrel politics (log rolling), 73, 145, 200

Porter, Eliot, 227

Possum Kingdom Dam (Morris Shepard Dam), 51–52, 91, 163–65

Potomic River, 61–62

poverty: decline in, 345; in eastern Williamson county, 340; politics and economic class, 363; rural interstates and, 299

"power centers," 140

preservation: county parks and, 351–52; Heritage Roads and Heritage Trees, 351; historic preservation, 356

property values: Army and prices paid for condemned property, 174–76, 184–87, 192, 195, 196–97, 353; Black Waxy soil and, 197; development and, 3, 236; I-35 and increase in land prices, 146–47, 295, 300, 312, 324, 326, 330, 358; law suits and increased prices for condemned lands, 196–97; "market price" and eminent domain, 174–76, 197; political clout and, 197; recreational opportunities and increase in real estate values, 177; in Round Rock, 341; subdivisions and increase in, 236; unearned increment and, 310, 313–14, 325–30

public works projects, 49

Q

Quick, Don, 287

Quíntanilla, Pablo L. and family, 43

racism, 345, 363

railroads, 18, 19, 21, 162; 1957 flood damage to Katy bridge, *100*; airdropped warnings during 1957 flood, 96, 100; bus/train accident, 248–49; damage caused by 1921 flood, *44*; M-K-T railroad, 18; Taylor as transportation hub, 250–51; towns established along routes, 19

ranching: as culture, 27–28, 195; displacement of ranchers by North Fork Dam, 176–77; on Edwards Plateau, 4, 28, 29; federal farm policies, 291; I-35 and, 327; longhorn cattle and early ranching, 26–27

Rayburn, Sam, 125–26

Reagan, Ronald, 74, 234

recreation: Army Corps of Engineers and, 90; Brazos River Authority resistant to recreational use, 230; canoeing on North Fork San Gabriel, 211–13, 350; dam construction and new recreation opportunities, 50; Granger Lake and recreational use, 220, 351; increase in real estate values and, 177; Lake Georgetown and, 220, 222, 228–30; parks, county parks, 351–52; San Gabriel River, 350; on undammed San Gabriel, 210–13

Reisner, Marc, 73–74

Republican Party, 234, 342–44, 363; dams as political issue for, 126–27; "quality of life" issues, 348n42

reservoirs: "dry" reservoir dams for flood control, 148–49; Granger Lake (Laneport Dam reservoir), 71, 83, 119–21, 201–2, 220, 222, 237–38n16; Lake Belton, 165–66; Lake Georgetown (North Fork reservoir), 173, 194, 220, 222, 228–30, 238, *240, 241,* 336, 348; Lake LBJ, 189, 194; Lake Stillhouse Hollow, 348–49; optimal size of, 201; recreational use of, 220, 222, 228–30; volume of dams on upper San Gabriel, 83; water conservation and, 146, 156, 217; water sales as purpose of, 142–43; water-storage volumes, 71, 83, 119–21, 144, 155, 222–23

residential development, 3, 204, 219, 334, 336, 348

Reston, Virginia, 315–16

resurvey of San Gabriel River watershed, 81–82, 84, 93–94, 96–98; differing expectations regarding, 118; funding of, 97–98; Poage on results of, 120–22; as political issue, 123; popular support for, 117–18; Thornberry and, 116

retail development, 10, 140, 345, 358–59

Road, River, and Ol' Boy Politics

COLOPHON

The typeface used for the text is Adobe Bembo.
The display face is Torino Bold.
Printed at Sheridan Books, Chelsea, Michigan,
on 50 lb. Natural.

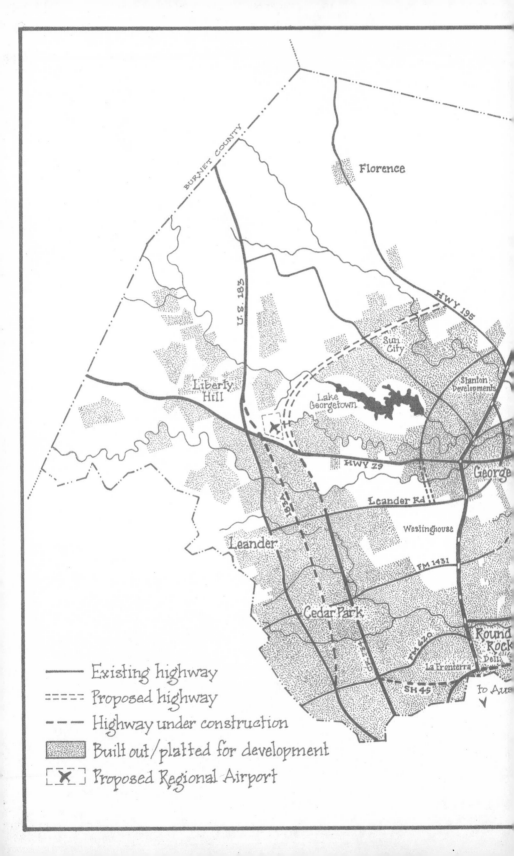

Florence

BURNET COUNTY

HWY 195

U.S. 183

Sun
City

Stanton
Developments

Liberty
Hill

Lake
Georgetown

HWY 29

George

Leander Rd

Westinghouse

Leander

FM 1431

Cedar Park

FM 620

Round
Rock
Dell

La Fronterra

SH 45

To Aus

—— Existing highway

===== Proposed highway

– – – Highway under construction

▓▓▓ Built out/platted for development

[X] Proposed Regional Airport